Energy and Climate
in the
Urban Built Environment

Energy and Climate
in the
Urban Built Environment

Edited by
M. Santamouris,
University of Athens, Greece

Authors
D.N. Asimakopoulos, V.D. Assimakopoulos,
N. Chrisomallidou, N. Klitsikas, D. Mangold,
P. Michel, M. Santamouris, A. Tsangrassoulis

Published by James & James (Science Publishers) Ltd,
35–37 William Road, London NW1 3ER, UK

A catalogue record for this book is available from the British Library

ISBN 1 873936 90 7

Printed in the UK by The Cromwell Press

Contents

બ

Part 1

1

———

On the built environment – the urban influence

M. Santamouris

Department of Applied Physics, University of Athens

ℛ

SOME BASIC NOTES ON THE BUILT ENVIRONMENT

Buildings provide shelter and places of retreat for human beings, while also defining our well being and helping to determine our quality of life. As Winston Churchill said, 'We shape our dwellings and afterwards our dwellings shape our lives'. The same is true of the streets, estates, villages, towns and cities where we live.[1] In addition to the social impact of buildings, the economic impact is also very important. The building industry has a pivotal place in the economy and is one of the biggest economic sectors. In Europe alone, business related to building represents a yearly turnover of the order of US $460 billion.[2]

The built environment is not just the collection of buildings; it is also the physical result of various economic, social and environmental processes, which are strongly related to the standards and needs of society. Economic pressures determine the built environment in which we live,[1] and these in turn are influenced by:

- the property and labour markets, investment and equity, household income and the production and distribution of goods;
- social aspects related to culture, security, identity, accessibility and basic needs;
- environmental influences related to the use of land, energy and materials.

Social, economic and environmental parameters should not be seen as isolated influences, but viewed in an integrated way and with consideration of the strong interrelationship with the other factors mentioned. As stated by Brugmann,[3] cities are integrated systems that facilitate the delivery of a wide range of services and activities. Synergies among these issues generate stress in the built environment and, in most cases, the solution to one problem is the cause of another. This integrated approach become clearer when the built environment is considered in terms of stocks, flows and patterns, as defined by the Expert Group on the Urban Environment.[4] Stocks include buildings, land, open spaces, streets and other tangible features; patterns involve all spatial and temporal patterns in urban and rural forms, neighborhood design and street layouts; while flows include all pressures of urbanization, pressures on rural communities, household trends, demands for energy, transport, materials, waste, etc.

The interrelated nature of almost all of the above aspects is evident, and perturbation of just one parameter may affect the other parts of the system in a way that is not easy to predict.

In particular, development of the urban environment has serious effects on the global environmental quality. Major concerns are the quality of air, increase in temperature, acoustic quality and traffic congestion. Buildings are related to global changes in the increase of urban temperatures, the rate of energy consumption, the increased use of raw materials, pollution and the production of waste, conversion of agricultural to developed land, loss of biodiversity, water shortages, etc.

Population growth in countries under development and improved standards of life in the developed world intensify environmental problems. As stated by the World Resources Institute and United Nations Environmental Programme:[5]

> In the wealthiest cities of the developed world, environmental problems are related not so much to rapid growth as to profligate resource consumption. An urban dweller in New York consumes approximately three times more water and generates eight times more garbage than does a resident in Bombay. The massive energy demand of wealthy cities contributes a major share of greenhouse gas emissions.

Cities are increasingly expanding their boundaries and populations, and 'from the climatological point of view, human history is defined as the history of urbanization'. The increased industrialization and urbanization of recent years have dramatically affected the number of the urban buildings, with major effects on the energy consumption of this sector. It is expected that 700 million people will have moved to urban areas during the final decade of this century. The number of urban dwellers rose from 600 million in 1950 to 2 billion in 1986 and, if this growth continues, more than half of the world's population will live in cities by the end of the century, whereas a hundred years ago, only 14% lived in cities and even in 1950, less than 30% of the world's population was urban.[6] Current and projected urban populations, by region as reported by the United Nations,[7] are given in Figure 1.1.

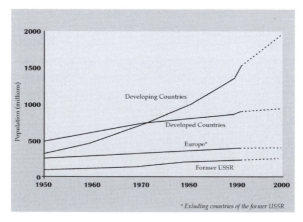

Figure 1.1. Current and projects urban populations by region
(source: *United Nations*[7])

Improving living standards increase the space requirements per person. It is characteristic that in the USA, between 1950 and 1990, the floor-space requirements per person doubled.[8] Very important variations in housing floor space per person also exist in Europe because of social and economic differences. Moscow has 11.6 m² net living space per person, while Paris has 28.2, Oslo 47.2 and Zurich 50.6 m² per person.[9] Today, at least 170 cities each have more than one million inhabitants. It is estimated that, in the USA, 90% of the population will be living in, or around, urban areas by the year 2000,[10] while other estimates show that urban populations will form 80% of the total world population in 2100. In Europe, more than 70% of the population is living in urban areas. Table 1.1 gives the percentages of the urban population of selected European countries in 1950 and 1990,[11] while Figure 1.2 shows the population growth of selected European cities.[12] In Africa, while in 1925 the population living in cities of more than 100,000 inhabitants was close to 500,000, this number increased to 20 million in 1972, and 72 million in 1988.[13]

Table 1.1. Urban populations of selected European countries – in order of the 1990 percentage (source: UNEP[11])

Country	Proportion of urban population in 1950 (%)	Proportion of urban population in 1990 (%)
Portugal	19	33
Albania	20	35
Romania	25	55
Former Yugoslavia	22	57
Ireland	41	57
Austria	49	59
Finland	32	61
Poland	39	62
Switzerland	43	62
Greece	38	63
Hungary	39	64
Former USSR	48	66
Bulgaria	24	68
Italy	53	69
Lithuania	29	70
Latvia	50	71
Estonia	49	72
France	62	74
Norway	50	75
Former Czechoslovakia	38	77
Spain	50	78
Sweden	65	82
Luxembourg	59	82
Germany	73	83
Denmark	68	83
Malta	69	84
The Netherlands	81	87
UK	82	87
Iceland	73	89
Belgium	90	95

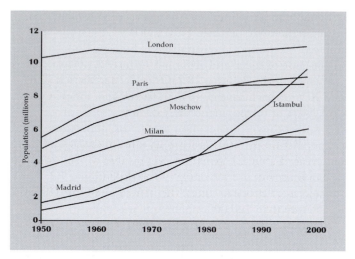

Figure 1.2. Population growth in selected European cities (source: UNEP/WHO[12])

Expansion of our cities means that more and more land is required to support them. According to Stanners and Bourdeau,[9] in Europe, 2% of agricultural land is lost to urbanization every ten years. Hahn and Simonis report that, during the last century, the surface of urban land per capita in Europe has increased ten times.[14] It is characteristic that an average European city of one million inhabitants consumes every day 11,500 tonnes of fossil fuels, 320,000 tonnes of water and 2,000 tonnes of food. It produces also 300,000 tonnes of wastewater, 25,000 tonnes of CO_2 and 1,600 tonnes of waste.[9] The direct and indirect needs for land are well represented by the notion of the 'ecological footprint', defined as the land required to feed a city, supply it with timber products and reabsorb its carbon dioxide emissions in areas covered with growing vegetation.[15] This concept helps to set the limits to the activities that an area can absorb in a sustainable way. Based on the calculations of the Sustainable London Trust,[16] London's ecological footprint is close to 50 million acres, which is 125 times its actual surface area. Calculations by Wackernagel and Rees[17] show that the mean ecological footprint in the world is close to 1.8 ha/person and, while in India it is close to 0.4, in Canada it is 4.3 and in the USA close to 5.1 ha/person. In addition to the above, the increased urbanization associated with the loss of agricultural land, wilderness and green areas adds an important additional cost because new infrastructure has to be developed and the existing infrastructure in older parts of a city is used less and thus not well amortized.[18] This is well supported by recent studies in the UK that show that in urban areas almost 60,000 hectares of land are vacant and 15–20% of office space is empty.[19]

URBANIZATION AND CLIMATE

Increasing urbanization and industrialization has caused the urban environment to deteriorate. Deficiencies in development control have important consequences for the urban climate and the environmental efficiency of buildings. The size of housing plots has been reduced, thus increasing densities and the potential for traffic congestion. The increasing number of buildings have crowded out vegetation and trees. It has been reported that New York has lost 175,000 trees, or 20% of its urban forest, in the last ten years.[20]

As a consequence of changes in the heat balance, air temperatures in densely built urban areas are higher than the temperatures of the surrounding country. The phenomenon known as the 'heat island' is due to many factors, the most important of which are summarized by Oke *et al.*[21] and deal with:

- the canyon radiative geometry that contributes to the decrease in long-wave radiation loss from within the street canyon due to the complex exchange between buildings and the screening of the skyline;
- the thermal properties of materials, which increase storage of sensible heat in the fabric of the city;
- the anthropogenic heat released from combustion of fuels and animal metabolism;
- the urban greenhouse, which contributes to the increase in the incoming long-wave radiation from the polluted and warmer urban atmosphere;
- the canyon radiative geometry, which decreases the effective albedo of the system because of the multiple reflection of short-wave radiation between the canyon surfaces;
- the reduction of evaporating surfaces in the city, which means that more energy is put into sensible heat and less into latent heat;
- the reduced turbulent transfer of heat from within streets.

Studies of the urban heat island refer usually to the 'urban heat-island intensity', which is the maximum temperature difference between the city and the surrounding area. Data compiled from various sources show that heat-island intensity can be as high as 15°C.[22] Extensive studies on the heat-island intensity in Athens, involving more than 30 urban stations, show that urban stations record temperatures that are between 5 and 15°C higher than temperatures recorded at reference suburban stations.

As well as the temperature increase, the urban environment affects many other climatological parameters. Global solar radiation is seriously reduced because of increased scattering and absorption. As mentioned by Landsberg,[23] the sunshine duration in industrial cities is reduced by 10 to 20% in comparison with the surrounding countryside and similar losses are observed in energy received.

In a general way, the wind speed in the canopy layer is seriously decreased compared to the undisturbed wind speed and its direction may be altered. This is mainly due to the specific roughness of a city, to channelling effects through canyons and also to the heat-island effect.

In parallel, the urban environment affects precipitation and cloud cover. The exact effect of urbanization depends on the relative place of a specific city with respect to the general atmospheric circulation. In Budapest, for example, mainly because of industrialization, the cloud cover has increased by 3% during the winter period. As mentioned by Escourrou,[13] urbanization causes a proportional increase in precipitation in cities like London, which, because of its geographic location, is more often in a perturbation zone, than in cities like Paris. Studies in Bombay, India, have shown that the development of an industrial zone close to the city has increased precipitation by 15%.[24]

ENERGY USE

Buildings are the most important energy-consuming economic sector. The world-wide primary energy consumption of buildings is close to 19 millions barrels of oil per day and represents almost the entire daily production of OPEC countries.[25]

Energy consumption in the building sector is characterized by a dynamic evolution. Although needs and consumption rates are very different in the various areas of the planet, absolute consumption figures are mainly determined by living standards, economic growth rates, actual energy prices, technological developments, weather conditions and increased population.

Predictions of trends in, and scenarios for, future energy needs indicate a significant increase of the absolute energy consumption, together with a considerable change in the actual structure of demand. Thus, definition of the technological priorities for the building sector requires a deep knowledge of the characteristics that determine, in the major geographic regions, environmental quality in buildings and the corresponding energy consumption.

Buildings in European Union countries have a primary energy consumption close to 740 Mtoe,[26] or a final energy consumption close to 357 Mtoe, which is close to 40% of the total energy consumption.[27] Buildings are responsible for about 18% of the total CO_2 emissions, 10% of CO emissions, 6% of SO_2 emissions and 4% of NO_x emissions,[28] while in selected countries buildings account for the 7.5% of the annual use of CFCs.[29] Most of the energy spend in the building sector is for space heating, 70% of the total energy consumption for domestic buildings and 55% for commercial and office buildings, followed by electrical appliances and hot water production.[30]

In the European Union, the energy consumption of buildings has remained quite constant, with an average yearly growth of 0.1% over the period 1985–1994. Although, for some countries, primary energy consumption of buildings has been reduced during the last few years as a result of intensive energy conservation measures, primary energy consumption in Southern Europe continues to grow, mainly as a result of the significant spread of air-conditioning equipment.[31] However, the energy standards for European buildings are far from ideal. In the UK alone, it is estimated that there are at least eight million homes that are inadequately heated, most being the dwellings of poor households.[32,33]

A study of the trends regarding the energy intensity of the domestic and tertiary sectors in the European Community over the last decade shows a significant increase,[34] although the total energy intensity index shows a small decrease. At the same time, evolution rates in central and eastern Europe have shown a continuous increase in consumption in the former USSR and a significant fall in central European countries, which is related to the decline in economic output. In these countries, the contribution of buildings to CO_2 emissions ranges from 4% (Estonia) to 32% (Czech Republic). Future trends are highly dependent on the pace of economic recovery in these areas. Projections show that energy consumption is expected to fall over the next few years and then start to grow again.[9] If the energy conditions of local buildings are taken into account, for example in Russian cities every fifth house is without central heating,[35] a significant increase in the energy consumption of buildings should be expected in the future.

The primary energy consumption of buildings in the USA is close to 700 Mtoe per year[36] and represents a value close to $200 billion,[37] or 36% of the total consumption. Energy in residential buildings covers mainly space and water heating, while in commercial buildings lighting and space heating account for fully 47% of the energy used. Buildings are responsible for a very high proportion of the pollution produced, but it is important to note that about 30% of the CFCs used in the country are related to buildings.[37] Evolution of building energy consumption shows a continuous increase and, according to official statistics,[37] given the historical growth rate in the number of homes and commercial floor space, the primary energy consumption of the building sector could increase to more than 1190 Mtoe annually by 2030.

Improved living standards and increased populations in developing countries are parameters that may contribute to a dramatic increase of building energy consumption. Although detailed energy consumptions for buildings are only available for some countries, reports from eight major developing countries, representing a very high percentage of the total energy consumption, show that the building sector absorbs almost 21% of the energy needs, even without electricity being taken into account.[38] The high variability of the requirements, combined with specific social development characteristics and a diversified technological potential, define for these countries some very specific regional priorities regarding the type of technological developments to adopt.

Today it is well accepted that urbanization leads to a very high increase in energy use. A recent analysis showed that a 1% increase in the per capita GNP leads to an almost equal (1.03%) increase in energy consumption.[39] However, an increase of the urban population by 1% has been reported to increase energy consumption by 2.2%, i.e. the rate of change in energy use is twice the rate of change in urbanization. Comparison of the energy consumption per capita for the inner and outer parts of selected cities shows that the consumption in the inner part is considerably higher. For example, inner London has a 30% higher energy consumption per capita than the outer part of the city.[40]

Buildings are the largest consumer of energy in cities. Data on the energy consumption of various European cities[40–42] show that the consumption of the residential sector varies from 48% of the end-use energy consumption in Copenhagen to 28% in Hanover. At the same time, buildings in the commercial sector absorb something between 20 and 30% of the final energy consumption of the cities (Table

Table 1.2. End-use energy consumption in selected European cities

City	Residential (%)	Commercial (%)	Industrial (%)	Transport (%)	Total (GJ/capita)
Berlin	33	29	15	23	78.10
Bologna	36	21	11	32	67.30
Brussels	43	29	5	23	94.70
Copenhagen	48	26	6	20	78.16
Hanover	28	25	26	21	112.43
Helsinki	34	23	9	34	89.50
London	36	24	11	29	89.10

Sources: ICLEI,[41] LRC[40] and IBGE[42]

1.2), while the electricity consumption in cities varies up to 23,000 kWh per capita with a mean value close to 5,700 kWh.[9]

Higher urban temperatures have a serious impact on the electricity demand for air conditioning of buildings and increase smog production, while also contributing to increased emission of pollutants from power plants, including sulphur dioxide, carbon monoxide, nitrous oxides and suspended particulates. The heat-island effect in warm-to-hot climates exacerbates cooling energy use in summer. The Environmental Protection Agency[4] has reported that, for US cities with populations larger than 100,000, the peak electricity load will increase 1.5 to 2% for every 1°F increase in temperature. Since urban temperatures during summer afternoons in the USA have increased by 2 to 4°F during the last 40 years, it can be assumed that 3 to 8% of the current urban electricity demand is used to compensate for the heat-island effect alone. Comparisons of high ambient temperatures with utility loads for the Los Angeles area have shown that an important correlation exists. It is found that the net rate of increase of electricity demand is almost 300 MW per °F. As there has been a 5°F increase in the peak temperature in Los Angeles since 1940, this is translated into an added electricity demand of 1.5 GW due to the heat-island effect. Similar correlations between temperature and electricity demand have been established for other selected utility districts in the USA. Based on these rates of increase, it has been calculated that for the USA the electricity cost for the summer heat island alone could be as much as $1 million per hour, or over $1 billion per year.[4] Computer studies for the whole country have shown that the possible increase in the peak cooling electricity load due to the heat-island effect could range from 0.5 to 3% for each 1°F rise in temperature.

Heat-island studies in Singapore show a possible increase in the urban temperature close to 1°C.[43] If there were to be similar changes in temperatures 50 years from now, the anticipated increase in building energy consumption, mainly for air conditioning, would be of the order of 33 GWh per annum for the whole island. Studies by Watanabe et al.,[44] in which the land temperature distribution and the thermal environment of the Tokyo metropolitan area are analysed, have shown that there is a much higher energy consumption in the central Tokyo area. Other studies on the Tokyo area conclude that during the period between 1965 and 1975, the cooling load of existing buildings increased by 10–20 % on average because of the heat-island phenomenon.[45] If this increase continues at the same rate, it will mean an increment of more than 50% by 2000.

Calculations of the spatial-cooling load distribution in the major Athens area, based on experimental data from 30 stations, have been reported by Santamouris et al.[22] It has been found that the cooling load of reference buildings at the centre of the city is about the double that in the surrounding Athens area. It is also reported that high ambient temperatures increase peak electricity loads and put a serious stress on the local utilities. Almost double the peak cooling load has been calculated for the central Athens area than for the surroundings of the city. Finally, a very important decrease in the efficiency of conventional air conditioners, because of the temperature increase, is reported. It has been found that minimum COP values are lower by about 25% in central Athens, obliging designers to increase the size of the installed air-conditioning systems and thus intensify peak electricity problems and energy consumption for cooling.

Increases in the energy consumption in urban areas put a high stress on utilities that have to supply the necessary additional load. Construction of new generating plants may solve the problem but this is not a sustainable solution; it is expensive and the plants take a long time to construct. Adoption of measures to decrease the energy demand in urban areas, such as the use of more appropriate materials, increased planting, use of sinks etc., in association with a more efficient use of energy, involving demand-side management techniques, district cooling and heating, etc. seems to be a much more reasonable option. Such a strategy, adopted by the Sacramento Municipal Utility District (SMUD), has proved to be very effective and economically profitable.[46] It has been calculated that a megawatt of capacity is actually eight times more expensive to produce than to save. This is because energy-saving measures have low capital cost and no running cost, while construction of new power plants involves high capital and running costs.

Although climatic issues have a major influence on the energy consumption of urban buildings, other aspects of the urban environment, such as transport, also play a very important role. Urban transportation is a major contributor to urban energy consumption, as well as to the emission of air pollutants, noise and traffic congestion. LRC reports that transport is responsible for almost 30% of the energy consumption in London and about 25% in the western part of Berlin,[40] while a mean value for most European cities is close to 30%. Road transport is the major contributor to energy consumption, because it accounts for more than 85% of transport's energy consumption.[9] Browning et al. modelled the energy consumption of two contrasting households, one living in a specially designed house with high energy-efficiency standards located in a typical suburban area, the other living in an old 'energy-hog' house of 1988 located in a traditional urban area.[47] Although the urban building used more energy for heating, overall it used ten times less energy because use of a car was much lower. The model assumed that access to services was easier in the urban area and that the level of public transport was good. Very similar results were reported by Goodacre for Lancaster in the UK.[48] Energy spend on urban transport has almost doubled in the last two decades[9] and may increase in the future as the number of cars increases and as mobility patterns constantly change.

USE OF MATERIALS

Each year almost three billion tonnes of raw materials are used worldwide to produce construction materials. This represents almost 40% of the total flow into the global economy,[49] while the building sector is responsible for the 50% of the material resources taken from nature;[50] thus, appropriate selection of materials may have an important ecological and financial impact. Recent research has shown that a high proportion of the material is wasted during the construction phase. Harland has calculated that about 20% of the total material is wasted as a result of damage or off-cutting at the installation stage, spoilage during transport or storage and inaccuracies in ordering and specifying.[51] According to Roodman and Lenssen,[49] a typical North American construction generates 2–35 kg of solid waste per square metre of building, while new ecological constructions generate only 5 kg/m^2.

In parallel, the embodied energy of construction materials accounts for a high part of global energy use. Connaughton has calculated that in the UK the embodied energy of construction materials accounts for 10% of total energy use,[52] while Tucker and Treloar estimate that this percentage in Australia is close to 19.5%.[53]

The technical characteristics of the materials used determine to a high degree the energy consumption and comfort conditions of individual houses, as well as of open spaces. In particular, the optical characteristics of materials used in urban environments and especially the albedo to solar radiation and emissivity to long-wave radiation have a very important impact on the urban energy balance.

The use of appropriate materials to reduce heat islands and improve the urban environment has attracted increasing interest during recent years. Much research has been carried out to identify the possible energy and environmental gains when light-coloured surfaces are used. In some recent research, Rosenfeld *et al.* have shown that the use of white roofs and shade trees in Los Angeles would lower the need for air conditioning by 18%, or 1.04 billion kWh, for the buildings directly affected by the roofs and shaded by the trees.[54] Indirect benefits may also be very important. If the temperature of the entire community drops by a degree thanks to lighter roofs and pavements and to the evapotranspiration from trees, then the air-conditioning load of all buildings may be reduced, even if these buildings are not themselves shaded or if they have dark-coloured roofs. This indirect annual saving would total an additional 12% or 0.7 billion kWh.

AIR QUALITY AND NOISE

In addition to the increased energy demand for cooling, increased urban temperatures affect the concentration and distribution of urban pollution because heat accelerates the chemical reactions in the atmosphere that lead to high ozone concentrations. Other sources such as transport, industry, combustion processes etc. contribute to increased pollution levels in urban areas. In Europe it is estimated that in 70 to 80% of European cities with more than 500,000 inhabitants, the levels of air pollution by one or more pollutants exceed the WHO standards at least once in a typical year.[9]

Urban pollution is linked to climatic change, acidification and photochemical smog. Comparison of daily peak temperatures in Los Angeles and in 13 cities in Texas show that, as temperatures rise, ozone concentrations reach dangerous levels.[55] Also, polluted days may increase by 10% for each 5°F increase.[4] Urban geometry plays an important role in the transport and removal of pollutants. The roughness of urban buildings and landscapes increases air turbulence, thus enhancing the dispersion of pollutants. Furthermore, if pollutants land in sheltered areas like street canyons, they may stay there longer than they would in a windy rural environment.

The roughness of buildings and urban structures affects wind within the city and slows down wind speeds, thus increasing pollutant concentration. Increased industrialization and urbanization have created important pollution problems in urban areas. Sulphur dioxide, particulate matter, nitrogen oxides, carbon monoxide, etc. directly affect both human health and historic monuments and buildings. It is calculated that the cost of damage to buildings and construction materials by sulphur dioxide alone may be of the order of €10 billion per year for the whole of Europe.[56]

Damage from increased pollutants is evident. Analysis of the relationship between hospital admissions and sulphur oxide levels in Athens found that a 'threefold increase in air pollutants doubles hospital admissions for the respiratory and cardiovascular

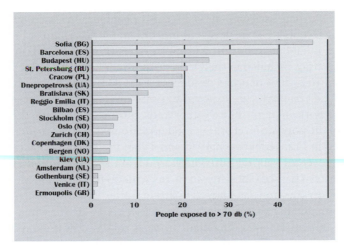

Figure 1.3. Acoustic quality in selected European Cities (Source: *Stamners and Bourdeau*[9])

disorders' and that 'acute respiratory illness shows the highest correlation for the SO_2 variable'.[57] Levels of nitrogen oxide are particularly high in urban environments. NO_2 levels in San Francisco and New York exceed 200 mg m^{-3}, while in Athens the corresponding concentration is close to 160 mg m^{-3}.[58]

Health problems associated with the urban environment are mainly associated with the increased use of cars. This has been acknowledged recently by the British Medical Association.[59] Pollution from diesel and petrol has been proved to be partly responsible for heart diseases. Poloniecki *et al.* have shown that in London, one in 50 heart attacks treated in hospitals are strongly linked with carbon monoxide, which is mainly derived from motor vehicle exhausts.[60]

The role of outdoor conditions and their impact on the indoor climate, as well as the relation between outdoor and indoor pollution, are obvious concerns of building physics and do not need to be discussed here. However, the intensive urbanization and the deterioration of the outdoor air quality observed during the last few years has created a new situation with serious consequences for indoor environmental quality. In fact, outdoor pollution is one of the sources of the so-called 'sick building syndrome'; the other is related to indoor sources. Numerous studies reported during recent years have shown the serious impact of the outdoor environment on indoor air quality.[61] Solutions to indoor air pollution problems are source control, avoiding or attenuating the emission of contaminants, air cleaning and appropriate use of ventilation.

Outdoor pollution and inadequate ventilation may be the primary causes of poor indoor air quality in buildings. Monitoring of 356 buildings with public access has shown that in approximately 50% of the buildings incorrect ventilation rates were the primary cause of illness complaints and poor air quality.[62] Increased outdoor concentrations seriously affect the indoor concentration of pollutants. Measurements of nitrogen oxide concentrations in a hospital in Athens showed high concentrations for indoor pollution, rising from 33 to 67 mg m^{-3}.[63] These results are supported by the findings of the European Audit Project,[64] which has developed assessment procedures for and guidance on ventilation and source control.

Noise in the urban environment is a serious problem. As stated by Stanners and Bourdeau,[9] unacceptable noise levels of more than 65 dB(A) affect between 10 and 20% of urban inhabitants in most European cities (Figure 1.3). The same authors report that, in cities included in the Dobris Assessment, unacceptable levels of noise affect between 10 to 50% of urban residences. OECD has calculated that 130 million people in OECD countries are exposed to noise levels that are unacceptable.[65] The same study reports that in The Netherlands during the decade 1977–1987, the proportion of the population claiming moderate noise disturbance increased from 48 to 60%, while in France, between 1975 and 1985, the urban population exposed to noise levels between 55 and 65 dB(A) increased from 13 to 14 million. UBA reported that in the western part of Germany, in towns with up to 5,000 inhabitants, 14–16 % of the population are strongly annoyed by street noise.[66] In towns with between 5,000 and 20,000 inhabitants this percentage increases to 17–19%, while in cities with between 20,000 and 100,000 inhabitants it rises to 19–25%. Finally, in cities with more than 100,000 inhabitants the percentage rises to between 22 and 33%.

FUTURE PRIORITIES

The continuous increase in urbanization, combined with the degradation of the urban climate and the recent upsurge of concern for the environment, as well as recent technological developments in the field of new energy technologies, define the major priorities and considerations for urban buildings. Thus, the main priorities are:

- Reconsideration of the architectural and planning priorities for the urban environment. Ideas like those developed by the New Urbanism movement,[67,68] based on mixed land use, greater dependence on public transport, cycling and walking, decentralization of employment locations, etc., may be further developed and applied to creating a more sustainable urban environment. In parallel, addressing urban environmental and energy problems, instead of treating their symptoms, is an absolute priority for improving the quality of the urban environment. All these are combined in a major goal that aims to achieve sustainability in urban areas. As defined by Stanners and Bourdeau:[9]

    Sustainable cities are cities that provide a livable and healthy environment for their inhabitants and meet their needs without impairing the capacity of the local, regional and global environmental systems to satisfy the needs of future generations.

    According to the same authors making cities sustainable entails:
    – minimizing the consumption of space and natural resources;
    – rationalizing and efficiently managing urban flows;
    – protecting the health of the urban population;
    – ensuring equal access to resources and services;
    – maintaining cultural and social diversity.

    Strategies that should be considered in moving towards more sustainable urban environments have been defined in the Green Paper on the Urban

Environment produced by the European Commission.[69] Another very important report, *Sustainable Cities*,[70] has explored in detail the application of sustainable development in urban areas.

- The environmental quality of indoor spaces is a compromise between building physics applied during the building's design, energy consumption and outdoor conditions. As buildings have a long life, of several decades or even centuries, all decisions made at the design stage have long-term effects on the energy balance and the environment. Thus, the adaption of existing and new urban buildings to the specific environmental conditions of cities in order to incorporate efficient solar and energy-saving measures and to counterbalance the radical changes and transformations in the radiative, thermal, moisture and aerodynamic characteristics of the urban environment seems to be of very high priority. It is notable that more than 70% of the building-related investment in Western Europe is channelled to urban renewal and building rehabilitation.

- The improvement of indoor environmental quality in urban areas can be seen as a combination of acceptable indoor air quality with satisfactory thermal, visual and acoustic comfort conditions. As the outdoor environment is the main source of indoor pollution and noise, improvements in indoor environmental quality should be considered together with possible improvements in outdoor urban environmental quality.

- The reduction of the energy consumption of urban buildings by combining techniques to improve the thermal quality of the ambient urban environment with the use of up-to-date alternative passive heating, cooling and lighting techniques. These strategies and techniques have already reached a very high level of architectural and industrial acceptance.

None of the above should be seen as isolated areas of concern. The interrelated nature of the parameters that define the efficiency of urban buildings requires that theoretical, experimental and practical actions undertaken at the various levels should be part of an integrated approach.

REFERENCES

1. Smith, M., Whitelegg, J. and Williams, N. (1998). *Greening the Built Environment*. Earthscan, London.
2. Palz, W., Caratti, G. and Zervos, A. (1992). *Renewable Energy Development in Europe. Proceedings of Euroforum on New Energy Sources*, Paris.
3. Brugmann, J. (1992). *Managing Human Ecosystems : Principles for Ecological Municipal Management*. International Council for Local Environmental Initiative, ICLEI, Toronto, Canada.
4. Expert Group on the Urban Environment (1996). *European Sustainable Cities*. Commission of the European Communities, Luxembourg.
5. World Resources Institute (1996). *World Resources : A Guide to the Global Environment 1996–1997*. Oxford University Press, New York.
6. Botkin, B.D. and Keller, E.A. (1995). *Environmental Science. Earth as a Living Planet*. John Wiley and Sons, New York and Chichester.
7. United Nations (1991). *World Population Prospect*. United Nations, New York.
8. Owens, M. (1994). 'Building Small, Thinking Big'. *New York Times*, 21 July.

9. Stanners, D. and Bourdeau, P. (eds) (1995). *Europe's Environment – The Dobris Assessment*. European Environmental Agency, Denmark.
10. Brown, L.R. (1987). *State of the World : A World Watch Institute Report on Progress Toward a Sustainable Society*, Chapter 3. Worldwatch Institute, New York.
11. UNEP (1993). *Environmental Data Report 1993–1994. United Nations Environment Programme*. Blackwell, Oxford.
12. UNEP/WHO (1992). *Urban Air Pollution in Megacities of the World*. Blackwell, Oxford.
13. Escourrou, G. (1991) *Le Climat et la Ville*. Nathan University, Paris.
14. Hahn, E. and Simonis, U.E. (1991) 'Ecological Urban Restructuring'. *Ekistics*. Vol. 58, May–June/July–August, pp. 348–349.
15. Rees, W. (1992). 'Ecological Footprints and Appropriated Carrying Capacity: What Urban Economics Leaves Out'. *Environment and Urbanization*, Vol.4, Part 2 (October), pp. 121–130.
16. Sustainable London Trust (1996). *Sustainable London*. Sustainable London Trust, London.
17. Wackernagel, M. and Rees, W. (1996). *Our Ecological Footprint*. New Society Publishers, British Columbia.
18. Bank of America (1995). *Beyond Sprawl: New Patterns of Growth to Fit the New California*. Bank of America, San Francisco.
19. CPRE (1997). *Planning More to Travel Less*. CPRE, London.
20. Environmental Protection Agency (1992). *Cooling Our Communities. A Guidebook on Tree Planting and Light Colored Surfacing*. US Environmental Protection Agency.
21. Oke, T.R, Johnson, G.T., Steyn, D.G. and Watson, I.D. (1991). 'Simulation of Surface Urban Heat Islands under "Ideal" Conditions at Night – Part 2: Diagnosis and Causation'. *Boundary Layer Meteorology*, Vol. 56, pp. 339–358.
22. Santamouris, M., Papanikolaou, N., Livada, I., Koronakis, I., Georgakis, C., Argiriou, A. and Assimakopoulos, D. N., (1999). 'On the Impact of Urban Climate on the Energy Consumption of Buildings'. *Solar Energy*. In press.
23. Landsberg, H. E. (1981). *The Urban Climate*. Academic Press, New York.
24. Rao, T.R. and Murty Bh. V. R. (1973). 'Effect of Steel Mills on Rainfall at Distantly Located Stations'. *Indian Journal of Meteorology and Geophysics*, Vol. 24, Part 1, pp. 15–26.
25. Flavin, C. and Durning, A. (1988). *Raising Energy Efficiency. State of the World. Worldwatch Institute Report*. Worldwatch Institute, New York.
26. CEC (1990). *Passive Solar Energy as a Fuel*. The Commission of the European Communities. Directorate General XII for Science Research and Development, Brussels.
27. CEC (1996). *Energy in Europe, Annual Energy Review*, European Commission, Brussels.
28. CEC (1995). *Europe's Environment. Statistical Compendium*. Eurostat, Brussels.
29. Butler, D.J.G. (1989). *CFC and the Building Industry*. BRE Information Paper IP 23/89. BRE, Garston (UK).
30. CEC (1995). 'For a European Union Energy Policy – Green Paper'. European Commission, Brussels.
31. Santamouris, M. and Argiriou, A. (1994). 'Renewable Energies and Energy Conservation Technologies for Buildings in Southern Europe'. *International Journal of Solar Energy*, Vol. 15, pp. 69–79.
32. Boardman, B. (1991). *Fuel Property : From Cold Homes to Affordable Warmth*. Belhaven/ Wiley, London.
33. Jacobs, M. (1996). *The Politics of the Real World*. Earthscan, London.
34. CEC (1993). *Energy in Europe, Annual Energy Review*. Special Issue, European Commission, Brussels.
35. Kosareva, N.B. (1993). 'Housing Reforms in Russia. First Steps and Future Potential'. *Cities*, Vol. 10, Part 3, pp. 198–207.
36. OECD (1990). *Energy Information Administration. Annual Energy Review*, OECD, *Paris*.
37. DOE (1991). *Conservation and Renewable Energy Technologies for Buildings*. Solar Energy Research Institute for the US Department of Energy.
38. World Energy Council (1993). *Energy for Tomorrow's World*. St. Martin's Press, *Vienna*.

39. Jones, B. G. (1992). 'Population Growth, Urbanization and Disaster Risk and Vulnerability in Metropolitan Areas: A Conceptual Framework'. In Kreimer, Alcira and Mohan Munasinghe (eds), *Environmental Management and Urban Vulnerability*. World Bank Discussion Paper No, 168, World Bank, Washington.

40. LRC (1993). *London Energy Study Report*. London Research Centre, London.

41. ICLEI (1993). 'Draft Local Action Plans of the Municipalities in the Urban CO_2 Reduction Project'. International Council for Local Environmental Initiatives, Toronto.

42. IBGE (1993). *Constitution d'un Système d'Eco–Geo-information Urbain pour la Région Bruxelloise Integrant l'Energie et l'Air*. Convention Institut Wallon et Mens en Ruimte, Institut Bruxellois pour la Gestion de l'Environment. Brussels.

43. Tso, C.P. (1994). 'The Impact of Urban Development on the Thermal Environment of Singapore'. *Report of the Technical Conference on Tropical Urban Climates*. World Meteorological Organization, Dhaka.

44. Watanabe, H., Yoda, H. and Ojima H. (1990/91). 'Urban Environmental Design of Land Use in Tokyo Metropolitan Area'. *Energy and Buildings*, Vol. 15–16, pp 133–137.

45. Ojima, T. (1990/91). 'Changing Tokyo Metropolitan Area and its Heat Island Model'. *Energy and Buildings*, Vol. 15–16, pp. 191–203.

46. Flavin, C. and Lenssen, N. (1995). *Power Surge: A Guide to the Coming Energy Revolution*. Earthscan, London.

47. Browning, R, Helou, M. and Larocque, P.A (1998). 'The Impact of Transportation on Household Energy Consumption'. *World Transport Policy and Practice*, Vol. 4, Part 1.

48. Goodacre, C. (1998). 'An Evaluation of Household Activities and their Effect on End-user Energy Consumption at a Social Scale'. PhD Thesis, Geography Department, Lancaster University.

49. Roodman, D.M. and Lenssen, N. (1995). *A Building Revolution: How Ecology and Health Concerns are Transforming Construction*. Worldwatch Institute, Washington, DC.

50. Anink, D., Boonstra, C. and Mak, J. (1996). *Handbook of Sustainable Buildings*. James and James Science Publishers, London.

51. Harland, E. (1993). *Eco-Renovation*. Green Books, Dartington, UK.

52. Connaughton, J.N. (1993). 'Real Low-energy Buildings: The Energy Costs of Materials'. In S. Roaf (ed), *Energy Efficiency*, pp. 87–100. Oxford University Press, Oxford.

53. Tucker, R.K. and Treloar, G.J. (1994). 'Energy Embodied in Construction and Refurbishment of Buildings'. *First International Conference on Buildings and the Environment*, CIB and BRE, Watford.

54. Rosenfeld, A., Romm, J., Akbari, H. and Lioyd, A. (1998). 'Painting the Town White and Green'. Paper available through the Web site of Lawrence Berkeley Laboratory (http://www.pbl.gov).

55. Argento, V.K. (1988). 'Ozone Non-attainment Policy vs the Facts of Life'. *Chemical Engineering Progress*, pp. 50–54.

56. Kucera, V., Henriksen, J.F., Knotkova, D. and Sjostrom, C. (1993). 'Model for Calculations of Corrosion Cost Caused by Air Pollution and its Application in 3 Cities'. Paper 084, 10th European Corrosion Congress, 5–8 July 1993, Barcelona.

57. Plessas, D. (1980). *The Social Cost of Air Pollution in the Greater Athens Region*. KEPE, Athens.

58. OECD (1983). *Control Technology for Nitrogen Oxides in the Atmosphere*. OECD, Paris.

59. British Medical Association (1997). *Road Transport and Health*. British Medical Association, London.

60. Polionecki, J.D., Atkinson, R.W., de Leon, A.P. and Anderson, H.R. (1997). 'Daily Time Series for Cardiovascular Hospital Admissions and Previous Day's Air Pollution in London, UK'. *Occupational and Environmental Medicine*, Vol. 54, pp. 535–540.

61. Godish, T. (1989). *Indoor Air Pollution Control*. Lewis Publishers, New York.

62. Wallingford, K.M. and Carpenter, J. (1986). 'Field Experience Overview: Investigating Sources of Indoor Air Quality Problems in Office Buildings'. *Proceedings of the IAQ 86*. ASHRAE, Atlanta.

63. Argiriou, A., Asimakopoulos, D., Balaras, C., Daskalaki, E., Lagoudi, A., Loizidou, M., Santamouris, M. and Tselepidaki, I. (1994). 'On the Energy Consumption and Indoor Air Quality in Office and Hospital Buildings in Athens, Hellas. *Journal of Energy Conversion and Management*, Vol. 35, Part 5, pp. 385–394.

64. Clausen, G., Pejtersen, J. and Bluyssen, P. (1994). *Manual of the European Audit Project to Optimise Indoor Air Quality and Energy Consumption in Office Buildings*. Commission of the European Commission, Directorate General for Research and Development.

65. OECD (1991). *Fighting Noise in the 1990's*. OECD, Paris.

66. UBA (1988). *Larmbekampfung 88, Tendenzen, Probleme, Lasungen*. Umweltbundesamt, Berlin.

67. Calthorpe, P. (1993). *The Next American Metropolis*. Princeton Architectural Press, New York.

68. Katz, P. (1994). *The New Urbanism: Toward an Architecture of Community*. McGraw Hill, New York.

69. CEC (1990). 'Green Paper for the Urban Environment', COM(90), 218. European Commission, Brussels.

70. CEC (1994). *European Sustainable Cities*. Report of the EC Expert Group on the Urban Environment, Sustainable Cities Project, European Commission, Brussels.

2

———

Climate and climate change

D. N. Asimakopoulos
Department of Applied Physics, University of Athens

℘

It is recognized that the climate is continuously being changed and that this happens in several ways. Even a small variation in the solar radiation would be sufficient to affect the radiation and energy budget of the Earth's surface, owing to the resulting melting or increasing of the ice cover. Changes in the orbit of the Earth or movements of the continents give climate changes with time scales of more than 10,000 years. Eruptions of volcanoes have definite global impact, because of the emission into the atmosphere of a large number of particles that cause absorption or refraction into space of the solar radiation. Any change in one of the subsystems of the climate system can affect the behaviour of other subsystems, resulting in effects that may amplify or reduce the original change (feedback mechanisms).

There is also no doubt that man is capable of influencing climate through human activities of many different kinds. Thus, owing to the increase of the atmospheric content of carbon dioxide (industrialization), climatic change may occur as a result of global warming (the greenhouse effect) as there is increased absorption of infrared radiation in the atmosphere. Changes in ground cover (deforestation) or on the surface of the Earth (buildings, highways) are also important on a local scale, since they modify the albedo, the surface roughness, thermal and moisture behaviour etc.

Since the number of possible mechanisms related to climatic changes is rather large, over the same period the climate can be changed in different ways and at different rates in various parts of the world.

THE CLIMATE SYSTEM

In general terms the climate system consists of five physical components (see Figure 2.1):

- The *atmosphere*, which comprises the Earth's gaseous envelope. It has a characteristic response (thermal adjustment) time of the order of 10–100 days.
- The *hydrosphere*, which comprises the surface water (oceans, lakes, rivers) as well as the water beneath the Earth's surface. The upper layer of the ocean is coupled with the atmosphere with time scales of the order of months, while the deep ocean waters have a characteristic response time of the order of 1000 years.

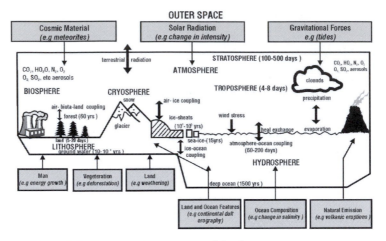

Figure 2.1. The physical components of the climatic system

- The *cryosphere*, which comprises the world's ice masses and snow deposits (glaciers, snow cover, ice sheets and sea-ice). The changes of the snow cover on the land are seasonal, while the glaciers and ice sheets change over periods ranging from hundreds to millions of years.
- The *lithosphere*, which includes the land masses over the surface of the Earth (mountains, ocean basin, surface rock and soil) with a relatively long characteristic response period.
- The *biosphere*, which includes the plant cover on land and in oceans as well as all the animals. Their response characteristics vary since they are sensitive to climate, but they can also influence climatic changes. Thus there are changes in surface vegetation resulting from changes in temperature and precipitation, but these changes in turn alter the surface albedo and roughness as well as rates of evaporation.

Solar radiation provides almost all of the energy that is needed from the climate system. Thus the rate at which heat is added to the system is of primary importance for the circulation of the atmosphere and oceans. This rate depends on the distribution of temperature and moisture in the atmosphere just as much as on the influence of the clouds on the solar and terrestrial radiation.

According to recent measurements, the solar constant (So), which is the solar energy per unit time per unit area perpendicular to the mean Earth–Sun distance, is about 1365–1372 W m^{-2}. Any change at this constant is an external force on the climatic system. For example, solar irradiance varies with sunspot cycle, being lowest at sunspot maximum.

Sunlight is mainly in the visible region of the spectrum, to which the atmosphere of the Earth is almost transparent.[1] If it is assumed that the Sun radiates as a 6,000 K black body, the wavelength of maximum emission is 0.5 mm in the spectral range from 0.1 to 10 μm. According to Figure 2.2, 7% of the Sun's radiation is in the ultraviolet ($\lambda \leq 0.4$ μm), 44% is in the visible region ($0.4 \leq \lambda \leq 0.7$ μm) and 37% is in the near-infrared. On the other hand, the cooler Earth with a surface temperature close to 300

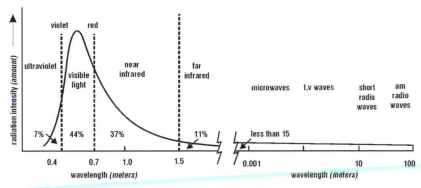

Figure 2.2. The electromagnetic spectrum of the Sun

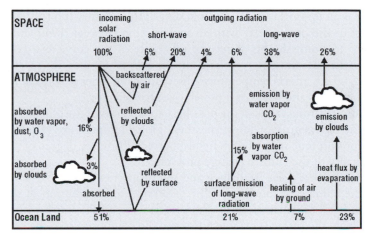

Figure 2.3. The energy balance of the Earth

K emits maximum radiation at wavelengths near 10 μm. Thus the Earth's radiation is called long-wave radiation, whereas the Sun's energy is referred to as short-wave radiation. Figure 2.3 gives the energy balance of the Earth if it is assumed that 100% incoming solar radiation represents 342 W m^{-2}. Thirty percent (105 W m^{-2}) of the incoming radiation is reflected, 20% by clouds, 6% by particles of the air and 4% by the Earth's surface. The other 70% (237 W m^{-2}) is absorbed by the Earth's surface (51%) and by water vapour, dust, ozone and clouds in the atmosphere. This gives a value of 0.3 to the planetary albedo (*a*), which is the ratio of the reflected to incoming radiation. In order to maintain the energy balance of the Earth, long-wave radiation has to be emitted from the Earth to space, equivalent to the 70% (237 W m^{-2}) of solar radiation that is absorbed. Thus there is an upward long-wave emission from the Earth's surface (390 W m^{-2}), of which 20 W m^{-2} escapes into space through the so-called atmospheric window (a region of the spectrum with low absorption by water vapour).

On the other hand, the atmosphere emits 327 W m^{-2} to the Earth's surface and 217 W m^{-2} to space, while absorbing 370 W m^{-2} in the infrared region and 68 W m^{-2} in the visible region (20% of the incoming solar radiation). Thus, the atmosphere is losing

106 W m^{-2} of radiating energy (radiative cooling), which is balanced by the corresponding heating given by sensible (16 W m^{-2}) and latent (90 W m^{-2}) heat transfer.

CLIMATE CHANGE

When we study the subject 'climate change', we are using temperature and precipitation as indicators of climate evolution. Of course, there are other changes that also have to be considered, such as changes in sea level or glacier extent. Thus, many climatic elements have to be used and these cause the characteristic variations over specific scales in time and space, which be the result of either internal or external effects.

As an example, Figure 2.4 gives the time series for the global mean temperature for the period 1861–1988. It can be seen that there are variations from year to year but overall there is an upward trend in the first part of this century (up to 1940) and this continues up to the present day with an interruption between 1940 and 1970. Of course, there are some doubts about the reliability of the curve as a result of the differences in the observations over space and time. Furthermore, the so-called 'heat-island' effect, which affects temperature observations in cities as a result of the heat capacity of buildings and roads, is a possible source of error.

Greenhouse warming is one of the possible explanations of the global temperature increase of the past century.[2,3] However, these global temperature variations, which are greater for different areas of each hemisphere, indicate the degree of uncertainty and the difficulty of the climate change problem.

The extraterrestrial solar irradiance and the properties of the Earth subsystems determine the climate of the Earth. The geographical distribution of the climate depends mainly on the inclination of the rotational axis of the Earth with respect to the orbital plane, the surface characteristics and the composition of the atmosphere.

There is a clear distinction between weather and climate. Weather is related to detailed instantaneous states of the atmosphere and their evolution from day to day. In contrast, climate is the total sum of the weather experienced over a long enough time period for the pattern to be established. Traditionally, the period for determining the 'normal climate' is 30 years. Thus, the climate is an ensemble average of climate states, together with some measure of its variability over a specific period of time. Differences between such climate states of the same kind over monthly, seasonal or annual time scales are referred to as climatic variations, including changes of the average or of the variability. Climate change, however, is a shift of the normal climate according to some factor, which may be known, lasting for many years. According to

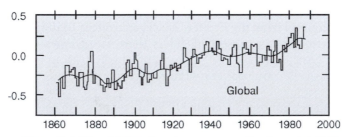

Figure 2.4. The global surface air temperature for the period 1861–1988

the above, a season-to-season or year-to-year effect, such as extremely wet or extremely dry conditions, can be described as a climatic variation if it is within the expected variations of the normal climate. The gradually melting of sea ice, however, as a result of warming may finally be interpreted as a climate change.

POSSIBLE CAUSES OF CLIMATE CHANGE

There are many theories that try to explain climate change, but none of them can give a complete explanation of all the changes. The major problem is the relationships and interactions among the physical components of the climate system. Thus, although we give here some possible causes of climate changes, only knowledge and under-standing of the majority of them can give a full understanding of the problem.

Variations in solar output

Recent measurements using radiometers on satellites suggest that solar energy, which is an input to our system, can vary considerably. Changes of the order of 0.1% of the total solar energy reaching the Earth have already been measured, within a period of less than 20 months. This kind of change could be linked to sunspot activity, which has a periodicity of 11 years. Sunspots are magnetic storms giving (or showing) cooler regions on the Sun's surface. Thus a sunspot maximum corresponds to a minimum of received solar energy. According to measurements during the period 1976 to 1980, the Sun's surface cooled by about 6°C corresponding to an increase in the number and the size of sunspots. These changes may alter the Earth's climate since, according to numerical climate models, a 0.5% change in solar output could be enough to change the climate. In addition, a decrease in solar energy of the order of 1% could lead to a decrease in the Earth's average temperature by 1.0°C. Of course, more satellite data are needed, so that the solar energy variations can be monitored to provide a better understanding of how solar energy is connected with climate change.

Variations in the Earth's orbit

Milankovitch's theory is the theory relating variations in the Earth's orbit to climate change. The basic concept of this theory is that changes in the Earth's orbit produce variations in the solar energy reaching the Earth's surface.

The first cause is changes in the shape (eccentricity) of the Earth's orbit. As shown in Figure 2.5, the Earth's orbit changes from nearly circular to elliptical and then back to circular. This cycle takes about 100,000 years to complete. If the eccentricity of the

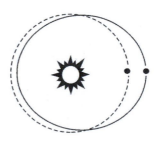

Figure 2.5. The change in the Earth's orbit from nearly circular (dotted line) to elliptical (solid line)

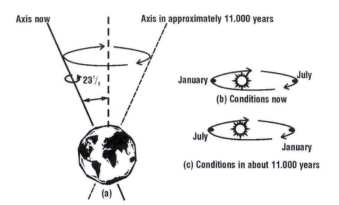

Figure 2.6. The Earth
wobbling like a spinning top

orbit is high, then the difference between the solar energy received at the top of the atmosphere at the two positions, closer and further away from the Sun, is also high.

We are now in a period of low eccentricity and the difference in distance between the closest and farthest distances of the Earth from the Sun is 3%, giving a 7% difference in solar energy between July and January. If the difference in distance were 9%, which is the case with high eccentricity, then the difference in solar energy could be 20%. A more eccentric orbit will also change the length of the seasons.

The second cause is the fact that the Earth wobbles, like a spinning top, as it rotates on its axis. According to the Milankovitch theory, this wobble has a period of about 22,000 years. As shown in Figure 2.6, at present the Earth is closer to the Sun in January and farther away in July. After about 11,000 years the Earth will be closer to the Sun in July, when the Northern hemisphere exhibits summer.

The third cause, according to the Milankovitch theory, is the fact that there are changes in the tilt of the Earth with a period of the order of 41,000 years. In Figure 2.6 the angle of the Earth's orbital tilt is currently 23.5°, but it can vary from about 22.5° to 24.5°. If the Earth's tilt has a smaller value, then the seasonal variations between summer and winter will be less. Thus summers will be cooler and winters will be milder.

Surface modifications

Modifications of the Earth's surface may cause climate changes. The major surface modification is the slow shifting of the continents and the ocean floors. According to the theory of plate tectonics, the outer shell of the Earth is composed of plates that are moving in relation to one another. The rate of motion is extremely slow, a few centimetres per year. This theory can explain facts such as the past existence of glacial features close to Africa and different paths of ocean currents because the continents have been arranged differently. The rearrangements change the transport of heat and the global wind system. Furthermore, if large areas of open water are covered with ice, the magnitudes of the sensible and latent heat fluxes are reduced. Finally, as will be shown in later sections, modification of the surface may influence the microclimate of certain areas. For example, overgrazing of grasslands or deforestation of large areas may increase the surface albedo and cause an increase in desert conditions (a process known as desertification).

Feedback mechanisms

Positive and negative feedback mechanisms reflect the ability of the Earth's atmosphere system to check and balance any forces influencing the system, in order that it can readjust to a new equilibrium. For example, if we assume that the Earth's temperature is exhibiting a slow increase over a long period, this results in an increase in the air's capacity to hold water vapour. More water is therefore evaporated and the increased amount of water vapour absorbs more of the infrared energy emitted from the Earth. This results in an increase in the greenhouse effect, increasing in turn the air temperature, which increases the evaporation, and so on. This mechanism is known as a positive feedback mechanism since the initial positive change in temperature results in a strengthening of this temperature increase as a result of a chain reaction.

It should be noted that this example is intended to give a general idea about the mechanism and we are not interested here in how the initial temperature increase started. It should also be mentioned that there is always a negative feedback mechanism that balances the positive feedback mechanism. For example, more moist warm air could lead to an increase in global cloudiness. This increase gives a higher albedo and thus a higher fraction of solar energy is reflected and there is less energy to warm the Earth's surface, so that a cooling of the surface begins. This is a negative feedback mechanism since the initial positive change in temperature finally results in a temperature decrease. Additionally, this cooling allows for more snow cover in middle and high latitudes and, since snow has a higher albedo, the solar energy reflected back into space will be greater, resulting in a further drop in surface temperature. This will create more snow cover and therefore a further temperature decrease. This is another positive feedback mechanism of the Earth–atmosphere system, which of course is counteracted by another negative feedback mechanism.

Variations within the atmosphere

Climate changes attributed to the constituents of the atmosphere either block out some of the incoming solar radiation or trap a fraction of the Earth-radiated infrared energy.

The greenhouse effect

If the energy balance of the Earth is considered, the main factor that determines the global temperature is solar radiation. The Earth re-emits long-wave radiation proportional to the fourth power of its absolute temperature. Thus, there is compensation between the emitted infrared radiation and the absorbed solar radiation. If we take into account the incident solar radiation that is absorbed (70%) and the emitted planetary long-wave radiation, then an effective planetary temperature can be estimated, which is about −18°C (225 K). However, since the actual surface temperature is about 1.5°C, the difference between surface and upper atmosphere is due to the fact that the infrared radiation emitted by the surface is absorbed in the lower levels of the atmosphere, mainly by water vapour and carbon dioxide. This absorbed energy is re-emitted back to the surface and thus the surface is heated to a temperature higher than the effective planetary temperature. This phenomenon is called the 'greenhouse effect'.

The greenhouse effect on the Earth can be identified in terms of the difference between the energy emitted by the Earth's surface and the energy emitted back into space by the upper atmosphere. Thus, it is the long-wave energy, which is trapped in the atmosphere and by feedback mechanisms, that responds to climate changes. This effect is attributed to the property of greenhouse gases that absorb strongly in the infrared region of the electromagnetic spectrum.[4]

It is not only CO_2, but also the other greenhouse gases, such as CH_4, N_2, O, O_3 and FCCs, that contribute to this phenomenon. These gases absorb infrared energy at wavelengths corresponding to the water-vapour spectral window (9–15 μm). Thus, an increase in atmospheric CO_2 decreases the transmissivity of the atmosphere in this spectral window and reinforces the greenhouse effect. In other spectral regions an increase of CO_2 also leads to higher radiation absorption, back-radiation and a higher surface temperature.

Climate models indicate that for twice the CO_2 concentration, the downward infrared flux will increase by 4 W m^{-2}. Of course, in order to estimate surface temperature changes in relation to increased atmospheric CO_2 concentration, one also has to consider a number of feedback mechanisms. For example, a significant increase in atmospheric CO_2 content leads to an increase in the temperature of the air at the Earth's surface that is due to the following feedback mechanisms:

- *Water-vapour feedback.* The increase in the surface temperature increases evaporation and therefore the water vapour content of the atmosphere and the water vapour absorbs infrared radiation even more intensely than CO_2. Finally the re-radiation of absorbed energy is resulting in an increase of air temperature.
- *Ice albedo feedback.* At high latitudes the increase in surface temperature causes ice and snow to melt, resulting in a decrease in the albedo, which in turn increases the absorption of solar radiation by the surface and thus an increase in surface temperature.
- *Cloud feedback.* Clouds influence the planetary albedo and the absorption of infrared radiation by the water and water vapour contained within them. The temperature increase influences the amount and the height of the clouds, leading to positive or negative feedback; which type of feedback it is very difficult to ascertain. For example, if we assume that the albedo effect dominates the greenhouse effect (infrared radiation), then increased cloudiness will result in an overall cooling and vice-versa.

Calculations that are part of climate models indicate that a doubling of the CO concentration will lead to an increase in the average global temperature by 1.5 to 4°C. Also, it has been estimated that for the next 30 years temperature changes due to other greenhouse gases may be about the same as those due to CO_2. Other models predict greater than the average warming and precipitation in high latitudes. It is obvious that the large number of possible estimates and feedback mechanisms lead to very complicated solutions and many uncertainties.

Particles in the atmosphere

Small particles of dust (most smaller than 5 μm in diameter), resulting from human activities, may remain suspended in the troposphere for several days. If there is a constant rate of emission of such particles into the atmosphere, a climate change may result. Depending on the size, shape, colour and vertical distribution of the particles and on the surface albedo, there are different actions that may take place. For example, if the particles are bright relative to the Earth's surface, more solar energy will be scattered back into space, causing a decrease in air temperature. However, if they are relatively dark compared to the Earth's surface, the higher absorption by the particles will increase the air temperature. Furthermore, the same particles may absorb infrared radiation emitted from the Earth's surface. This amplifies the greenhouse effect, leading to a warming of the air. Thus it is obvious that the mechanisms related to particles in the atmosphere are very complicated and more research has to be done.

Volcanic eruptions

During volcanic eruptions, ash particles, dust and gases are ejected into the stratosphere. If these eruptions are rich in sulphur gases, they can have a great influence on climate change. While, volcanic ash falls out of the stratosphere quickly and has no serious effect, the sulphur gases react with water vapour, producing bright sulphuric acid particles that gradually form a dense haze layer. This layer, which may remain for a long period, absorbs and reflects back into space a fraction of the solar energy. According to mechanisms mentioned previously, this could lead to a decrease in the temperature of the Earth's surface. Although the calculation is very complex, mathematical models predict that large volcanic eruptions may give an average temperature drop of the order of 0.2 to 0.5°C over a period of three years.

Variations of land surface albedo due to vegetation

The surface albedo is the ratio of the reflected to the incident solar radiation. Changes in surface albedo are caused either by the extent of snow cover and ice or by a surface change due to desertification, deforestation or greenhouse warming. These changes introduce a variability into the climatic system, with the appearance of corresponding feedback mechanisms.

It should be mentioned that there is a difference between incoming solar flux reaching the top of the atmosphere and the flux reaching the surface. This is attributed to the absorption scattering and reflection mechanism involving water vapour, ozone, dust, air and clouds. Thus the solar radiation reaching the ground is mainly direct radiation plus diffuse radiation, generally about 75% of the incident solar radiation at the top of the atmosphere. Changes in the surface albedo modify the planetary albedo, which is the ratio of radiation reflected from the system made up of the Earth and its atmosphere to the incident solar radiation.

Climate changes may therefore be caused by changes in the land surface albedo, which, although it represents only 29% of the total surface of the Earth, is the part that varies most over both time and space. The surface albedo depends on the wavelength of the radiation, the elevation of the Sun and the nature of the Earth's surface. Figure 2.7 gives the spectral albedo over various areas, with and without vegetation. From this figure it is evident that the albedo of soil depends to a high degree on the type, colour, roughness and moisture content of the soil. There is a range from less than 0.1 to greater than 0.5. The albedo of snow and ice similarly shows a high variability, ranging from 0.8 (fresh dry snow) to 0.1, depending on the density, thickness and contamination of the snow and the stage of the ice.

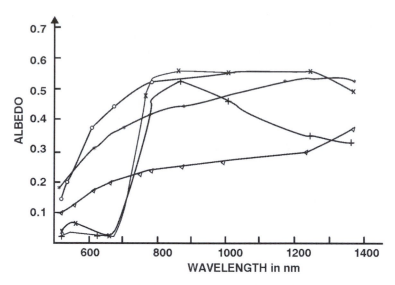

Figure 2.7. The spectral albedo over various surfaces: yellow dune (o),
bright sand (♦), dark sand (Δ), grassland (x) and wheat (+)

The results of climate models and observations seem to indicate that increased albedo resulting from removal of vegetation can decrease cloudiness and precipitation since there is a net radiative loss. On the other hand, a decrease in albedo due to increased vegetation gives a positive feedback with increased rainfall. To distinguish the feedback mechanisms that involve surface vegetation and climate, we have to consider the effect of climate change on vegetation and soil, the sensitivity of the climate to land surface changes and possible feedback mechanisms between the climate system and the above-mentioned changes. There are numerous climate scenarios concerning the changes in vegetation due to climatic changes, but it is very difficult to take into account either the possible feedback resulting from these changes or the adaptation of plants to the changing environment. The two most important quantities are connected with the land surface and the changes in vegetation; these are the surface albedo and the ratio of heat flux to the net solar radiation at the surface.

According to climatic models, if we assume an average land surface albedo of 0.25, with an albedo change of +1%, the global temperature effect will be –0.032 K. This is, of course, a very small change, but we have to take into account the fact that a greater increase in albedo is quite possible and also that there may be geographical differences in the relative insolation in different areas.

If we now consider the absorbed solar energy that is converted into heat, we have to separate this into three different heat fluxes. One part of the heat is directed into the soil, heating the ground. This fraction of the energy is stored during the day and, during the night, is released to the atmosphere by radiative cooling. The remaining energy is converted into sensible and latent heat fluxes, leaving a very small portion of energy to be used for photosynthesis. The sensible heat flux transports heat by turbulence from the Earth' surface to upper layers of the atmosphere. The magnitude of the sensible heat flux depends on the thermal gradient between the Earth's surface

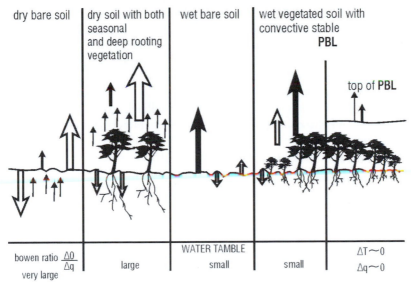

Figure 2.8. The three types of heat flux over different terrains. The heat flux into the soil is shown by the stippled arrows, the sensible heat flux by hatched arrows and the latent heat flux byfull arrows

and the atmosphere, as well as on the wind speed, which is an additional source of energy for turbulence generation. The latent heat flux (water-vapour flux) penetrates the atmospheric boundary layer without any heating effect and releases energy at the condensation level, where clouds are formed. In Figure 2.8 the relative magnitudes of the three types of heat flux are shown over different terrains. The surface is heated by the absorbed solar radiation until the different types of flux are in equilibrium.

In dry climates the latent heat flux is very small compared with the sensible heat flux. However, if vegetation is present, then the water-vapour flux is high, even over dry soil, as a result of the transpiration mechanism through the leaves of the vegetation. In the case of a moist surface, the latent heat flux is the largest of the energy fluxes. Furthermore, because wet surfaces have a lower albedo than dry surfaces, more solar radiation is absorbed and converted into heat. The surface temperature does not increase much because most of this heat is latent heat.

Figure 2.8 also shows the difference between convective and stable planetary boundary layers, in relation to heat fluxes. In the second case, where small vertical gradients of temperature and moisture exist, the values of fluxes are very small and only heating of the surface or an increased wind speed can change this situation.

With a climatic change, the processes at the surface can be influenced in different ways. Changes in cloudiness or precipitation can affect the hydrological cycle as well as the albedo. With less precipitation, vegetation will be reduced and albedo will be increased, resulting finally in less solar short-wave radiation being converted into (mainly) sensible heat flux. Changes in surface temperature affect the vegetation but not the magnitudes of the fluxes, since they depend upon temperature gradients. An increase in surface albedo leads to a reduction of the solar input, which in turn leads to the decrease in the sensible heat flux.

If the vegetation of an area is removed or is changed, then the supply of water to the atmosphere through the latent heat flux will change, resulting in a change of the area's precipitation. This may then lead to a local climate change, which may produce drying of the surface, the final stage being desertification. Changes of vegetation may change the roughness of the surface, which is important in calculating heat fluxes that depend on the roughness parameter Z_0.

From these scenarios, it is evident that there is feedback on different scales between climate change and the changes in land surface characteristics and that it is very difficult and uncertain to predict these changes quantitatively.

The most important climatic variables for the structure of natural vegetation are:

- *Radiation*: the intensity and extreme values of both short- and long-wave radiation; day length; seasonal changes.
- *Temperature*: mean and extreme values; daily fluctuations; seasonal changes.
- *Precipitation*: mean annual total; seasonal changes; state (rain, snow or ice).
- *Atmospheric properties*: air moisture; CO_2 level; the levels of other gases. In addition, the impact of vegetation on climate depends mainly on relevant biospheric properties, such as the structure of ecosystems, water use, water storage and metabolism mechanisms such as gas exchange (CO_2).

Climate modification in the urban environment

When we examine the climate of a small area, several square kilometres in size, then we are looking at the mesoclimate of the area. Such areas could be valleys, forests, beaches or towns. On this scale, human activities are the major influence on climate change, for example, creating the urban environment, which may differ considerably from the rural regions around it.

Urban climate

Most cities are sources of heat and pollution and the thermal structure of the atmosphere above them is affected by the so-called 'heat-island' effect.[5] The heat that is absorbed during the day by the buildings, roads and other constructions in an urban area is re-emitted after sunset, creating high temperature differences between urban and rural areas. The greatest temperature differences are observed during the night since the heat island is attributed mainly to urban–rural cooling, rather than to heating differences, especially in the period around sunset (see Figure 2.9). Additional city heat is given off by vehicles and factories, as well as by industrial and domestic heating and cooling units. It has been observed that 'cities' with populations of thousands have maximum temperature differences from the surrounding rural area of 2 to 3K, while cities with a population of one million can have temperature differences of 8 to 12K. Sometimes during a calm night with large temperature excesses, a light breeze, called a 'country breeze', due to the formation of a low-pressure area over the city, blows from the rural to the urban area. This wind may transport more pollutants into the city if there are big industries in the surrounding countryside.[6]

Thus an urban boundary layer (UBL) is created in the lower atmosphere above an urban area, with micrometeorological characteristics determined by the city.[7] During the day, the UBL increases in depth as a result of the city warming, and reaches heights

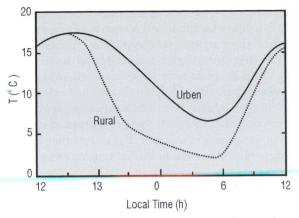

Figure 2.9. Idealized diurnal temperature cycle for urban and rural areas

that are about 25% greater than the mixing height over rural areas. At night, the UBL retains a surface mixed layer, giving rise to layering above this level, since the surrounding rural area is characterized by a strong surface-based radiation inversion (Figure 2.10). Thus with weak winds and intense inversions that suppresses vertical mixing, episodes of severe air pollution can occur (Figure 2.11).

Pollution influences the climate of a city. The particles reflect solar radiation, leading to a decrease in the solar energy that reaches the surface. Some particles also serve as nuclei upon which water vapour condenses, even at 70% relative humidity, forming haze and increasing the frequency of city fog. Precipitation and cloudiness may be greater in cities than in rural areas. This is due to the added nuclei and to the excess heat that reduces air stability and enhances vertical air motions. Thus, in an urban area, we expect to have higher levels of pollution than in a rural area, higher values of temperature, precipitation, cloudiness and frequency of fog and lower values of solar energy reaching the surface, relative humidity, wind speed and visibility.

Cities also influence the climate of areas that are downwind from them. The annual precipitation in such areas is increased as a result of the additional transport of particles or moisture emitted by industries close to this area.

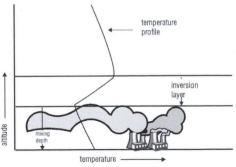

Figure 2.11. The inversion layer prevents vertical mixing, leading to increased pollutant concentrations

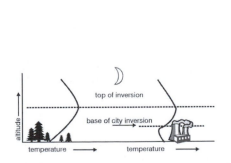

Figure 2.10. Night-time temperature profile

Acid deposition

Air pollutants, such as oxides of sulphur and nitrogen, can be transported over great distances downwind of their release position. These pollutants either reach the surface in dry form (dry deposition) or are removed from the air and are carried to the ground by means of rain or snow (wet deposition or acid rain). Acid deposition contains both dry and wet deposition. Sulphur dioxide and oxides of nitrogen that are on the ground are transformed into acids when they react with water. Similarly, air pollutants that remain aloft may be transformed into drops of sulphuric acid (H_2SO_4) and nitric acid (HNO_3) and fall to the Earth. Because of this, precipitation in many different areas of the world is becoming acidic, causing many problems. It is well known that Sweden suffers from acid precipitation that comes from factories in the eastern part of England. Acid deposition may affect plants and water resources, as is the case in Germany, where more than 200,000 acres of woodland have been seriously damaged. In addition, thousands of lakes in different countries have been affected by acid deposition. The foundations of structures, building surfaces, monuments and other structures in many cities have also been seriously affected. This is thus a major environmental problem that will become even more serious if adequate precautions are not taken.

REFERENCES

1.   Ahrens, C.D. (1985). *Meteorology Today*. West Publishing Company, New York.
2.   Duplessy, J.C.,Pons, A. and Fantechi, R. (1990). *Climate and Global Change*. Proceedings of the European School of Climatology and Natural Hazards course, Arles/Rhone, 4–12 April. *European Commission, DG XII*.
3.   Fantechi, R., Maracchi, G. and Almeida-Teixeira, M.E. (1988). *Climatic Change and Impacts. A General Introduction*. Proceedings of the European School of Climatology and Natural Hazards course, Florence, 11–18 September. *European Commission, DG XII*.
4.   Oppenheimer, M. (1989) 'Greenhouse Gas Emissions Environmental Consequences and Policy Responses'. *Climatic Change Journal*, Vol. 15, Special issue, pp. 335.
5.   Oke, T.R. (1982). 'The Energetic Basis of the Urban Heat Island'. *Quarterly Journal of the Royal Meteorological Society*, Vol. 108, pp. 1–24.
6.   Plate, E.J. (1982). *Engineering Meteorology*. Elsevier Scientific, Oxford.
7.   Stull, R.B. (1988). *An Introduction to Boundary Layer Meteorology*. Kluwer Academic, Dordrecht and London.

3

Wind patterns in urban environments

M. Santamouris,
Department of Applied Physics, University of Athens

ℰᴐ

WIND DISTRIBUTION IN THE URBAN ENVIRONMENT

The urban wind field is complicated. Small differences in topography can cause irregular air flows. As the air flows from the rural environment to the urban environment, it must adjust to the new boundary conditions defined by the cities (Figure 3.1). This adjustment results from the higher-level development flow field and the uniqueness of local effects, such as topography, building geometry and dimensions, streets, traffic and other local features, like trees. In a general way, the wind speed in the canopy layer significantly decreases relative to the undisturbed wind speed. This was first observed and documented by Kremser,[1] and then by many authors,[2-6] for a two-layer vertical structure. Oke characterized the wind variation with height over cities by defining two specific sublayers,[7] the so called 'obstructed sublayer', or urban canopy sublayer, which extends from the ground surface up to the height of the buildings, and the so-called 'free surface layer', or urban boundary layer, which exists above the roof tops. The obstructed or canopy sublayer has its own flow field, driven and determined by the interaction with the local features. A very detailed discussion of the problems related to the air flow in the urban canopy layer is given by Landsberg.[8]

Particular intention has to be given to the air flow during the night. Heat islands in cities create an unstable rate of temperature drop and thus a convergence of air flow into urban areas is observed. This flow is rarely centripetal, because the irregularity

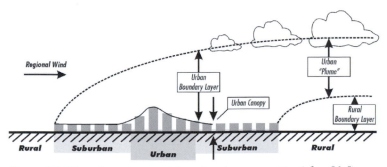

Figure 3.1. The urban canopy and boundary layer over cities (after Oke[7])

of the urban area and the differences with the surroundings prevent such a flow.[8,9] Night wind flow is not even stable because the temperature in the urban environment changes abruptly over short distances. This leads to short bursts of wind from the country to the urban area.[10] In theory, this type of flow may result in uplift above the city core and a high-level counterflow from the city to the rural areas.[7] Studies in Paris,[11] concerning this country breeze, show a correlation between the wind speed measured at the top of the Eiffel tower and the temperature difference between the city centre and the surrounding rural areas. It has been found that when the wind speed is lower than 2 m/s, the temperature difference between the centre and the periphery increases compared to the other days.

This chapter is devoted to the general characteristics of the air flow in the canopy layer. It also serves as an introduction to Chapter 7, where the particular air-flow patterns in urban streets are discussed in detail.

WIND SPEED CHARACTERISTICS

The Earth's surface has a significant influence on the wind speed. Classical fluid-flow theory states that, under ideal conditions, the influence of the surface causes the wind speed vectors in successive layers, up to the layer where friction ceases, to have an effect on the wind. This wind that is not influenced by friction is called the geostrophic wind and is given by:

$$V_G = -1/(\rho f)\ \partial p/\partial x, \tag{3.1}$$

where V_G is the geostrophic wind, r is the atmospheric density, f is the Coriolis parameter equal to 2Ω sin, where Ω is the angular velocity and φ the latitude, and $\partial p/\partial x$ is the horizontal pressure gradient from higher to lower pressure. The relationship between the geostrophic wind and the wind speed u_z at any height z, in the layers below is given by:

$$u_z - V_G = 1/f\ \ d(\tau_z/\rho)/dz, \tag{3.2}$$

where τ_z is the shear stress.
In the obstructed sublayer, the variation of wind with height is described by the exponential law used to describe air flow beneath forest canopies:[12–14]

$$u = U_0 e^{z/Z_0}, \tag{3.3}$$

where U_0 is a constant reference speed, (see Figure 3.1), and Z_0 is the roughness length of the obstructed sublayer calculated from the expression:

$$Z_0 = h_b D^*/z_0, \tag{3.4}$$

where D^* is an effective diameter of air space between obstacles and can be tentatively approximated for the city by:

$$D^* = 0.1h_b.$$

The variation of mean wind speed u with height z in the free surface layer above the roof tops is given by the following logarithmic law:[13]

$$\hat{u} = u^*/k\left[\ln(z+d+z_0)/z_0\right]. \tag{3.5}$$

This assumes a unidirectional mean flow with only vertical shear u_z so that $u_x = u_y = v = w = 0$. In equation (3.5), k is the Karman constant ($k = 0.38$), d is the zero-plane displacement and u^* is the frictional velocity given by:

$$u^* = C_g V_G, \tag{3.6}$$

where V_G is the geostrophic wind speed and C_g is the geostrophic drag coefficient given by:

$$C_g = 0.23/\log R_\infty, \tag{3.7}$$

where R_∞ is the surface Rossby number for thermally neutral atmospheres given by:

$$R_\infty = G/z_0\, F, \tag{3.8}$$

where z_0 is the aerodynamic roughness length and f is the Coriolis parameter. The aerodynamic roughness z_0 may be calculated from the expression proposed by Lettau:[15]

$$z_0 = 0.5h_b\left(A^*/A'\right), \tag{3.9}$$

where A^* and A' are the typical silhouette area and the average lot area (number of roughness elements divided by the city area), respectively.

Typical values of z_0, as given by Oke,[7] are shown in Table 3.1.

The zero-plane displacement parameter d is calculated from:

Table 3.1. Typical roughness length z_0 of urbanized terrain (from Oke[7])

Terrain	z_0 (m)
Scattered settlement (farms, villages, trees, hedges)	0.2–0.6
Suburban	
Low-density residences and gardens	0.4–1.2
High density	0.8–1.8
Urban	
High density, < five-storey row and block buildings	1.5–2.5
Urban high density plus multistorey blocks	2.5–10

$$d = z_0 - (h_b + z_0), \tag{3.10}$$

where:

$$x \ln x = 0.1(h_b)^2 / (z_0)^2. \tag{3.11}$$

Both d in equation (3.5) and U_0 in equation (3.3) are determined by the requirement that the log law and the exponential law must give the same u and $\hat{u}$ at height $z = h_b$.

Both the logarithmic and exponential profiles are mathematical idealizations and do not hold at a particular point in the city. Instead, they represent an average for roughness z_0 over the entire city or city sector (Figure 3.2).

Estimation of the wind speed in a city is of vital importance for passive cooling applications and especially for the design of naturally ventilated buildings. The above discussion has indicated that wind speeds measured above the buildings or at airports differ considerably from the speed at an urban monitoring site. As the roughness length z_0 is greater in an urban area than in the surrounding countryside, the wind speed u at any height z is lower in the urban area, and much lower within an obstructed area.

Simultaneous measurements of the wind speed in various canyons in Athens, Greece, are reported by Santamouris *et al.*[17] Data for the measured wind speed in the Ippokratous canyon, as well as the simultaneous measurements at the airport, are given in Figure 3.3, while simultaneous measurements of the wind speed in and above the Mavromihali canyon are given in Figure 3.4. As shown, although the wind speed above the canyon may reach values up to 5 m/s, within the canyon the air speed never exceeds 1 m/s.

Analysis of the frequency distribution, given in Figure 3.5, of the wind speed measured at the Ippokratous and at the airport shows clearly that although the wind speed at the airport may vary up to 6 m/s the wind speed within the canyon never exceeds 1 m/s.

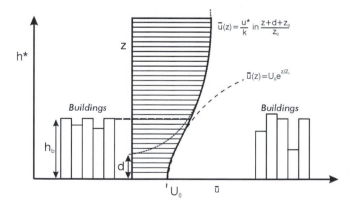

Figure 3.2. Logarithmic and exponential wind profiles in the surface layer, above and below building height (after Nicholson[16])

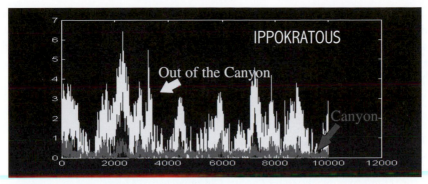

Figure 3.3. Wind speed in and above the Ippokratous canyon, Athens, Greece (source Santamouris et al.[17])

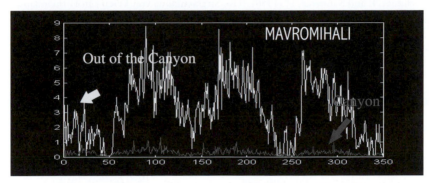

Figure 3.4. Wind speed in and above the Mavromihali canyon, Athens, Greece (source Santamouris et al.[17])

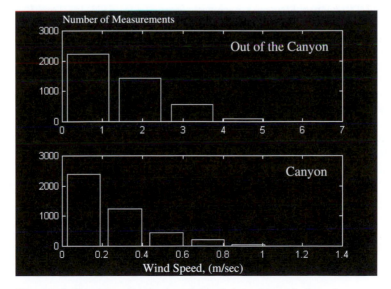

Figure 3.5. Frequency distribution of the wind speed in and above the Ippokratous canyon, Athens, Greece (source Santamouris et al.[17]).

Air speed in urban areas may be higher than in rural areas under two specific conditions. The first is when the high-speed air layers are either deflected downwards by tall buildings or channelled as 'jets' along canyons in the same direction as the flow. The second is when the induced low-level flow from the country, discussed previously, is strong enough to overcome the frictional drag of the canyon walls.[7]

REFERENCES

1. Kremser, V. (1909). 'Ergebnisse vieljahriger Windregistrierungen in Berlin'. *Meteorologische Zeitschrift*, Vol. 26, pp. 238–252.
2. Rubinstein, E.C. (1979). *Odnorodnost meteorologicheskikh Ryadov vo Vremeni I Prodstanstve v Svya zi c Issledovanien izmeneniya Klimata.* Gidrometeozdat, *Moscow.*
3. Zanella, G (1976). 'Il Clima Urbano di Parma'. *Rivista Meteorologia e Aeronautica*, Vol. 36, pp. 125–146.
4. Dirmhim, I. and Sauberer, F. (1959). 'Das Strassenklima von Wien'. In F. Steinhauser, Eckel, O. and Sauberer, F. (eds), *Klima und Bioklima won Wien*, Part III, Ch. 4, pp. 122–135. *Vienna.*
5. Steinhauser, F., Eckel, O. and Sauberer, F. (1959). *Klima und Bio Klima von Wien*, Part III. *Welter Leben. Vienna.*
6. Frederick, R.H. (1961).'A Study of the Effect of Tree Leaves on Wind Movement'. *Monthly Weather Review*, Vol. 89, pp. 39–44.
7. Oke, T.R. (1987). *Boundary Layer Climates.* Cambridge University Press.
8. Landsberg, H. (1981). *The Urban Climate.* Academic Press, New York and London.
9. Stummer, G. (1939). 'Klimatische Untersuchungen in Frankfurt am Main und seinen Vororten'. *Berichte Meteoroligische und Geophysikalische Institut, Universität Frankfurt*, Vol. 5.
10. Schmauss, A. (1925). 'Eine Miniaturpolarfront'. *Meteorologische Zeitschrift*, Vol. 42, p. 196.
11. Escourrou, G. (1991). *Le Climat et la Ville.* Nathan University Editions, Paris.
12. Cionco, R. (1965). 'A Mathematical Model for Air Flow in a Vegetative Canopy'. *Journal of Applied Meteorology*, Vol. 4, pp. 517–522.
13. Cionco, R. (1971). 'Application of the Ideal Canopy Flow Concept to Natural and Artificial Roughness Elements'. Technical report ECOM-5372, US Army Electronics Command, Fort Monmouth, NJ.
14. Inoue, E. (1963). 'On the Turbulent Structure of Air Flow within Crop Canopies'. *Journal of the Meteorological Society of Japan*, Ser. II, Vol. 41, pp. 317–326.
15. Lettau, H.H. (1970). 'Physical and Meteorological Basis Mathematical Models of Urban Diffusion Processes'. *Proceedings of Symposium on Multiple Source Urban Diffusion Models*, pp. 2.1–2.26, Chapel Hill, NC. *University of North Carolina Publications, North Carolina.*
16. Nicholson, S.E. (1975). 'A Pollution Model for Street-level Air'. *Atmospheric Environments*, Vol. 9, pp. 19–31.
17. Santamouris, M., Papanikolaou, N., Livada, I., Koronakis, I., Georgakis, C., Argiriou, A. and Assimakopoulos, D. N. (1999). 'On the Impact of Urban Climate on the Energy Consumption of Buildings'. *Solar Energy*, in the press.

4

Thermal balance in the urban environment

M. Santamouris,
Department of Applied Physics, University of Athens

℘

The thermal balance in the urban environment differs substantially from that of rural areas. The anthropogenic heat released by cars and combustion systems, the higher amounts of solar radiation stored and the blockage of the emitted infrared radiation by urban canyons makes the global thermal balance more positive and contributes to the warming of the environment.

In this chapter and those that follow urban energy fluxes are discussed in detail. The specific characteristics of the anthropogenic heat released in cities, as well as the characteristics of the solar and infrared radiation received and emitted by cities, are analysed. Specific information and details on the thermal balance of urban canyons is given in Chapter 6.

URBAN ENERGY FLUXES

The energy balance of the 'Earth surface's–ambient air' system in the urban environment is governed by the energy gains and losses, as well as by the energy stored in the opaque elements of the city, mainly buildings and streets. In general:

Energy gains = Energy losses + Energy storage.

Energy gains involve the sum of the net radiative flux Q_r, in the form of both solar radiation and long-wave radiation emitted by the opaque elements (building, streets, etc.), as well as the anthropogenic heat Q_T related to transportation systems, power generation and other heat sources.

Energy losses take the form of sensible heat Q_E or latent heat Q_L, resulting from heat convection between the opaque surfaces and the air as well evapotranspiration. Losses can also occur because of the advective heat flux between the urban and the surrounding environments. Thus the energy balance of the surface–air system can be written as (Figure 4.1):

$$Q_r + Q_T = Q_E + Q_L + Q_s + Q_A, \tag{4.1}$$

where Q_s is the stored energy and Q_A is the net energy transferred to or from the system through advection in the form of sensible or latent heat. The advective term can be

(a)

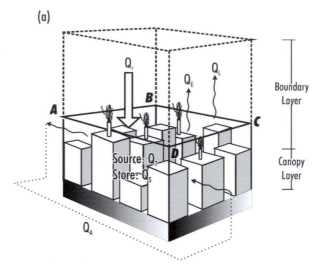

Figure 4.1. Schematic presentation of the energy fluxes in the urban environment (after Oke[1])

ignored in central urban areas surrounded by an almost uniform building density, but may be important in the boundary areas between the urban and the rural environments.

Specific information and data on the thermal balance of urban canyons are given in Chapter 6.

Both the radiation and the anthropogenic terms of equation (4.1) are discussed in the following chapters. A comparison betwen the energy balances in suburban and rural environments is provided by Cleugh and Oke for suburban and grassland sites in and around Vancouver in Canada, for a period of 30 summer days.[2] Table 4.1 gives the daily totals of the more important terms of equation (4.1) as measured or estimated in both environments.

As shown in the table, the radiative flux is the most important term and is slightly higher in the rural environment. During the daytime, the net radiative input is higher in the suburban environment, although night losses are much higher and compensate daytime gains. The sensible heat flux is the main mechanism for dissipating the daytime radiative surplus in the suburban environment. Sensible heat remains positive during the late afternoon hours, contributing significantly to the heating of the

Table 4.1. Energy balance of suburban and rural areas in Vancouver, Canada during a period of 30 summer days (after Cleugh and Oke[2]). Values are in MJ/m²/day

Term	Suburban	Rural
Q_r	11.8	13.0
Q_E	6.4	4.0
Q_L	4.2	8.7
Q_s	1.2	0.3

urban atmosphere. The higher stored energy in the suburban environment is mainly due to there being a greater surface area for absorption than in the rural environment, where rural vegetation cover provides a type of insulation.[2] Problems related to heat storage in the urban environment, as well as those concerned with the properties of the materials used and their impact on the energy balance of cities, are considered in Chapter 11. Problems related to evapotranspiration are discussed in Chapter 10. It should be noted, however, that the latent heat in urban areas is seriously reduced. The energy content of evapotranspiration is close to 597 kcal/g of evaporated water at 0°C, and 575 kcal/g at 40°C. In urban areas the total latent heat is reduced according to the morphology and the characteristics of the city and in particular is a function of the existing green areas. Escourrou reports that the reduction of evaporation is a function of the 'waterproofness' of the urban soil.[3] Thus, evaporation is reduced by 19% when 25% of the urban soil is waterproof, by 50% when the figure is 38% of the surface and by close to 75% when 59% of the soil is waterproof.

THE ROLE OF ANTHROPOGENIC HEAT

Anthropogenic heat is mainly related to transportation systems, power generation and other heat sources. Anthropogenic heat in urban areas may significantly affect the ambient temperature and increase the heat-island intensity. It is characteristic that in New York anthropogenic heat is almost twice the solar radiation input,[3] while in Barcelona human-generated energy is equivalent to about a fifth of the total solar radiation.[4]

Escourrou, on the basis of Miller's research, gives indicative values for the strength of the various sources of anthropogenic heat.[3] In particular, heating and lighting produces 25 kcal/cm^2/day, urban circulation 9 kcal/cm^2/day and industry close to 8 kcal/cm^2/day, while human metabolism produces 1 kcal/cm^2/day. Other estimates of the anthropogenic heat rate for some US city centres range from 20 to 40 W/m^2 in summer and from 70 to 210 W/m^2 during winter.[5] Oke reported that the mean annual magnitude of the heat source for typical temperate-latitude cities lies between 15 and 50 W/m^2 for a horizontal unit area of the city.[6] Gutman and Torrance suggested a range between 84 and 167 W/m^2 for mid-latitude cities in winter,[7] while Atwater suggested constant fluxes of 92 and 17 W/m^2 for the winter and summer period respectively.[8] Kerschgens and Drauschke have reported measurements from Bonn, Germany.[9] For the winter period, they found a diurnal pattern comprising two maximum inputs of 45 W/m^2 at 08:00 and 17:00 and a minimum of 30 W/m^2 from midnight until 06:00. Torrance and Shum reported an artificial heat flux with a daily mean magnitude of 54 W/m^2 resulting from transportation and domestic electrical use in summer.[10] Taha gives the anthropogenic heating rate for certain US, European and Asia cities (Table 4.2) and also includes net radiation at all wavelengths.[11]

There are contrasting estimates of the effect of artificial heat release on the urban temperature. Based on a diurnal anthropogenic heating profile involving heat rejection from buildings and motor vehicles, Taha et al. estimated the impact of the anthropogenic heat on ambient temperature.[18] Their simulations show that the anthropogenic heat in the centre of a large city can create a heat island of up to 2–3K during both day and night.

Table 4.2. *Anthropogenic heating and net radiation at all wave-lengths in selected cities. These are annual average values within the urban limits of the cities. The data do not include suburban or rural surrounding areas. (Source: Taha,[11] with primary data taken from Hosler and Landsberg,[5] Oke,[6] Chandler,[12] Flohn,[13] Gay and Stewart,[14] McNaughton and Black,[15] Dabberdt and Davis,[16] and Mayer and Noack[17])*

City	Anthropogenic rate (W/m²)	Net radiation at all wavelengths (W/m²)
Chicago	53	–
Cincinnati	26	–
Los Angeles	21	108
Fairbanks	19	18
St. Louis	16	–
Manhattan, NewYork City	117–159	93
Moscow	127	–
Montreal	99	52
Budapest	43	46
Osaka	26	–
Vancouver	19	–
West Berlin	21	57

Oke *et al.* simulated the impact on the maximum heat-island intensity of anthropogenic heat coming from the heating of buildings.[19] For various street-canyon geometries they found that the heat-island intensity is much higher at low ambient temperatures and can reach values between 2 and 8K.

Gutman and Torrance have reported that, for a theoretical city, an invariant flux of 125 W/m² during the night in winter produces a heat island of 2.9 K,[7] while Oke and East have reported for Montreal, a city most nearly comparable to the physical characteristics of Gutman and Torrance's theoretical city, a heat island of 4K.[20] Atwater, on the basis of simulations, revealed urban temperature excesses of, respectively, 2.6 and 9.7 K for daily maximum and minimum temperatures in winter and 0.2 and 0.1 K in summer.[21] Finally, Swaid,[22] on the basis of the heat addition rate proposed by Grimmond,[23] has simulated the hourly air-temperature increase due to heat addition and found that this causes an almost invariant diurnal rise in temperature of 0.7 K.

SOLAR, ATMOSPHERIC AND TERRESTRIAL RADIATION

The radiation balance R in the urban environment is the sum of the incoming absorbed short- and long-wave radiation minus the long-wave radiation emitted by all the components of the Earth's surface:

$$R = (I_b + I_d)(1 - a) - I_l \qquad (4.2)$$

where I_b and I_d are the beam and diffuse solar radiations incident on the surface of a city, while a is the mean albedo of the city to the solar radiation. The term I_l refers to the long-wave radiation balance:

$$I_l = I_l\uparrow - I_l\downarrow \qquad\qquad (4.3)$$

where $I_l\uparrow$ is the long-wave radiation emitted by the city surfaces and $I_l\downarrow$ is the long-wave atmospheric radiation absorbed by the city structure.

Solar radiation and sunshine duration in the urban environment are seriously reduced as a result of the increased scattering and absorption by particulates in the urban atmosphere. As mentioned by Landsberg,[24] the sunshine duration in industrial cities is reduced by 10 to 20% in comparison with the surrounding countryside and similar losses are observed in received energy. Oke mentioned that on especially polluted days and at times of low solar elevation, the reduction in solar radiation may be in excess of 30%.[6]

Urban pollution drastically affects both the spectral composition and the direction of the incoming solar radiation, while, because of the increased scattering and of the characteristics of the scattering agents, diffuse radiation increases while visibility is reduced and the colour of the sky changes. In particular ultraviolet wavelengths are much more affected than the visible and the infrared. Measurements by Maurain have shown that in Paris ultraviolet radiation is reduced by a factor of ten, while in the visible and the infrared wavebands the reductions are only 7% and 4%, respectively.[25] Results from Leicester, UK[26] and Los Angeles[27] show that ultraviolet radiation is reduced by 30% and 50%, respectively, compared to the surrounding countryside. Important reductions in the visible waveband have also been reported by Steinhauser et al. for Vienna.[28] Losses are about 10% in summer and about 18% in winter.

Several studies have shown that solar radiation and sunshine duration are seriously reduced by the urban atmosphere. According to Hufty,[29] Liege in Belgium loses 55 minutes of sunshine per day because of the increased pollution. In the inner part of London, the sunshine duration is about 16% less than in the surrounding countryside.[30] Unsworth and Monteith reported that in Britain the solar radiation flux in urban areas is only 82% of the lowest value found in rural areas.[31] East has reported that solar radiation losses for Montreal and Toronto are 9% and 7% respectively.[32] Similarly, Nishizawa and Yamashita have reported that solar radiation losses in Tokyo are between 12% and 30%.[33] Studies in the USA have shown that the direct solar radiation decreases between 30% and 50%, according to the atmospheric turbidity, while the diffuse solar radiation increases between 40% and 70%.[34] In total, the global solar radiation decreases between 10% and 20%.

As shown in equation (4.1), absorbed solar radiation is a direct function of the mean albedo of a city. Use of high-albedo materials reduces the amount of solar radiation absorbed through building envelopes and urban structures and thus keeps their surfaces cooler. More information and data on the albedo and the spectral characteristics of materials used in the urban environment are given in Chapter 11.

The atmosphere emits thermal radiation as a result of the vibrational and rotational transitions of the asymmetric molecules of the various constituents of the air, such as water vapour, carbon dioxide and ozone. Radiation is emitted in the infrared spectrum, except in the spectral region from 8 to 13 μm, known as the 'atmospheric window', and has a spectral distribution very similar to that of a black body at a temperature equal to the dry-bulb temperature of the air close to the ground. The deviations from black-body radiation are due to the emissivity of the sky, $\varepsilon_s(\lambda)$.

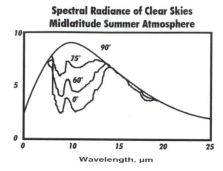

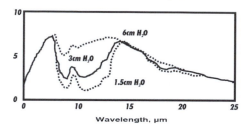

Figure 4.2. Emission spectrum of a clear atmosphere; data from Clark[38]

Figure 4.3. Effect of water vapour on the infrared emission spectrum depth of the atmosphere (source: Clark[38])

Atmospheric radiation received in urban environments has been extensively studied and the role of pollutants has been investigated.[8,35–37] Under standard clear-sky conditions during summer, the spectral distribution of the atmosphere has the distribution shown in Figure 4.2. This spectrum corresponds to a dry-bulb temperature of 21°C and a wet-bulb temperature of 16°C. The spectrum is given for four zenith angles, 0°, 60°, 75° and 90°. The emission within the atmospheric window is due to the water molecules. The effect of the water vapour on the infrared emission spectrum depth of the atmosphere is important, as illustrated in Figure 4.3.

Atmospheric pollutants including aerosols 'close' the atmospheric window, because they absorb and emit at these wavelengths, and thus increase the long-wave flux from the atmosphere. The sensitivity of the emission spectrum of the atmosphere to the urban and maritime aerosols is demonstrated in Figure 4.4.

Knowledge of the atmospheric radiation incident on the urban environment comes either from direct measurements performed by a few stations or from using the Stefan–Boltzmann law. If it is taken into account that the atmosphere is not characterized by a unique temperature, the expression:

$$I\downarrow = \sigma T_{sky}^4 \tag{4.4}$$

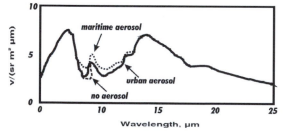

Figure 4.4. Sensitivity of the emission spectrum of the atmosphere to the atmospheric aerosol (source: Clark[38])

cannot be applied directly. In equation (4.4), T_{sky} is the sky temperature and is defined as the black-body temperature that gives the same spectral distribution as the sky. To overcome this problem, the sky emissivity is defined as:

$$\sigma T_{sky}^4 = \varepsilon_s \sigma T_a^4, \tag{4.5}$$

where σ is the Stefan–Boltzmann constant and T_a is the absolute ambient temperature close to the ground.

To calculate the sky emissivity under clear skies ε_{sc}, Berdhal and Martin have proposed the following expression:[39]

$$\varepsilon_{sc} = 0.711 + 0.56(T_{dp}/100) + 0.73(T_{dp}/100)^2, \tag{4.6}$$

where T_{dp} is the dew-point temperature.

The sky emissivity under cloudy sky conditions can be calculated from the expression proposed by Clark:[38]

$$\varepsilon_s = \varepsilon_{sc}(1 + 0.0224n - 0.0035n^2 + 0.00028n^3), \tag{4.7}$$

where n is the total amount of opaque cloud, ($n = 0$ for clear skies and $n = 1$ for completely overcast skies).

An empirical model to calculate the downward infrared radiation emitted by the atmosphere (W/m^2) has been proposed by Aida and Yaji:[40]

$$I\!\downarrow = \sigma T_a^4 \big[0.127 + (-0.114\Gamma^3 - 0.168\Gamma^2 - 0.173 + 0.603) \\ \times (0.0000438e^3 - 0.001e^2 + 0.123e + 0.05)^{0.107} \big], \tag{4.8}$$

where e is the surface water vapour pressure in mbar and Γ is the vertical temperature gradient (K/mbar).

The long-wave radiation emitted by buildings, streets and all emitting surfaces in the canopy layer may be calculated using the Stefan–Boltzmann law:

$$I\!\uparrow = e\sigma T^4, \tag{4.9}$$

where T is the temperature of the specific surface. A detailed discussion of the emissivities of the materials used in the urban environment, as well as of the surface temperature distribution in cities, is given in Chapter 11.

REFERENCES

1. Oke, T.R. (1988). *Boundary Layer Climates*. Routledge, London.
2. Cleugh, H.A. and Oke, T.R. (1986). 'Suburban–Rural Energy Balance Comparisons in Summer for Vancouver, B.C.'. *Boundary Layer Meteorology*, Vol. 36, pp. 351–169.
3. Escourrou, G. (1991). *Le Climat et la Ville*. Nathan University Editions, Paris.

4. Pares, M. *et al.* (1985). *Descombrir el Medi Urba. Ecologia d' Una Ciutat: Barcelona*. Centre del Medi Urba, Barcelona.
5. Hosler, C.L. and Landsberg, H.E. (1977). 'The Effect of Localized Man-made Heat and Moisture Sources in Mesoscale Weather Modification'. *Energy and Climate*. National Academy of Sciences, Washington, DC.
6. Oke,T.R (1988). 'The Urban Energy Balance'. *Progress in Physical Geography*, Vol. 12, p. 471.
7. Gutman, D.P. and Torrance, K.E. (1975). 'Response of the Urban Boundary Layer to Heat Addition and Surface Roughness'. *Boundary Layer Meteorology*, Vol. 9, pp. 217–233.
8. Atwater, M.A. (1971). 'The Radiation Budget for Polluted Layers of the Urban Environment'. *Journal of Applied Meteorology*, Vol. 10, pp. 205–214.
9. Kerschgens, M.J. and Drauschke, R.L. (1986). 'On the Energy Budget of Wintry Mid-latitude City Atmosphere', *Beitrage zur Physik der Atmosphare*, Ser. B, Vol. 59, pp. 115–125.
10. Torrance, K.E. and Shum, J.S.W. (1976). 'Time Varying Energy Consumption as a Factor in Urban Climate'. *Atmospheric Environment*, Vol. 10, pp. 329–337.
11. Taha, H. (1997). 'Urban Climates and Heat Islands : Albedo, Evapotranspiration, and Anthropogenic Heat'. *Energy and Buildings*, Vol. 25, pp. 99–103.
12. Chandler, T.J. (1964). 'City Growth and Urban Climates'. *Weather*, Vol. 19, pp. 170–171.
13. Flohn, H. (1971). 'Saharization: Natural Causes or Management'. *Environmental Report No 2*, WMO Rep. No 312, pp. 101–106. World Meteorolgical Organization, Geneva.
14. Gay, L.W. and Stewart, J.B. (1974). *Energy Balance Studies in Coniferous Forests*. Rep. No. 23, Institute of Hydrology, Natural Environment Research Council, Wallingford, Berks, UK.
15. McNaughton, K. and Black, T.A. (1973). 'A Study of Evaportanspiration from a Douglas Forest using the Energy Balance Approach'. *Water Resources Research*, Vol. 9, p. 1579.
16. Dabberdt, W.F. and Davis, P.A. (1978). 'Determination of Energetic Characteristics of Urban – Rural Surfaces in the Greater St-Louis Area'. *Boundary Layer Meteorology*, Vol. 14, p. 105.
17. Mayer, H and Noack, E.M. (1980). 'Einfluss der Schneedecke auf die Strahlungsbilanz im grossraum Munchen'. *Meteorologische Rundschau*, Vol. 33, p. 65.
18. Taha, H., Akbari, H., Sailor, D. and Ritschard, R. (1992). 'Causes and Effects of Heat Islands: Sensitivity to Surface Parameters and Anthropogenic Heating'. Lawrence Berkeley Laboratory, Rep. 29864, Berkeley, CA.
19. Oke, T.R., Johnson, G.T., Steyn, D.G. and Watson, I.D. (1991). 'Simulation of Surface Urban Heat Islands under "Ideal" Conditions at Night – Part 2 : Diagnosis and Causation'. *Boundary Layer Meteorology*, Vol. 56, pp. 339–358.
20. Oke, T. R. and East, C. 91971). 'The Urban Boundary Layer in Montreal'. *Boundary Layer Meteorology*, Vol. 1, pp. 411–437.
21. Atwater, M.A. (1972). 'Thermal Effects of Urbanization and Industrialization in the Boundary Layer: A Numerical Study'. *Boundary Layer Meteorology*, Vol. 3, pp. 229–245.
22. Swaid, H. (1993). 'Urban Climate Effects of Artificial Heat Sources and Ground Shadowing by Buildings'. *International Journal of Climatology*, Vol. 13, pp. 797–812.
23. Grimmond, C.S.B. (1992). 'The Suburban Energy Balance: Methodological Considerations and Results for a Mid Latitude West Coast City under Winter and Spring Conditions'. *International Journal of Climatology*, Vol. 12, pp. 481–497.
24. Landsberg, H. E. (1981). *The Urban Climate*. Academic Press, New York and London.
25. Maurain, C.H. (1947). *Le Climat Parisien*. Presses Universitaires, Paris.
26. Department of Scientific and Industrial Research (1947). 'Atmospheric Pollution in Leicester: A Scientific Survey'. Atmospheric Pollution Research Technical Paper, p. 161, London.
27. Peterson, J.T., Flowers, E.C. and Rudisill, J.H. (1978). 'Urban–Rural Solar Radiation and Atmospheric Turbidity Measurements in the Los Angeles Basin'. *Journal of Applied Meteorology*, Vol. 17, pp. 1595–1609.

28. Steinhauser, F., Eckel, O. and Sauberen, F. (1955). 'Klima und Bioklima von Wien. 1. Teil'. *Wetter Leben*, Special issue, p. 17.

29. Hufty, A. (1970). 'Les Conditions de Rayonnement en Ville'. In *Urban Climates*, WMO Techn. Note. No. 108, pp. 65-69. World Meteorogical Organization, Geneva.

30. Chandler, T.J. (1965). *The Climate of London*, p. 122. Hutchinson, London.

31. Unsworth, M.H. and Montheith, J.L. (1976). 'Aerosol and Solar Radiation in Britain'. *Quarterly Journal of the Royal Meteorological Society*, Vol. 98, pp. 778–797.

32. East, C. (1968). 'Comparison du Rayonnement Solaire en Ville et à la Campagne'. *Cahiers de Geographie du Quebec*, Vol. 12, pp. 81–89.

33. Nishizawa, T. and Yamashita, S. (1967). 'On Attenuation of the Solar Radiation in the Largest Cities'. *Japanese Progress in Climatology, Tokyo*, pp. 66–70.

34. Davenport, A.G. (1968). 'The Dependances on Wind Loads on Meteorological Parameters'. In *Procedings of Symposium on Wind Effects on Structure*, National Physical Laboratory, pp. 53–102. HMSO, London.

35. Bergstrom, R.W. Jr and Viskanta, R. (1973). 'Modeling the Effects of Gaseous and Particulate Pollutants in the Urban Atmosphere, Part I : Thermal Structure'. *Journal of Applied Meteorology*, Vol. 12, pp. 901–912.

36. Viskanta, R., Bergstrom, R.W. and Johnson, R.D. (1977). 'Radiative Transfer in a Polluted Urban Planetary Boundary Layer'. *Journal of Atmospheric Science*, Vol. 34, pp. 1091–1103.

37. Viskanta, R. and Daniel, R.A. (1980). 'Radiative Effects of Elevated Pollutant Layers on Temperature Structure and Dispersion in an Urban Atmosphere'. *Journal of Applied Meteorology*, Vol. 19, pp. 53–70.

38. Clark, G. (1981). 'Passive/Hybrid Comfort Cooling by Thermal Radiation'. In Bowen, A., Clark, E. and Labs, K. (eds), *Proceedings of the International Passive and Hybrid Cooling Conference*, Miami Beach, FL, pp. 682–714.

39. Martin, M. (1989). 'Radiative Cooling'. In J. Cook (ed.), *Passive Cooling*, Ch. 4, pp. 138–196. MIT Press, Cambridge, MA.

40. Aida, M. and Yaji, M. (1979). 'Observations of Atmospheric Downward Radiation in the Tokyo Area'. *Boundary Layer Meteorology*, Vol. 16, pp. 453–465.

5

Heat-island effect

M. Santamouris,
Department of Applied Physics, University of Athens

ॐ

DEFINITIONS AND PHYSICAL PRINCIPLES

Air temperatures in densely built urban areas are higher than the temperatures of the surrounding rural country (Figure 5.1). The phenomenon known as the 'heat island' was first noticed by meteorologists more than a century ago,[1] and is the most well documented phenomenon of climatic modification. According to Landsberg, the heat island is present in every town and city and is the most obvious climatic manifestation of urbanization.[2]

Higher urban temperatures have a serious impact on the electricity demand for air conditioning of buildings and increase smog production, as well as contributing to increased emission of pollutants from power plants, including sulphur dioxide, carbon monoxide, nitrous oxides and suspended particulates.

The phenomenon of the heat island is characterized by an important spatial and temporal variation related to climate, topography, physical layout and short-term weather conditions. A detailed description of the more important factors influencing heat islands is summarized by Oke *et al.*:[3]

- *Canyon radiative geometry* contributes to decreasing the long-wave radiation loss from street canyons as a result of the complex exchange between buildings and the screening of the skyline. Infrared radiation is emitted from the various building and street surfaces within the canyons. Buildings replace a fraction of the cold sky hemisphere with much warmer surfaces, which receive a high portion of the infrared radiation emitted from the ground and radiate back an even greater amount.
- *Thermal properties of materials* may increase storage of sensible heat in the fabric of the city during the daytime and release the stored heat into the urban atmosphere after sunset. Furthermore, the replacement of natural soil or vegetation by the materials, such as concrete and asphalt, used in cities reduces the potential to decrease ambient temperature through evaporation and plant transpiration.
- *Anthropogenic heat* is released by the combustion of fuels from either mobile or stationary sources, as well as by animal metabolism.

- *The urban greenhouse effect* contributes to the increase in the incoming long-wave radiation from the polluted urban atmosphere. This extra radiative input to the city reduces the net radiative drain.
- *Canyon radiative geometry* decreases the effective albedo of the system because of the multiple reflection of short-wave radiation by the canyon surfaces.
- *The reduction of evaporating surfaces* in the city puts more energy into sensible heat and less into latent heat.
- There is *reduced turbulent transfer* of heat from within streets.

The heat-island phenomenon may occur during the day or during the night. The intensity of the heat island is mainly determined by the thermal balance of the urban region and can result in a temperature difference of up to 10 degrees. IPCC has compiled data from various cities so as to be able to assess the impact of the heat island.[4] The data show that the effect is quite strong in large cities. The temperature increase due to the heat island varies between 1.1 K and 6.5 K (see Table 5.1).

Table 5.1. Heat-island effects in some cities[4]

City	Temperature increase, (K)
30 US cities	1.1
New York	2.9
Moscow	3–3.5
Tokyo	3.0
Shanghai	6.5

Regarding the variation of ambient temperature with distance between the rural area and the city centre, it is well known that for a large city with a cloudless sky and light winds just after sunset, the boundary between the rural and the urban areas exhibits a steep temperature gradient to the urban heat island (Figure 5.1), and then

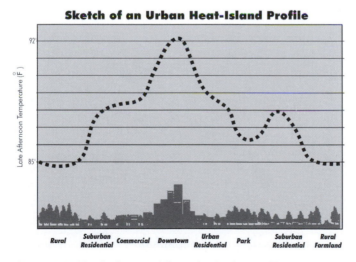

Figure 5.1. Sketch of a typical heat-island urban profile

the rest of the urban area appears as a 'plateau' characterized by a weak gradient of increasing temperatures and finally a peak at the city centre, where the urban maximum temperature is found. However, it should be pointed out that heat-island patterns are significantly determined by the unique characteristics of each city. The difference between the maximum urban temperature and the background rural temperature is defined as the urban *heat-island intensity*.[5]

Numerous studies have been performed to analyse and understand the heat island. Most of the studies concentrate on night heat islands during the winter period, but a few of the studies analyse the daytime temperature field and summer heat islands.

HEAT-ISLAND PARAMETRIC MODELS

Heat-island studies are mainly aimed at understanding the role of the main parameters influencing the temperature increase in cities. Studies have concentrated on the role of city size and population, weather conditions such as cloud cover, wind speed, humidity, urban canyon characteristics, etc.

Oke has correlated the heat-island intensity to the size of the urban population.[6] He proposed two different regression lines for North American and European cities (Figure 5.2). As shown, the expected heat-island intensity for a city of one million inhabitants is close to 8 K in Europe and 12 K in the USA. Oke explains that the higher values for the American cities are because the centres of North American cities have taller buildings and higher densities than typical European cities.[6]

Based on the above data for North American cities, Oke suggested the following formula for calculating the heat-island intensity, near sunset under cloudless skies, as a function of the population and the regional wind speed:[6]

$$dT = P^{0.25}/(4U)^{0.5},\qquad\qquad(5.1)$$

where dT is the heat island intensity in K, P is the population and U is the regional non-urban wind speed in m/s at a height of 10 m. The influence of wind on the heat-island

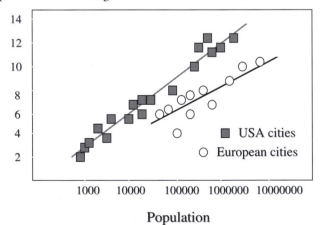

Population

Figure 5.2. The maximum difference between urban and rural temperatures for US and European cities[6]

intensity was studied by Escourrou for the city of Paris.[7] He found that the heat-island intensity decreases for increasing wind speeds. Table 5.2 gives the heat-island intensity as a function of the wind speed. For winds higher than 5 m/s the temperature difference was not important.

Table 5.2. *Heat-island intensity and corresponding wind speeds in the greater Paris area*[7]

Wind speed in rural areas (m/s)	Heat-island intensity (K)
1	4.5
2	3.4
3	3.4
4	2.6
5	2.2

Jauregui has supplemented the work of Oke by providing data from various cities located in South America and India (Figure 5.3).[8] As shown in the figure, the heat island in these cities is weaker. According to Jauregui, this phenomenon can be attributed in part to the difference in morphology (physical structure) between South American and European cities. Park has updated Oke's figures by including data from Korea and Japan, Figure (5.4).[9] As shown in this figure the heat-island intensity is much lower in these countries.

Ludwig has suggested a formula for calculating the urban–rural temperature difference dT as a function of the corresponding lapse rate (in K/millibar) over the rural area Y:[10]

$$dT = 1.85 - 7.4Y. \tag{5.2}$$

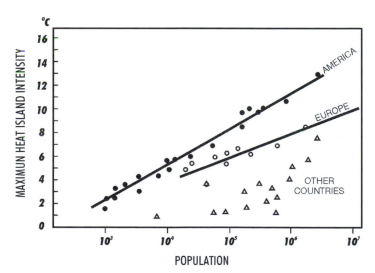

Figure 5.3. *The maximum difference between urban and rural temperatures for US and European cities*[6] *together with data for tropical and mid-latitude cities*[8]

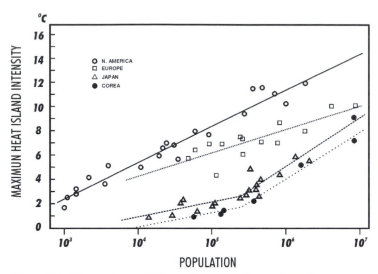

Figure 5.4. The maximum difference between urban and rural temperatures for US and European cities[6] together with data for Korea and Japan[9]

This equation is based on statistical analysis of measurements.

Bornstein[11] cited another statistical model, developed by Dundborg, which relates the nocturnal heat island in Uppsala, Sweden, to various meteorological elements: cloudiness (N), wind speed (U), temperature (T) and specific humidity (q). The suggested formula is:

$$dT = 2.8 - 0.1N - 0.38U - 0.02T + 0.03q. \tag{5.3}$$

Summers, using data from Montreal, correlated wind speed with the heat-island intensity and proposed the following expression:[12]

$$Dt = \left(2r\frac{\theta T}{\theta z}Qu\right)\Big/\rho c_p u, \tag{5.4}$$

where r is distance from the upwind edge of the city to the centre, $\theta T/\theta z$ is the potential temperature increase with height z, Qu is the urban excessive heat per unit area, ρ is the air density, c_p is the specific heat and u is the wind speed.

Oke,[13] in addition to equation (5.1) correlating the heat-island intensity with local wind speed and population, has also suggested another formula correlating the maximum heat-island intensity with the geometry of the 'urban canyon', as expressed by the relationship between a building's height (H) and the distance between buildings (W), namely the ratio (H/W):

$$dT = 7.54 + 3.97\ln(H/W) \tag{5.5}$$

or, in terms of the sky view factor of the middle of the canyon floor, Y_{sky}:

$$dT = 115.27 = 13.88Y_{sky}. \qquad (5.6)$$

These formulae express the concept that the urban heat island is caused by reduced radiant heat loss to the sky from ground level in densely built urban centres as a result of the restricted view of the sky.

HEAT-ISLAND STUDIES IN EUROPE

Various studies on the intensity of heat islands have been carried out for many European cities. Lyall,[14] reporting on the heat-island effect in London in June–July 1976, noted that the magnitude of the nocturnal heat island averaged over the two months was of the order of 2.5 K. This is not far below the daily upper decile limit of 3.1 K found by Chandler in a comparison of Kensington and Wisley from 1951–60.[15]

Urban–rural temperature traverses in Goteborg, Sweden showed a well developed urban heat island of magnitude 5 K.[16] The data show an urban heat island, ranging from 3.5 K in winter and 6 K in summer. During the summer season for nearly all the night hours, the heat-island intensity was greater than 0.5 K and, for 40% of the night hours, it was greater than 1 K. Measurements in Malmo, Sweden were performed during the winter and spring periods and a mean heat-island intensity close to 7 K was found.[17]

Limited data on the heat-island intensity in Essen, Germany were reported for September 1986.[18] The observed heat-island intensity was between 3 and 4 K for both day and night. Heat-island studies for the small German town of Stolberg, with 60,000 inhabitants and located in a valley,[19] reported high excess temperatures between the urban and rural areas, 6 K at night. These high temperature differences are because the heavily built-up town centre, which is situated in such a narrow valley, obstructs the flow of cold air at night, so that it cannot penetrate into the urban area. Measurements of the heat island in Fribourg, Germany, have demonstrated that the heat-island intensity is close to 10 K.[20]

Escourrou recorded a horizontal thermal differential between Paris, France and its suburbs of close to 14 K.[21]

For Basle and Berne in Switzerland, the heat-island intensity was close to 6 K, while for Biel and Fribourg the value was 5 K, and for Zurich close to 7 K.[22] Finally, Abbate,[23] using satellite data for Rome, Italy, has reported important temperature differences between high-density urban areas and low-density urban and agricultural areas.

HEAT-ISLAND STUDIES IN NORTH AMERICA

Several studies have been performed in the USA, aimed at investigating the heat island in urban environments. The most recent are those performed by NASA and the Lawrence Berkeley Laboratory (LBL).[24,25]

Akbari *et al.* of LBL have presented trends in absolute urban temperatures in several cities in California and elsewhere in the USA.[26] In the absence of summer data, the overall analysis is based on the use of average annual and maximum annual temperatures. Table 5.3 summarizes the measured temperature trends and the type of temperature data used.

The observed increase of temperature in some North American cities becomes much more apparent when the cooling degree days at urban and rural stations are compared. Taha has given the increase of the cooling and heating degree days due to urbanization and heat-island effects for selected North American locations (Tables

Table 5.3. Measured temperature trends in selected cities[26]

City	Trend (°F/decade)	Type of temperature data
Los Angeles	1.3	Highs
Los Angeles	0.8	Means
San Francisco	0.2	Means
Oakland	0.4	Means
San Jose	0.3	Means
San Diego	0.8	Means
Sacramento	0.4	Means
Washington	0.5	Means
Baltimore	0.4	Means
Fort Lauderdale	0.2	Means

5.4 and 5.5).[27] As can be seen in the tables, the difference in the cooling degree days can be as high as 92%, while the minimum difference is close to 10%. For the heating degree days, the maximum difference is close to 32%, while the minimum difference is close to 6%. The increase in the cooling degree days has a tremendous impact on the energy consumption of buildings for cooling. The possible increase in the cooling load is discussed in Chapter 7.

Table 5.4. The difference in the cooling degree days due to urbanization and heat-island effects – averages for selected locations for the period 1941–1970[27]

Location	Urban	Airport	Difference (%)
Los Angeles	368	191	92
Washington DC	440	361	21
St Louis	510	459	11
New York	333	268	24
Baltimore	464	344	35
Seattle	111	72	54
Detroit	416	366	14
Chicago	463	372	24
Denver	416	350	19

Table 5.5. The difference in the heating degree days due to urbanization and heat-island effects – averages for selected locations for the period 1941–1970[27]

Location	Urban	Airport	Difference (%)
Los Angeles	384	562	−32
Washington DC	1300	1370	−6
St. Louis	1384	1466	−6
New York	1496	1600	−7
Baltimore	1266	1459	−14
Seattle	2493	2881	−13
Detroit	3460	3556	−3
Chicago	3371	3609	−7
Denver	3058	3342	−8

Information on the increase in surface temperature in cities is usually obtained using satellites. Recent pictures of New York (Figure 5.5), Tallahassee and Washington show clearly that urban areas are of much higher temperature than the surrounding rural spaces. Unfortunately, no more quantitative data are available.

NASA, in collaboration with other agencies, has launched a campaign to study extreme heat in urban environments. The project, known as the Urban Heat Island Pilot Project (UHIPP), is based on images obtained with heat-sensitive instruments, such as the Airborne Terrestrial and Land Acquisition Sensor (ATLAS), aboard a NASA Lear 23 jet operating from Stennis Space Center. At the time that this book was in preparation, data were available for four locations: Atlanta, Salt Lake City. Sacramento and Baton Rouge.

Figure 5.6 shows a picture of the surface temperature distribution in and around the city of Atlanta, taken with the thermal channels of the GOES 8/9 geostationary

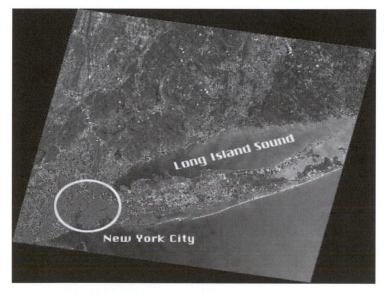

Figure 5.5. Satellite infrared picture of New York City, USA (photo: S.W. Stetson, http://www.RSAT.com.sws@rsat.com)

Figure 5.6. The surface temperature distribution in and around Atlanta, obtained with the GOES 8/9 satellite (photo: NASA's Marshall Space Flight Center, Space Science Laboratory)

satellites at 10:30 p.m. EDT on 6 May 1997. A very clear surface-temperature increase in the city and in particular in the city centre can be seen. The temperature in the city centre corresponds to about 59°F, while the rest of the city is close to 57°F and the city surroundings at about 55°F.

Recent measurements using the ATLAS sensor over Salt Lake City, taken on 21 July 1998 at 12:34 p.m. have shown that the city has rooftops and other structures that reach a blistering 71°C. Roads and buildings are at about 71°C, while vegetation is at approximately 29 to 36°C. Similar data provided by NASA for Sacramento, California, are based on measurements taken on 29 June at 1:00 p.m. by the ATLAS imager. Buildings are at about 60°C, while vegetation is at approximately 29 to 36°C. Finally, data, also based on measurements taken by the ATLAS imager, of Baton Rouge, LA, at 1:00 p.m. on 19 May 1998 show rooftops and other hot spots close to 65°C, while cooler areas, mostly with vegetation, are close to 25°C.

Several studies on heat-island conditions have been performed in Canada. Oke and East first studied heat-island characteristics in Montreal.[28] They reported a maximum heat-island intensity during winter nights close to 10.5 K at 22:00, which gradually dropped to 6.5 K by 05:00. Hage studied heat-island conditions at Edmonton, Alberta during both winter and summer periods.[29] During the cold period, a heat-island intensity of about 6.5 K was found to occur at about 21:00, while, during the summer, the heat-island intensity had a similar value but did not reach its maximum until midnight. Finally, Nkederim and Truch studied heat island conditions at Calgary, Alberta.[30] They reported a mean monthly heat-island intensity of about 10.1 K during the winter period and 6 K during the summer months.

HEAT-ISLAND STUDIES IN OTHER COUNTRIES

Numerous studies on the heat-island intensity in tropical cities were presented at the 1986 WMO Technical Conference on Urban Climatology and its application with Special Regard to Tropical Areas.[31] Givoni provided a comprehensive summary of the most important information presented at this conference.[32] More data were presented at a second WMO conference in 1994 on Tropical Urban Climates.[33]

According to the various presentations, in São Paolo, Brasil, the mean annual temperature has increased by about 2 K.[34] In Nigeria, the heat-island phenomenon in Lagos is mainly due to the dense traffic and is experienced at noon or in the late afternoon, ranging between 2 and 4 K.[35] In Ibadan, the heat-island effect 'is most marked at the height of the dry season in March when the rural/urban heat island ranges between 5 and 7 K in the city centre'. During the wet season the mean temperature differences are around 1–3 K.[35]

Data from India were reported by Padmanabhamurty for Delhi, Bombay and Calcuta,[36] the maximum heat-island intensity being 6 K in Delhi, 9.5 K in Bombay and 4 K in Calcutta. During the summer period, the heat-island intensity in Delhi varies between 2 and 5 K. Padmanabhamurty has also given data on the heat-island intensity for eight different Indian cities (Table 5.6) and, as can be seen, the heat island intensity reaches values up to 10 K.[37]

Data on the heat island in Shanghai, China, show that the urban area is always warmer than the countryside, and on calm clear nights the temperature difference is about 6 K.[38]

Table 5.6. Heat-island intensities in some Indian cities[37]

Station	Heat-island intensity (K)
New Delhi	6.0
Bhopal	6.5
Calcutta	4.0
Bombay	9.5
Pune	10.0
Visakhapatnam	0.6
Vijayawada	2.0
Madras	4.0

Table 5.7. The intensity of the heat island for selected urban centres in Malaysia[43]

Urban centre	Maximum heat island (K)
Kuala Lumpur – Petaling Jaya	6–7
Urban centres in the Klang Valley	2–5
Georgetown	4
Johor Bahru	3
Kota Kinabalu	3

Studies on the heat-island intensity in Singapore have shown that the intensity of heat island is close to 1 K.[39] Similar studies for Havana, Cuba, showed a heat-island intensity between 1 and 3 K,[40] while studies for Cairo, Egypt showed that a heat-island intensity close to 4 K occured during the night and early morning in the summer period.[41] Similar studies on the heat island in Dhaka, Bangladesh, showed an intensity between 0.5 and 6 K occuring during the night. The intensity during the summer period was relatively low (0.6 K), as a result of the high relative humidity and strong surface wind.[42]

Sani has published data on the heat-island intensity for selected cities in Malaysia (Table 5.7) and, as can be seen, the heat-island intensity ranges from 2 to 7 K.[43]

Heat island studies for Johannesburg, South Africa have been reported by Goldreich.[44] During the summer period, the heat-island intensity is found to be close to 1.9 and 2.0 K respectively during the night and day times.

Heat-island studies for some cities located in the Tama River Basin in Japan have been reported by Yamashita et al.[45] In all the cities, a heat island was observed to develop to some extent. The intensity was largely dependent on weather conditions. In Tachinawa City, the heat-island intensity amounted to 3.5 K during daytime in May. In Fussa City the heat island almost always appeared, except during the daytime in February and March. The temperature difference between the urban centre and rural areas was usually bigger in the daytime than at night. Kimura and Takahashi[46] and Kawamura[47] reported that in the greater Tokyo area the average summer night temperature is higher by 3 to 5 K than in the surrounding rural areas; furthermore, the temperature excess has been increasing in recent years.

Measurements of the heat-island intensity in Buenos Aires, Argentina, for a five-day period during June 1978 showed that the maximum value of the urban heat island was 7.4 K.[48]

HEAT-ISLAND STUDIES IN HOT MEDITERRANEAN CLIMATES

In the frame of the urban climate experiment carried out in Athens, partly financed by the POLIS project of the Directorate General for Science, Research and Technology of the European Commission, 20 automatic temperature and humidity stations were installed in the major Athens area during spring 1996; in a later phase the number of stations was increased to thirty. The instrumentation used was selected to satisfy several criteria, such as acceptable cost, in order to cover as many locations as possible and to provide satisfactory performance according to international meteorological standards, low maintenance, an internal power supply and high data-storage capacity. In order to protect the instruments from solar radiation and rain, white wooden boxes with lateral slots were constructed, approximating the Stevenson screen. The detailed and specific characteristics of used sensors are given by Santamouris *et al*.[49]

As the study had to cover the Athens basin, measurement points were selected to meet the following criteria:

- to obtain information about the boundary conditions around the basin;
- to study densely built areas with heavy traffic;
- to study densely built areas with less traffic;
- to study the conditions in the green areas of the city centre;
- to study medium-density built-up areas.

An attempt was therefore made to select areas that not only met these criteria, but also were located on the north–south and east–west axes of the city. A detailed map showing the relative position of the stations is given in Figure 5.7. In general terms, seven stations were placed in the very central area of Athens, 22 stations in urban areas and in a radial configuration around Athens, while one station was placed in an almost rural region ouside city so that it could be used as the reference station. A description of the stations is given in Table 5.8. The numbering of stations refers to Figure 5.7.

Measurements were taken on a minute-by-minute basis and all sensors were calibrated against one another and against high-precision sensors.

The collected data were analysed in detail in order to assess the heat-island characteristics in the city of Athens, as well as the specific distribution of the ambient temperature in the city. Specific and detailed statistical and climatological analysis was performed.

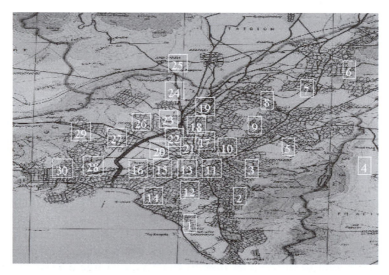

Figure 5.7. The positions of the measuring stations in the major Athens area

Table 5.8. Description of the meteorological stations placed in the major Athens area

Station	Description
1	South coastal area. Close to the sea and the airport. Very low traffic and building density.
2	South-eastern area. Very close to a mountain forest. Low traffic, medium building density.
3	Central area. Five km from the city centre. High traffic and building density.
4	North-eastern suburban area. Low traffic and building density.
5	Eastern area. Close to the mountain. High building density. Low traffic.
6	North suburban area. Low traffic and building density. Relatively green area.
7	Northern area. Low traffic. Medium building density. Relatively green area.
8	North-western area. Traffic and building density very high
9	Central area. Four km from the city centre. High traffic and building density
10	Central area. Close to the park. Traffic and building density very high
11	Central area. National Park.
12	South central area. Three km from the city centre. Four km from the sea. High traffic and building density.
13	Central area. Pedestrian street.
14	Southern area. Very close to the coast. Six km from the city centre. High traffic, medium building density.
15	Central area. Archaeological space. Bare soil. No buildings. Some trees.
16	Central area. On a green hill. Low building density. No traffic.
17	Central area. Very high traffic and building density.
18	Central area. Very high traffic and building density.
19	Central area. National Park.
20	Central western area. Moderate vegetation. High traffic. Low building density.
21	Central area. Very high traffic and building density.
22	Central area. Very high traffic and building density.
23	Western area. Very high traffic and building density.
24	Western area. Very high traffic and building density.
25	North-western area. Very low traffic and building density. Close to a park.
26	Western area. Very high traffic and building density.
27	Western area. Very high traffic and building density.
28	South-western area. High traffic, very high building density.
29	Western area. Low traffic. Medium building density. Close to a park.
30	South-western area. Very high traffic and building density.

During both summer and winter periods much higher temperatures were recorded in the central Athens area, especially during the daytime. Figure 5.8 plots the relative temperature difference between 12 urban stations located in and around the centre as a function of the temperature of the urban station. The plotted data refer to the summer of 1996.

As can be see in Figure 5.7, the temperature increase in the very central area may reach values up to 15 K. Daily heat-island intensity for most of the central urban stations is close to 10 K. The heat-island intensity is much lower in suburban areas and ranges between 6 and 2 K. At night, the heat island-intensity (Figure 5.9) varies between 2 and 5 K depending on the characteristics of each station. Urban green areas show 2–3 K lower than the reference station.

During the winter period the temperatures in the central Athens area are much higher than in the surrounding areas. The relative temperature differences, during the daytime period, between 18 urban and suburban stations located in and around the centre of Athens are plotted in Figure 5.10 as a function of the temperature of the

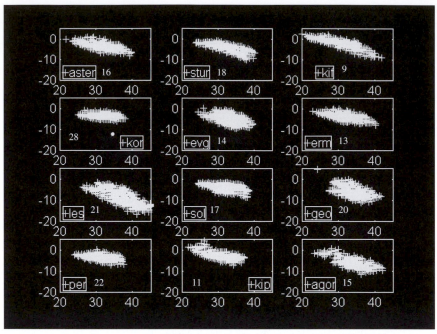

Figure 5.8. Temperature difference between urban stations and the reference station as a function of the absolute temperature of the urban station on summer days (for the numbering of the stations, refer to Figure 5.7)

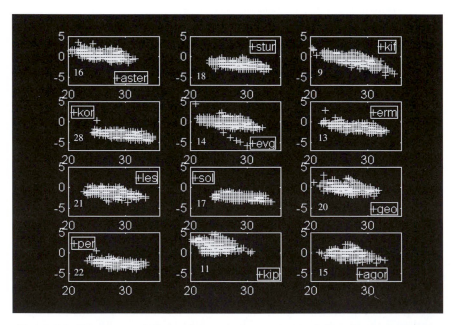

Figure 5.9. Temperature difference between urban stations and the reference station as a function of the absolute temperature of the urban station during summer nights (for the numbering of the stations, refer to Figure 5.7)

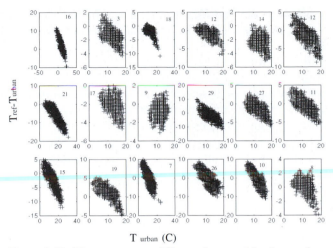

T_{urban} (C)

Figure 5.10. The temperature difference between 18 urban and suburban stations and the reference station as a function of the absolute temperature of the urban station during winter days in Athens; data refer to the winter period of 1997 (for the numbering of the stations, refer to Figure 5.7)

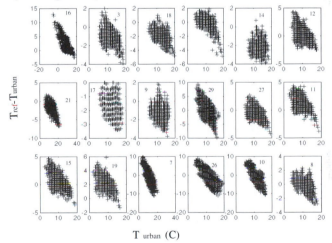

T_{urban} (C)

Figure 5.11. The temperature difference between 18 urban and suburban stations and the reference station as a function of the absolute temperature of the urban station during winter nights in Athens; data refer to the winter period of 1997 (for the numbering of the stations, refer to Figure 5.13)

urban station. The plotted data refer to the whole winter period of 1997. During the day, the mean heat-island intensity in the central Athens area is close to 8 K, while the absolute heat-island intensity in the city centre is about 13 K, corresponding to a station located in the central area of the city characterized by a very heavy traffic load. The heat-island intensity is reduced to 3–6 K in the surrounding suburban areas. During the night period (Figure 5.11) the heat-island intensity can be up to 10 K and varies as a function of the characteristics of the area. The maximum heat-island intensity corresponds to the Western Athens area, which is characterized by high building density and low vegetation. The heat-island intensity in the city centre varies between 3 and 6 K.

Tables 5.9 and 5.10 give data regarding the mean T_{aver}, the mean maximum T_{avmax}, the mean minimum T_{avmin}, the absolute minimum T_{absmin} and the absolute maximum T_{absmax} temperatures recorded at 20 selected stations during summer 1996 and winter 1997. Some comparisons with a reference station are also given. These are ΔT_{abmxdf}, the absolute positive maximum temperature difference recorded in the month between a particular station and the reference station, ΔT_{abmndf}, the absolute negative temperature difference between the stations, ΔT_{mndf}, the difference between the daily means, and ΔT_{mnmxdf}, ΔT_{mnmndf}, the differences between the mean maximum and minimum temperatures respectively. Finally, the heating, HDH, base 18°C, and the cooling, CDH, base 26°C, are given. The reference station used is station No 2, as shown in Figure 5.7.

Analysis of the ambient distribution during the summer period in Athens, as shown in Table 5.9, leads to the following conclusions:

- The cooling degree hours in the central area of the city are about 350% greater than in the suburban areas.
- The maximum heat-island intensity in the very central area is close to 16 K, while a mean value for the major central area of Athens is close to 12 K.
- The absolute maximum temperature in the central area is close to 15 K higher than in the suburban areas, while the absolute minimum temperature is up to 3 K higher in the centre.
- The West Athens area, characterized by low vegetation, high building density and a high anthropogenic emission rate, presents almost twice the number of cooling degree days than the northern or southern areas of Athens.
- The heat-island intensity in the central Athens park is close to 6.1 K, while the heat island intensity of nearby stations is close to 10 K. In addition, the park presents almost 40% fewer cooling degree hours than the other nearby urban stations. The park also presents the lowest recorded absolute minimum temperature of all the stations, but, as shown in Table 5.9, during the daytime the absolute temperature inside the park is higher than in the suburban areas.

Analysis of the temperature distribution during the winter period permits the following conclusions:

- Heating degree hours in the very central area of Athens are about 40–60% lower than in the surrounding suburban areas.
- The absolute heat-island intensity in the city centre is to about 13 K, while the mean value of the heat-island intensity for the city centre is close to 8 K. The maximum heat-island intensity corresponds to a station located in the central area of the city characterized by a very heavy traffic load.
- The heat island intensity corresponding to the centrally located urban park is close to 4.5 K. The park also presents the lowest absolute minimum temperature of all the stations.

Table 5.9. Characteristic temperatures for 20 selected stations during August 1996, together with comparisons with a reference station

Station	1	2	3	5	6	9	11	12	13	14	15	16	17	18	20	21	22	27	28	29
T_{aver}	29.2	27.2	28.7	27.3	27.0	27.6	27.3	29.6	29.5	29.6	30.3	28.4	32.1	30.0	30.2	30.1	30.3	29.7	30.3	28.4
T_{avmax}	33.5	29.9	32.8	30.4	30.0	31.9	33.2	32.8	34.9	34.8	37.6	34.3	41.9	33.1	36.9	33.7	36.3	33.9	33.1	32.2
T_{avmin}	25.4	24.8	25.5	23.7	24.1	23.8	21.9	27.1	25.6	24.3	24.0	23.7	25.6	26.9	24.3	27.2	26.4	26.3	27.5	24.4
T_{absmax}	22.4	21.5	21.7	19.6	20.9	18.8	17.7	23.8	21.5	20.9	20.7	19.9	22.4	23.6	20.5	23.3	23.0	22.1	23.9	20.3
T_{absmin}	49.6	33.7	36.7	35.9	33.8	36.5	37.1	36.7	39.3	37.9	42.1	38.4	46.6	36.8	41.5	37.6	39.8	38.3	36.8	36.8
$\Delta T_{absmxdf}$	12.8	0.0	4.5	7.9	2.7	5.4	6.1	4.2	8.2	8.8	11.6	6.8	16.7	7.2	11.2	7.0	10.2	5.9	5.0	6.7
$\Delta T_{absmndf}$	-3.1	0.0	-0.5	-3.4	-3.8	-4.4	-4.9	-0.5	-5.1	-4.5	-2.9	-5.1	-0.6	1.1	-5.2	-0.6	-0.2	-2.3	-0.4	-5.1
ΔT_{mndf}	2.0	0.0	1.5	0.0	-0.2	0.4	0.1	2.5	2.3	2.3	3.0	1.1	4.7	2.8	2.9	2.9	3.1	2.5	3.1	1.2
ΔT_{mnmxdf}	5.8	0.0	3.0	4.1	0.8	2.8	4.6	3.4	6.9	6.9	8.8	5.3	13.5	5.5	9.0	5.0	6.9	4.4	4.2	4.5
ΔT_{mnmndf}	-0.3	0.0	0.3	-1.9	-1.2	-1.4	-3.2	1.4	-0.5	-0.9	-1.1	-1.7	0.4	1.7	-1.2	1.57	1.3	1.16	1.9	-0.9
HDH (18°C)	0.0	0.0	0.0	0.0	0.0	0.0	0.0	0.0	0.0	0.0	0.0	0.0	0.0	0.0	0.0	0.0	0.0	0.0	0.0	0.0
CDH (26°C)	2422	1223	2080	1401	1183	1655	1777	2623	2622	2728	3263	2133	4351	2888	3173	2954	3116	2727	3064	2056

Table 5.10. Characteristic temperatures for 20 selected stations during March 1997, together with comparisons with a reference station

Station	1	2	3	5	6	9	11	12	13	14	15	16	17	18	20	21	22	27	28	29
T_{aver}	13.1	10.9	12.0	13.7	11.3	10.9	11.0	13.4	12.4	12.8	12.9	11.3	14.7	12.9	13.0	13.1	15.2	12.3	13.2	11.1
T_{avmax}	17.0	13.4	15.1	16.2	13.3	14.0	15.0	15.9	16.6	16.8	19.4	15.3	22.7	15.3	18.3	16.1	23.0	15.6	18.2	14.1
T_{avmin}	9.5	9.0	9.3	11.2	9.3	8.2	7.0	11.3	9.5	8.9	8.0	8.2	9.7	10.6	8.8	10.6	11.1	9.5	10.5	8.1
T_{absmax}	22.8	17.8	20.6	22.9	18.6	19.2	20.3	20.3	23.7	21.8	25.4	21.1	28.1	19.7	25.6	21.2	28.6	21.0	25.4	20.0
T_{absmin}	4.9	4.1	5.6	3.3	4.8	4.5	2.8	7.1	5.3	4.9	4.1	4.1	5.4	6.9	4.4	5.7	7.3	6.0	6.9	4.1
$\Delta T_{absmxdf}$	7.8	0.0	3.5	16.1	2.3	3.5	4.5	4.1	6.8	9.3	11.1	5.4	12.6	5.5	9.8	6.5	13.1	4.6	9.3	4.8
$\Delta T_{absmndf}$	-2.0	0.0	-1.2	-3.1	-1.7	-2.3	-4.4	0.9	-1.1	-2.7	-3.9	-3.1	-1.4	0.7	-3.2	-0.8	0.5	-1.6	-0.8	-3.5
ΔT_{mndf}	1.9	0.0	1.1	2.8	0.3	-0.1	-0.2	2.3	1.4	1.6	1.6	0.3	3.4	1.9	1.7	1.9	4.0	1.4	2.2	0.0
ΔT_{mnmxdf}	4.5	0.0	2.1	5.1	1.2	1.2	2.4	3.1	3.7	4.6	3.8	2.6	4.8	3.2	5.8	3.5	6.0	2.9	5.3	2.2
ΔT_{mnmndf}	-0.1	0.0	0.0	0.9	-0.6	-1.2	-2.6	1.5	0.1	-0.7	-1.6	-1.2	0.1	1.0	-0.8	0.7	1.5	0.1	0.9	-1.5
HDH (18°C)	3532	5271	3837	4064	5034	4543	4747	3123	4290	3692	3933	5093	3048	3872	3774	3490	3772	4298	3450	4847
CDH (26°C)	0.0	0.0	0.0	0.0	0.0	0.0	0.0	0.0	0.0	0.0	0.0	0.0	0.0	0.0	0.0	0.0	0.0	0.0	0.0	0.0

- The absolute maximum temperature is to about 6–7 K higher in the centre than in the surrounding areas, while the absolute minimum is about 1–2 K higher.
- In the western areas of Athens, characterized by low vegetation, high building density and a high anthropogenic emission rate, there are almost 30% fewer heating degree days than in the northern or southern areas of Athens.

The cooling degree hours determine to a high degree the cooling load of buildings,[50] and thus knowledge of their spatial and temporal distribution, as well as of their absolute value, in a city is of great interest to designers and climatologists.

The daily spatial distribution of cooling degree hours in a day in Athens (August 1996) is given in Figure 5.12. As shown, the central Athens area starts to be heated at about 10:00 a.m., gets the maximum difference compared to the surrounding area at about 14:00 to 15:00, while the phenomenon is amortized at about 19:00. Maximum cooling degree hours in the central Athens area are close to 450, while the corresponding value for the reference suburban station is close to 150. During night cooling degree hours present a maximum in western Athens characterized by high building density and low vegetation. Maximum cooling degree hours during night are close to 104 while the minimum value at the reference station is close to 26. These results indicate clearly the role of urban layout, existence of vegetation and type of materials used on the potential cooling energy demand of urban buildings.

To provide a better picture of the variability of cooling degree hours in Athens, their spatial distribution for 13:00 for the whole August 1996 is given in Figure 5.13. Reported cooling degree hours have been calculated for a temperature base of 26°C and are the sum of daily cooling degree days at 13:00 for all the days in August. As can be seen, close to noon, the centre of the city presents almost three times the cooling

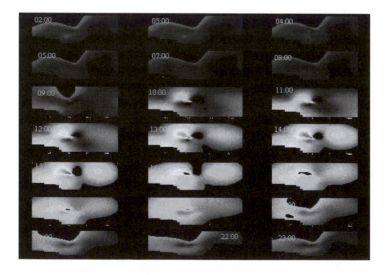

Figure 5.12. Spatial and hourly distribution of the cooling degree hours in Athens for August 1996

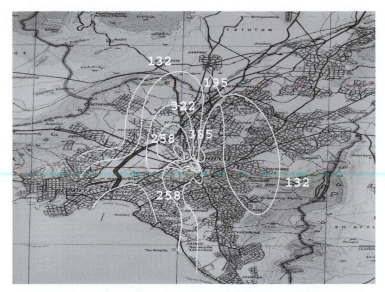

Figure 5.13. Iso-cooling degree hours lines in Athens for a 26°C temperature base, at 13:00 for August 1996

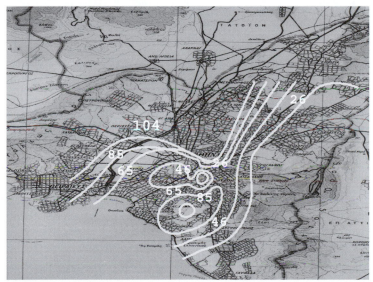

Figure 5.14. Iso-cooling degree hours lines in Athens for a 26°C temperature base, at 1:00 for August 1996

degree hours than the city surroundings. In particular, in the central Athens area the cooling degree hours are close to 385, while the corresponding value for the suburban areas is close to 132.

As already reported, a very interesting spatial distribution of cooling degree hours has been calculated for the night period. Figure 5.14 shows this distribution for 1:00 for the whole August 1996. As can be seen, higher cooling degree hours have been calculated in the western Athens area than in central one. In particular, while the

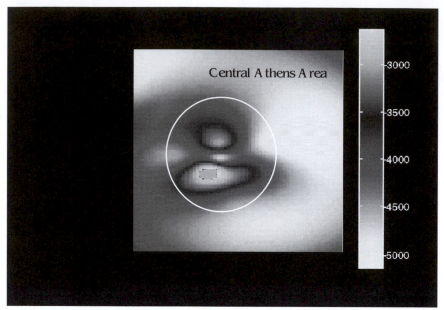

Figure 5.15. Spatial distribution of the heating degree days in Athens for December 1997

cooling degree hours in the reference area are close to 26, the corresponding values for the western and central Athens area are close to 104 and 85 respectively. This is mainly due to the high building density of western Athens, as well as to the lack of green spaces and other cool sinks.

Heating degree hours determine to high degree the heating load of building. An indicative spatial distribution for December 1997, base 18.3°C, is given in Figure 5.15. As shown, the heating degree hours in the central Athens area are about 3000–4000, while in the surrounding areas can be high as 5000, which corresponds to a decrease of about 40–60%. The higher heating degree hours in the centre of the city, as shown in Figure 5.15, correspond to stations located in urban parks or areas of high vegetation.

REFERENCES

1. Howard, L. (1833). *The Climate of London*, Vols. I–III. London.
2. Landsberg, H. E. (1981). *The Urban Climate*. Academic Press, New York.
3. Oke, T.R, G.T. Johnson, D.G. Steyn and I.D. Watson (1991). 'Simulation of Surface Urban Heat Islands under "Ideal" Conditions at Night – Part 2: Diagnosis and Causation'. *Boundary Layer Meteorology*, Vol. 56, pp. 339–358.
4. IPCC Working Group II (1990). *Climate Change – The IPCC Impacts assessment*, pp. 3–5. International Panel on Climate Change.
5. Oke, T.R. (1987). *Boundary Layer Climates*. Routledge, London.
6. Oke, T.R. (1982). 'Overview of interactions between settlements and their environments', WMO experts meeting on Urban and building climatology. WCP-37, WMO, Geneva.
7. Escourrou, G. (1991). *Le Climat et la Ville*. Nathan University Editions, Paris.
8. Jauregui, E. (1986). 'The Urban Climate of Mexico City'. In *Proceedings, Technical Conference in Mexico City: Urban Climatology and its Applications with Special Regard to Tropical Areas*, No. 652, WMO, pp. 63–86.

9. Park. H.S. (1987). 'City Size and Urban Heat Island Intensity for Japanese and Korean Cities', *Geographical Review of Japan*, A, Vol. 60, September, pp. 238–250.
10. Ludwig, F.L. (1970). 'Urban Temperature Fields in Urban Climates', WMO Technical Note No 108, pp. 80–107.
11. Bornstein, R.D. (1986). 'Urban Climate Models: Nature, Limitations and Applications'. In *Proceedings, Technical Conference in Mexico City: Urban Climatology and its Applications with Special Regard to Tropical Areas*, No. 652, WMO, pp. 237–276.
12. Summers, P.W. (1964). 'An Urban Ventilation Model Applied to Montreal'. Ph.D. Dissertation, McGill University, Montreal.
13. Oke, T.R. (1981). 'Canyon Geometry and The Nocturnal Urban Heat Island: Comparison of Scale Model and Field Observations', *Journal of Climatology*, Vol. 1, pp. 237–254.
14. Lyall, I.T. (1977). 'The London Heat-Island in June–July 1976', *Weather*, Vol. 32, No. 8, pp. 296–302.
15. Chandler, T.J. (1965). 'City Growth and Urban Climates', *Weather*, Vol. 19, pp. 170–171.
16. Eliasson, I. (1996). 'Urban Nocturnal Temperatures, Street Geometry and Land Use', *Atmospheric Environment*, Vol. 30, No. 3, pp. 379–392.
17. Barring, L., Mattsson, J.O. and Lindovist, S. (1985). 'Canyon Geometry, Street Temperatures and Urban Heat Island in Malmo, Sweden', *Journal of Climatology*, Vol. 5, pp. 433–444.
18. Swaid, H. and Hoffman, M.E. (1990). 'Climatic Impacts of Urban Design Features for High and Mid Latitude Cities', *Energy and Buildings*, Vol. 14, pp. 325–336.
19. Kuttler, W., Barlag, A.-B. and Robmann, F. (1996). 'Study of the Thermal Structure of a Town in a Narrow Valley', *Atmospheric Environment*, Vol. 30, No. 3, pp. 365–378.
20. Nubler, W. (1979). 'Konfiguration and Genese der Warmeinsel der Stadt Freiburg', *Freiburger Geographische Hefte*.
21. Escourrou, G. (1990/1991). 'Climate and Pollution in Paris', *Energy and Buildings*, Vol. 15–16, pp. 673–676.
22. Wanner, H. and Hertig, J.A. (1983).'Temperature and Ventilation of Small Cities in Complex Terrain (Switzerland)'. Study supported by Swiss National Science Foundation, Berne.
23. Abbate, G. (1998). 'Heat Island Study in the Area of Rome by Integrated Use of ERS-SAR and LANDSAT TM'. Report available from earthnet online, service provided by the European Space Agency.
24. NASA Climate News and Research. (1998). Information is available from URL: http://www.climatenews.com.
25. Lawrence Berkeley Laboratory (1998). 'Mitigation of Heat Islands'. Information is available from URL: http://www.lbl.gov/HeatIsland.
26. Akbari, H., Davis, S., Dorsano, S., Huang, J. and Winett, S. (1992). *Cooling our Communities – A Guidebook on Tree Planting and Light Colored Surfacing*. US Environmental Protection Agency. Office of Policy Analysis, Climate Change Division.
27. Taha, H. (1997).'Urban Climates and Heat Islands: Albedo, Evapotranspiration and Anthropogenic Heat', *Energy and Buildings*, Vol. 25, pp. 99–103.
28. Oke, T.R. and East, C. (1971). 'The Urban Boundary Layer in Montreal', *Boundary Layer Meteorology*, Vol. 1, pp. 411–437.
29. Hage, K.D. (1972). 'Nocturnal Temperatures in Edmonton, Alberta.' *Journal of Applied Meteorology*, Vol. 11, pp. 123–129.
30. Nkederim, L.C. and Truch, P. (1981). 'Variability of Temperatures Field in Calgary, Alberta' *Atmospheric Environment*.
31. WMO (1986). *Proceedings, Technical Conference in Mexico City : Urban Climatology and its Applications with Special Regard to Tropical Areas*, WMO, No 652.
32. Givoni, B. (1989). *Urban Design in Different Climates*. WMO Technical Report 346.
33. WMO (1986). *Report of the Technical Conference on Tropical Urban Climates*. WMO, Dhaka.
34. Monteiro, C.A.F. (1986) 'Some Aspects of the Urban Climates of Tropical South America: The Brazilian Contribution'. In *Proceedings, Technical Conference in Mexico City: Urban*

Climatology and its Applications with Special Regard to Tropical Areas, WMO, No 652, pp. 166–198.

35. Oguntoyinbo, J.S. (1986). 'Some Aspects of the Urban Climates of Tropical Africa'. In *Proceedings, Technical Conference in Mexico City : Urban Climatology and its Applications with Special Regard to Tropical Areas*, WMO, No 652, pp. 110–135.

36. Padmanabhamurty, B. (1986) In *Proceedings, Technical Conference in Mexico City: Urban Climatology and its Applications with Special Regard to Tropical Areas*, WMO, No 652.

37. Padmanabhamurty, B. (1990/91). 'Microclimates in Tropical Urban Complexes', *Energy and Buildings*, Vol. 15–16, pp. 83–92.

38. Chow, S.D. (1986). 'Some Aspects of the Urban Climate of Shanghai'. In *Proceedings, Technical Conference in Mexico City: Urban Climatology and its Applications with Special Regard to Tropical Areas*, WMO, No 652, pp. 87–109.

39. Tso, C.P. (1994). 'The Impact of Urban Development on the Thermal Environment of Singapore'. In *Report of the Technical Conference on Tropical Urban Climates*. WMO, Dhaka.

40. Estela, L.L., Poveda, M.N. and Castro, L.P. (1994). 'Investigations on Urban Climate in Cuba'. In *Report of the Technical Conference on Tropical Urban Climates*. WMO, Dhaka.

41. Fouli, R.S. (1994). 'Effect of Urbanization on Some Meteorological Elements in Greater Cairo Region', In *Report of the Technical Conference on Tropical Urban Climates*. WMO, Dhaka.

42. Ershad, M.H and Nooruddin, Md. (1994). 'Some Aspects of Urban Climates of Dhaka City', In *Report of the Technical Conference on Tropical Urban Climates*. WMO, Dhaka.

43. Sham, S. (1990/91). 'Urban Climatology in Malaysia: An Overview', *Energy and Buildings*, Vol. 15–16, pp. 105–117.

44. Goldreich, Y. (1985). 'The Structure of the Ground-level Heat Island in a Central Business District'. *Journal of Climate and Applied Meteorology*, Vol. 24, pp. 1237–1244.

45. Yamashita, S, Sekine, K., Shoda, M. Yamashita, K. and Hara., Y. (1986). 'On the Relation-ships between Heat Island and Sky View Factor in the Cities of Tama River Basin, Japan'. *Atmospheric Environment*, Vol. 20, No. 4, pp. 681–686.

46. Kimura, T and Takahashi, S. (1991). 'The Effect of Land Use and Anthropogenic Heating on the Surface Temperature in the Tokyo Metropolitan Area: A Numerical Experiment'. *Atmospheric Environment*, Vol. 25B, pp. 439–454.

47. Kawamura, T. (1979). *Urban Atmospheric Environment*. Tokyo University Press, Tokyo.

48. Mazzeo, N.A. and Camilioni, I. (1990/91). 'Buenos Aires Urban Meteorological Data Analysis of a Five Day Period'. *Energy and Buildings*, Vol. 15–16, pp. 339–343.

49. Santamouris, M., Argiriou, A. and Papanikolaou, N. (1996). 'Meteorological Stations for Microclimatic Measurements'. Report to the POLIS project, Commission of the European Commission, Directorate General for Science Research and Technology (available from the authors).

50. Santamouris. M. and Assimakopoulos, D. (eds) (1996). *Passive Cooling of Buildings*. James & James (Science Publishers), London.

6

The canyon effect

M. Santamouris,
Department of Applied Physics, University of Athens

∞

THE CLIMATE OF URBAN CANYONS

According to Oke's theory,[1] the air space above a city may be divided into the so-called urban air 'canopy' and the boundary layer over the city space called the 'urban air dome'. The urban air canopy is the space bounded by the urban buildings up to their roofs. The urban canopy layer includes an unlimited number of microclimates generated by the various urban configurations. The specific climatic conditions at any given point within the canopy are determined by the nature of the immediate surroundings and, in particular, their geometry, materials and properties. The upper boundary of the urban canopy varies from one point to another because of the variable heights of the buildings and the wind speed.

Oke has defined the air dome layer as 'that portion of the planetary boundary layer whose characteristics are affected by the presence of an urban area at its lower boundary';[2] the air dome layer is more homogeneous in its properties over the urban area at large than the urban canopy.

Studies focusing on the canopy layer aim, in general, to improve the understanding of the thermal and flow characteristics in similar and common urban structural forms, such as urban canyons. Thus, they are concentrated on the energy balance in the canyon, the temperature distribution of the air and of the surfaces and the air circulation in the canyon.

Air circulation and temperature distribution within urban canyons are significant for the energy consumption of buildings, pollutant dispersion studies, and heat and mass exchange between the buildings and the canyon air and are therefore of interest in studies on the energy potential of natural ventilation techniques for buildings, pedestrian comfort, etc.

TEMPERATURE DISTRIBUTION IN URBAN CANYONS

The temperature distribution in the urban canopy layer is greatly affected by the urban radiation balance. Solar radiation incident on urban surfaces is absorbed and then transformed to sensible heat. Most of the solar radiation impinges on roofs and on the vertical walls of buildings; only a relatively small part reaches ground level.

Walls, roofs and the ground emit long-wave radiation to the sky. The intensity of the emitted radiation depends on the view factor of the surface relative to the sky. Under urban conditions, most of the sky dome that could be viewed by walls and surfaces is blocked by other buildings and thus long-wave radiant exchange does not really result in significant losses.

The net balance between the solar gains and the heat loss due to emitted long-wave radiation determines the thermal balance of urban areas. Because the radiant heat loss is slower in urban areas, the net balance is more positive than in the surrounding rural areas and thus temperatures are higher.

As reported by Park,[3] who discusses the relationship between population and the sky view factor for different regions, the effects of a building's layout are different from region to region because different regions are characterized by different building heights and different canyon widths (Figure 6.1).

There have been various studies of the relationship between the canyon layout, and especially the sky view factor, and both the heat-island intensity in the canyon and the surface temperatures. Yamashita *et al.* reported a clear correlation of urban air temperature and sky view factor for some Japanese cities.[4] Barring *et al.* studied the relationship between the street surface temperature and the sky view factor in Malmø, Sweden.[5] They reported a strong correlation between the surface temperature pattern and the street geometry; the higher the sky view factor, the lower the surface temperature. Although high surface temperatures have also been recorded in canyons with low sky view factors outside the city, no clear correlation between the urban temperature and the sky view factor has been found. This clearly indicates that the average air temperature of the streets is governed by more complex and regional factors than their surface temperature, even if the local canyon geometry is of importance.

Very similar results were reported by Eliason for the city of Goteborg in Sweden.[6,7] During the winter period, the surface temperature was greatly influenced by the city

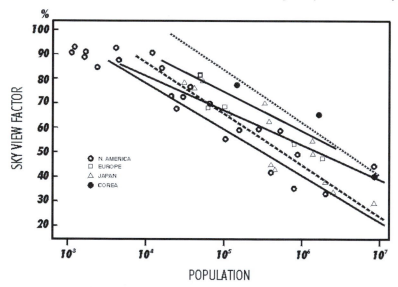

Figure 6.1. Relationship between the sky view factor and the urban population for Japanese, Korean, North American and European Cities[4]

structure and a clear correlation between the surface temperature and the sky view factor was found. However, no correlation was found between the urban temperatures and the sky view factor. On the contrary, it was found that the maximum surface temperature difference observed between urban sites of $H/W = 0.5$ and $H/W = 2.0$ (H = height, W = width) was 3.5 K. In a similar study, Arnfield reported a surface temperature difference of 4 K between urban sites of different density (H/W ratios of 0.5 and 2.0 respectively).[8] Both studies show clearly that important variations in surface temperature exist between urban sites of different geometry. However, the results of other studies clearly show that, in spite of the fact that the street surface is influenced by the canyon geometry, there is a weak connection between geometry and air temperature and this is because the air temperature is dependent upon the flux divergence per unit volume of air including the effects of the horizontal transport.

Surface temperature

The temperature of the external materials in a canyon is governed by its thermal balance. Surfaces absorb short-wave radiation as a function of their absorptivity and their exposure to solar radiation. They absorb and emit long-wave radiation as a function of their temperature, emissivity and view factor. And they transfer heat to or from the surrounding air and exchange heat via conduction procedures with the lower material layers.

The thermal balance of a surface in a canyon can be expressed as follows:[9]

$$Q^* = Q_H + Q_G,\qquad(6.1)$$

where Q^* is the net radiation, Q_H are the convective heat exchanges and Q_G the conductive heat exchanges with the substrate. The net radiation is the balance of the received beam, diffuse and reflected solar radiation K, as well as of the received and emitted long-wave radiation L:

$$Q^* = K\downarrow_S + K\downarrow_T + K\uparrow_r + L\downarrow_S + L\downarrow_T + L\uparrow_e,\qquad(6.2)$$

where the arrows represent directions to ($\downarrow$) and from ($\uparrow$) the surface, and subscripts T and s represent the sky and surrounding terrain sources respectively; r represents the reflected radiation and e the emitted radiation.

The optical and thermal characteristics of materials used in urban environments and especially the albedo for solar radiation and emissivity for long-wave radiation have a very important impact on the urban energy balance. Yap has reported that systematic urban–rural differences of surface emissivity have the potential to cause part of the heat island.[9]

The albedo of a surface is defined as its reflectivity integrated over the complete hemisphere and over wavelength. Use of high-albedo materials reduces the amount of solar radiation absorbed in building envelopes and urban structures and keeps their surfaces cooler. Materials with high emissivities are good emitters of long-wave energy and readily release the energy that has been absorbed as short-wave radiation. Lower surface temperatures contribute to a decrease in the temperature of the ambient air because heat convection intensity from a cooler surface is lower. Such temperature reductions can have significant impacts on consumption of cooling energy in urban areas, a fact of particular importance in cities with hot climates.

As explained above, surface temperature measurements have been performed in different canyons all over the world. Extensive measurements were performed, during the summer period, in 12 urban canyons in Athens, Greece.[10] A detailed analysis of the temperature characteristics for streets and pavements is given in the chapter devoted to materials (Chapter 11). In this chapter the surface distribution of the vertical walls in a canyon is discussed.

Experiments have shown that when the surface temperatures of opposite facades in a canyon are compared, then, as expected, surfaces that are almost south facing present much higher temperatures during the daytime than north-facing surfaces. The maximum daily simultaneous temperature difference between two facing panels can be as much as 19 K. The highest reported recorded difference in the daily maximum temperature of two facing surfaces is close to 14 K. There is a period during the daytime, when one of the two facades presents a higher temperature than the opposite one. This time period is a function of the canyon orientation, the layout and the characteristics of the materials forming the facades.

In N–S, NE–SW and SE–NW oriented canyons, where the solar surfaces are of south, south-east or south-west orientation, the instantaneous temperature difference between opposite surfaces during the day is lower at ground level and increases as a function of canyon height. This is because, in this type of canyon, the lower surfaces of the S, SW and SE facades receive much less radiation than the upper ones and thus the upper S, SW and SE floors present higher surface temperatures. This has been experimentally verified. Since the facing N, NE and NW facades receive very low levels of radiation, this explains the highest instantaneous temperature differences on the upper floors

During the night the temperature difference between the opposite surfaces in the canyon is generally not significant. The maximum recorded temperature difference is close to 2 K. In almost all cases where a significant temperature difference has been found, and the view factor of the opposite facades is almost the same, the S, SW and SE facades are warmer than the opposite surface in the canyon. Higher temperatures are only reported on non-south-facing facades during the night when these surfaces present higher temperatures during the daytime as well.

The temperature of a surface during the night is mainly governed by its radiative balance. The highest temperature differences are measured in asymmetric canyons. In these canyons the sky view factor and thus the radiative balance of the opposite surfaces is different. In general, surfaces with a higher sky view factor present a lower temperature during the night.

The vertical distribution of the external temperature of buildings in a canyon is of great interest because it defines the convective transfer from and to the ambient air. Measurements of the vertical distribution of surface temperatures in various urban canyons have shown the following:

- On S-, SW- and SE-oriented facades and during the daytime, the temperature difference varies between zero and 14 K. During the same period, the lowest surface temperature is presented at ground-floor level. This is mainly because the incident solar radiation is much lower at this level. It has also been observed that sometimes middle-height surfaces (intermediate floors) present the highest temperatures. This is because, in these canyons, intermediate floors receive almost the same solar radiation as the upper floors and more infrared radiation from the facing buildings, while convective fluxes are lower because of the lower wind speed at this level. Also, in most of the cases, the ambient temperature at this level is higher than the air temperature outside the canyon.

- On N-, NE- and NW-oriented facades and during the daytime, the temperature difference varies between 2 and 6 K. The lower surface temperatures are also presented on the ground floor for the reasons explained previously. However, in canyons with small H/W ratio, higher temperatures can be observed on the ground floor, especially when the ground-level surface is highly absorbent. The difference between the surface temperature of the middle and upper floors is not significant in this case. This is because, for these orientations, higher solar radiation falls on the upper-floor surfaces and thus counterbalances the additional radiative and convective gains of the middle floors.

- During the night, the temperature difference in the S, SW and SE facades varies between 0 and 5 K, with a mean value close to 3 K. In all cases the higher temperatures are presented on the ground floor. Then temperature falls as a function of canyon height. This is because of the radiative balance in the canyon. Lower surfaces have lower sky view factors and higher view factors for the other canyon buildings, and thus present a more positive radiative balance. However, no correlation between the vertical temperature intensity and the H/W ratio has been found. Moreover, the night vertical temperature intensity cannot be correlated with the corresponding daytime vertical intensity. This is because the temperature of the surfaces is not only governed by their radiative balance and the stored heat, but also by the convective and conductive losses.

- During the night, the temperature difference in the N, NW and NE facades varies between zero and 3 K. As in the case of south-oriented surfaces, the higher temperatures are presented on the ground floor and the temperature falls as a function of height.

Air temperature

The distribution of the ambient air in a canyon greatly influences the energy consumption of the buildings. Higher temperatures in a canyon increase the heat convection to a building and increases the cooling load due to ventilation. Therefore, it is very important to understand the mechanism that determines the distribution of the ambient temperature in a canyon.

The temperature in a canyon is influenced by the temperature of the canyon surfaces, because energy is transferred through convective process. However, as already noted, in spite of the fact that the street surface is influenced by the canyon geometry, there is a weak connection between geometry and air temperature and this is because the air temperature is flux divergence per unit volume of air including the effects of the horizontal transport.

Experiments that have focused on the distribution of the air temperature in canyons have mainly concluded that the air temperature stratification in the canyons during the daytime is not significant. The maximum daily temperature differences rarely exceed 2–3 K. No specific temperature distribution pattern as a function of canyon height has been observed. In most cases lower temperatures are measured at the ground levels and the temperature increases as a function of the canyon height. This matches the distribution of building surface temperatures in the canyon.

The air temperature distribution across a canyon is of great interest. It has been observed that, close to the facade of a building, an air film, which is a function of the temperature of the building surface and the rate of vertical air transport, is developed. In the middle of the canyon, and at ground level, the air temperature is more dependent upon the flux divergence per unit volume of air including the effects of the horizontal transport.[11,12] Thus, the temperature at the middle of the canyon is very different from the average of the temperatures of the two air films developed close to the facades of the buildings on the two sides.

As would be expected, the air temperature close to a S, SW or SE facade of a canyon will be higher. The measured air temperature differences between the two facades vary as a function of the canyon layout and the surface characteristics. The mean maximum reported temperature difference during the peak temperature period is close to 3 K. The absolute maximum measured temperature difference may be close to 5 K.

Measurements have shown that, in most cases, the temperature at the middle of the canyon is lower than the corresponding air-film temperature. In all these cases the air-film temperature is higher than the undisturbed air temperature measured above the buildings.

Comparison of the surface and air temperatures in canyons, measured during the summer period, shows clearly that in most cases, the surface temperature is higher than the corresponding air temperature. In particular, S, SW and SE facades show temperatures that are higher by up to 13 K. Lower temperature differences were observed on N, NW and NE facades, where the maximum observed temperature difference was close to 10 K. In all these cases the air temperature inside the canyon was higher than the undisturbed air temperature measured above the canyon. In some cases, the air temperature was higher than the corresponding surface temperature by about 3–4 K (in peak conditions). At the same time, the undisturbed ambient temperature was 4–8 K higher than the air temperature. Because of the vertical and horizontal air transport in these canyons, the air temperature close to the surface lies between the temperature of the surface and the undisturbed ambient temperature.

The temperature distribution in a canyon during the night is low. During the summer period, the maximum temperature difference between the different canyon levels never exceeds 1.5 K.[10]

In all cases, higher temperatures are measured at ground level, and the temperature is found to decrease as a function of the height. This matches the distribution of the surface temperature in the canyon during the night and is related to the radiative balance of the canyon surfaces. No significant air temperature differences have been measured between the air temperature close to S, SW and SE facades and N, NW and NE facades. In a general way, S, SW and SE facades have a higher air temperature, but the temperature difference rarely exceeds 0.5 K.[10] The temperature of the air in the middle of the canyon is found to be higher than that of the air film close to the facades of the canyon. In particular, the difference for S, SW and SE facades is close to 0.3 K, while the corresponding difference for N, NW and NE facades is close to 0.7 K.

During the night, the maximum difference between surfaces and the air is close to 2 K. In most canyons, the surface temperature is higher than the corresponding temperature of the air. For some canyons, however, air temperatures higher than the surface temperatures have been recorded. In these canyons it is found that the temperature of the asphalt on the ground is always higher than the air temperature by about 1 K. Thus, there is a convective flow from the street surface to the adjacent air that contributes to increasing its temperature.

The orientation of the streets determines the amount of solar radiation received by the canyon surfaces. The degree to which this influences the air temperature in a street is a very interesting problem. To investigate the effect of the street orientation on the canyon temperature, simultaneous measurements have been performed in three sets of canyons, each composed of two crossing streets.[10] The results indicate that the air temperature measured in the middle of the canyon is not influenced by the orientation of the street, either during the day or during the night. This reinforces the previous conclusions that air temperature in the canyon is not greatly influenced by the canyon configuration and is mainly controlled by the air-flow processes. It can therefore be concluded that the orientation of a street does not influence the bulk air temperature in the canyon, although it has a rather strong influence on the air film developed in the proximity of the facades. In all tested configurations, S or SW facades have been found to have the highest peak temperatures during the day. No significant temperature differences have been found during the night.

AIR FLOW IN THE URBAN CANYON

The pattern of air flow around isolated buildings is well known. It is characterized by a bolster eddy vortex due to flow down the windward facade, while behind there is a lee eddy drawn into the cavity of low pressure resulting from the flow separation at the sharp edges of the building at both top and sides. Further downstream is the building wake characterized by increased turbulence but also by lower horizontal speeds than the undisturbed flow.

As noted above, knowledge of the air flow characteristics in urban canyons is necessary for all studies related to natural ventilation of buildings, pollution studies, thermal comfort, etc.

The air flow patterns in urban canyons have received much attention during recent years. Urban canyons are characterized by three main parameters, as shown in Figure 6.2: H the mean height of the buildings in the canyon, W the canyon width and L the canyon length. Given these parameters, the geometrical descriptors are limited to three simple measures. These are the ratio H/W, the aspect ratio L/H and the building density $j = A_r/A_1$ where A_r is the plan of roof area of the average building and A_1 is the 'lot' area or unit ground area occupied by each building.

Knowledge of the air flow patterns in urban canyons results either from numerical studies or from field experiments in real urban canyons or with scaled physical models in wind tunnels. Most of the existing studies deal with the determination of the pollution characteristics within the canyon and put their emphasis on situations where

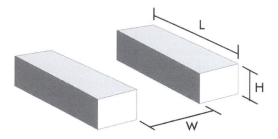

*Figure 6.2. Height, width and
length of a canyon*

the ambient flow is perpendicular to the canyon long axis, when the highest pollutant concentration occurs in the canyon.

The present chapter aims to present the actual state of knowledge concerning the air flow in urban canyons. The chapter presents the phenomena when the flow is perpendicular, parallel or at an angle to the canyon axis.

Perpendicular wind speed

The flow over arrays of buildings has been the subject of many studies. When the predominant direction of the air flow is approximately normal (say ±30°) to the long axis of the street canyon, three types of air flow regime are observed as a function of the building (L/H) and canyon (H/W) geometries (Figure 6.3).[13,14]

When the buildings are well apart, ($H/W > 0.05$), their flow fields do not interact. At closer spacing (Figure 6.3(a)), the wakes are disturbed and the flow regime is known as 'isolated roughness flow'. When the height and spacing of the array combine to disturb the bolster and cavity eddies, the regime changes to one referred to as wake

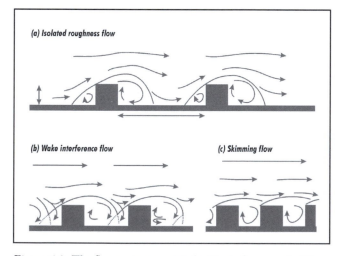

*Figure 6.3. The flow regime associated with air flow over building
arrays of increasing H/W[13]*

interference flow (Figure 6.3(b)). This is characterized by secondary flows in the canyon space, where the downward flow of the cavity eddy is reinforced by the deflection down the windward face of the next building downstream. At even greater H/W and density, a stable circulatory vortex is established in the canyon, because of the transfer of momentum across a shear layer of roof height, and a transition to a 'skimming' flow regime occurs, in which the bulk of the flow does not enter the canyon (Figure 6.3(c)). The existence of the vortex flow within the urban canyon was first measured by Albrecht,[15] and has been confirmed through numerous wind tunnel and field studies.

The transitions between these three regimes occur at critical combinations of H/W and L/W. Oke has proposed threshold lines that divide the flow into three regimes as functions of the building (L/H) and canyon (H/W) geometries.[13] The proposed threshold lines are shown in Figure 6.4.

Because high H/W ratios are very common in cities, the skimming air flow regime has attracted considerable attention. Numerous wind-tunnel and field experiments have been performed and some of the main conclusions are as described in the following paragraphs.

The air flow in the canyon can been seen as a secondary circulation feature driven by the above-roof imposed flow.[16] If the wind speed out of the canyon is below some threshold value, the coupling between the upper and secondary flows is lost,[16] and the relation between the wind speed above the roof and that within the canyon is characterized by considerable scatter. According to De Paul and Shieh,[17] who worked in an almost symmetrical canyon where $H/W = 1.4$, the threshold value is between 1.5 and 2 m/s. Similar values were reported by Nakamura and Oke,[16] who worked with a H/W ratio close to unity (1.06), also in a symmetrical canyon. McCormick also reported a threshold wind speed close to 2 m/s,[18] while Yamartino and Wiegand, who

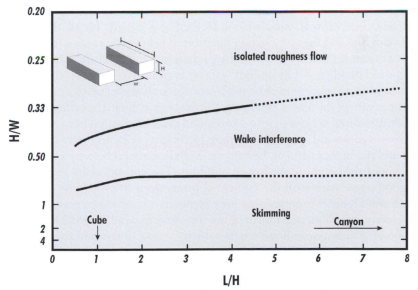

Figure 6.4. Threshold lines dividing flow into three regimes as functions of the building (L/H) and canyon (H/W) geometries[13]

worked on a $H/W = 1$ symmetric canyon, also reported that for almost perpendicular winds, the existence of a vortex in the canyon was the result of a sufficient cross canyon flow out of the canyon.[19] In all these studies, higher wind speeds were found to produce a stable vortex circulation within the canyon. For lower wind speeds, thermal as well as mechanical influences may play an important role in the canyon circulation.

Regarding the relation between the wind speed out of the canyon and the corresponding vortex velocity, De Paul and Shieh reported that, for wind speeds higher than the threshold value, the speed of the vortex increases with the speed of the cross-canyon flow. Yamartino and Wiegand found that the transverse vortex speed inside the canyon is proportional to the above-roof transverse component and is independent of the above-roof longitudinal component.[19] Arnfield and Mills,[20] who worked in an asymmetric canyon with a mean H/W value close to 1.52, found that there was no evidence of dependence between the vortex speed and the horizontal or total wind velocity in step-down and step-up configuration canyons.

Regarding the direction of the vortex, it has to be expected that, as the vortex is driven by a downward transfer of momentum across the roof-level shear zone, a flow normal to the canyon axis has to create a vortex in which the direction of the air flow near the ground should be directly opposite to the wind direction outside the canyon. This was verified for a symmetric canyon configuration by Nakamura and Oke[16] and for a step-up canyon configuration by Hoydysh and Dabbert.[21] Arnfield and Mills reported that in a step-up asymmetric canyon, the vortex direction was consistent with the mechanism described above, although a reversed vortex was detected in some cases with wind speeds below the threshold value of 2 m/s.[20]

As far as the air velocity inside the canyon is concerned, Nakamura and Oke reported that, for wind speeds up to 5 m/s, the general relation between the two wind speeds appears to be linear, $u_{in} = pu_{out}$.[16] For wind speeds normal to the canyon axis, and for a symmetric canyon with $H/W = 1$, p varies between 0.66 and 0.75 under the condition that winds in and out were measured to be about $0.06H$ and $1.2H$ respectively. For normal wind speeds DePaul and Shieih found that downdraft vertical velocities in a canyon a strong function of height and reach a maximum close to 95% of the ambient horizontal velocity, at heights near three quarters of the height of the upwind building.[17] The updraft velocity appeared to be relatively independent of height and had a maximum close to 55% of the ambient velocity, at a height of one half of the height of the upwind building. Vertical velocities in the centre of the canyon have been measured as close to zero. Horizontal velocities varied from zero up to 55% of the free-stream wind speed. The highest horizontal velocities were obtained at the bottom and the higher parts of the canyon. Zero horizontal velocity was measured at about 75% of the height of the lowest building. The authors claim that their results are in good agreement with the numerical predictions of Hotchkiss and Harlow.[22] Hoydysh and Dabbert reported that for a symmetric canyon the average circumferential velocity was about one fourth of the free-stream wind speed (2 m/s), while the corresponding mean ascending vertical velocity was 0.26 m/s and the mean vertical ascending vertical velocity was 0.24 m/s.[21] For a step-up canyon configuration, the mean circumferential velocity was between 1.02 and 1.07 m/s, while the corresponding mean ascending and descending vertical velocities were 0.34 and 0.32 m/s respectively. Measurements by Albrecht and Grunow show that as horizontal roof-level wind varies from 1 to 3 m/s, the ascending and descending currents vary

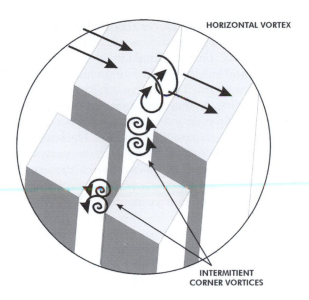

HORIZONTAL VORTEX

INTERMITTENT
CORNER VORTICES

Figure 6.5. Intermittent vortices at the building corner

proportionally from 0.1 to 1 m/s.[23] Numerical simulations by Lee *et al.*, in a symmetric canyon with an undisturbed wind speed of 5 m/s flowing perpendicular to the canyon, show that the strength of the vortex developed inside the canyon was less than the wind speed above the roof level by about an order of magnitude.[24]

Very few studies have been performed on the length of the vertical displacement of the vortex in a canyon. For normal wind speeds, Hoydysh and Dabbert reported that for a symmetric canyon the average vertical displacement of the vortex was of the same magnitude as the canyon width, while in a step-up canyon the vortex was smaller and the mean vertical displacement was equal to 0.61 of the canyon width.[21]

End effects, or finite-length canyon effects, play an important role in the air flow distribution in canyons. Yamartino and Wiegand reported that when $L/W \approx 20$, finite-length canyon effects begin to dominate over the vortex.[19] Hoydysh and Dabberdt reported that intermittent vortices are created at building corners.[21] These vortices are responsible for the mechanism of advection from the building corners to the mid-block position, so that a convergence zone is created in the mid-block region of the canyon, resulting in larger concentrations there, both at street level and higher up (Figure 6.5). Similar phenomena have been reported by Santamouris *et al.*[25]

Hoydysh and Dabbert reported that, when the free-stream winds are perpendicular to the canyon axis, for both symmetric and step-up canyons the pollutant concentrations are a factor of two or more greater for the leeward than the windward facade.[21] Almost same results have been reported by other field studies for symmetric canyons,[26–30] as well as by fluid modelling studies.[31–39] Lee *et al.* reported rather different results of numerical simulations in a canyon with $H/W = 1$, with a free-stream wind velocity of 5 m/s, perpendicular to the canyon.[24] They found that the maximum concentration coincides with the vortex centre, which is located at the windward side of the canyon at 0.7 of the building height. For a step-down configuration, Hoydysh and Dabbert reported that the concentration in the windward facade was slightly greater than to leeward.[21] Zhang and Huber reported that the concentration at the windward facade of a step-down configuration was 2.5 times higher that the concentration at the leeward facade.[39] They explained the flow pattern in such a configuration: in their case the upwind building was 1.5 times higher than the downwind one and the flow past it was downward. The downward flow had the tendency to form a large wake behind the building. As there was a second building within this potential wake, the flow hit the roof of the second building and tended to move up upstream, forming two counter-direction recirculating vortices. The larger and stronger clockwise one overlapped the step-down building and the canyon itself generated a secondary counterclockwise vortex at the bottom of the canyon. Hoydysh and Dabbert reported that concentrations are generally a factor of two lower in a step-up canyon than in symmetric and step-down canyons.[21]

Perpendicular wind speed in deep canyons
In deep canyons, wind tunnel research has shown that two vortices are developed, an upper one driven by the ambient air flow and a lower one driven in the opposite direction by the circulation higher up (Figure 6.6).[40] Santamouris *et al.* performed air-flow measurements in a deep SE–NW canyon in Athens with H/W = 2.5.[25] For perpendicular air flow, measured data indicate strongly the creation of a single vortex driven by the ambient air flow. When all wind speeds are considered, in almost two thirds of cases the air moved down the windward facade and in a direction from SE to NW along the canyon. The direction of the air speed measured across the canyon is almost always opposite to the wind direction out of the canyon, but measured values are small and less than 0.1 m/s. The mean measured downward vertical air speed was close to 12% of the horizontal ambient wind speed, 0.25 m/s, with a standard deviation

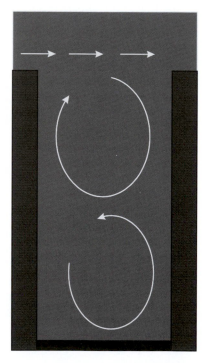

Figure 6.6. The two-vortex regime in deep canyons

close to 10%, 0.19 m/s. The downward air flow was found to create an angle relative to the vertical axis of between 25° and 40°.

Direct comparison of the wind speed out of the canyon with the secondary air-flow components inside the canyon shows a considerable scatter independent of the wind speed. The lack of a clear threshold wind-speed value for which the link between the wind speeds in and out the canyon can be established is mainly due to the following. The air circulation inside the canyon is not only due to the ambient wind flow, but is the resultant of three specific mechanisms, the ambient air flow above the canyon, the vertical stratification of the air inside the canyon, which can reach values up to 6 K, and the mechanism of advection from the building corners that results from end effects. Further analysis has shown that, when stratification does not contribute to the air circulation, the speed of the vortex increases with the transverse ambient wind component below a threshold cross-canyon wind speed of between 2 and 3 m/s.

Santamouris *et al.* also reported that for almost one third of cases a counterclockwise air motion is observed.[25] The measured air-flow characteristics strongly indicated the existence of a second, counter-clockwise, vortex inside the canyon. This is in agreement with the results of Chang *et al.*,[40] who observed such a double-vortex regime within the canyon. According to Chang *et al.*, the upper vortex rotates in a clockwise direction and transmits the energy that causes the lower part of the fluid to rotate in a counterclockwise vortex motion. The existence of the double vortex is considered to be essentially a phenomenon with a moderately low Reynolds number, although it has not been possible to determine a critical Reynolds number. Arnfield and Mills reported that in a step-up asymmetric canyon with low wind speeds below the threshold value of 2 m/s, a reversed vortex was detected in some cases.[20] However, in both of the above studies the role of the wind was the main determinant because temperature stratification in the canyon was not important.

In the results reported by Santamouris *et al.*,[25] counterclockwise motion occurred only when the temperature difference between the lower-level and mid-height air-temperature layers close to the south-oriented facade of the canyon was positive. This usually happened during the period around midday when first the NE facade and then the floor of the canyon were strongly irradiated. Two hours before noon, when the NE facade was irradiated, the surface temperature on the upper floors of the facade was to about 6–10 K higher than on the lower ones or at ground level. This resulted in the development of a downward air flow along the NE facade of the canyon. At midday, when the canyon floor was sunlit, the difference between the surface temperature of the canyon floor close to the SW facade and that of air above it reached 8–12 K. This stratification contributed to the warming of the air and resulted in an upward air flow along the SW wall. The role of the ambient wind speed was also important. It was found that an increase in the ambient wind speed, combined with an air-temperature stratification in the canyon, increased both the vertical wind speed and the wind speed along the canyon.

A possible explanation of the mechanism is that temperature stratification drives an upward air movement along the SW facade of the building and at the lower layers of the canyon it is not possible for the upper vortex to be extended downwards. Under these conditions, higher ambient winds contribute to the transmission of more energy from the upper to the lower vortex and thus increase its speed.

Figure 6.7. Air flow characteristics along the canyon axis

Flow along the canyon

Air-flow characteristics in an urban canyon with an air flow parallel to the canyon axis have been studied in many field and wind-tunnel experiments. As in the case of perpendicular winds, the air flow in the canyon has to be seen as a secondary circulation feature, driven by the above-roof imposed flow.[16] If the wind speed out of the canyon is below some threshold value, the coupling between the upper and secondary flows is lost,[16] and the relationship between wind speeds above and below roof level is characterized by a considerable scatter. For higher wind speeds, the main results and conclusions resulting from the existing studies are as discussed in the following paragraphs.

Parallel ambient flow generates a mean wind along the canyon axis,[16, 36] with possible uplift along the canyon walls as the air flow is retarded by friction at the building walls and the street surface (Figure 6.7).[41] This is verified by Arnfield and Mills,[20] who found that with no winds along the canyon the mean vertical canyon velocity is close to zero. Measurements performed in a deep canyon have also shown a flow along the canyon in the same direction.[25] The flow is characterized by a velocity along the canyon that is almost always parallel to the axis of the canyon and that has a downward angle of incidence relative to the canyon floor of between 0° and 30°.

Yamartino and Wiegand reported that the along-canyon component of the wind velocity v, within the canyon, is directly proportional to the along-canyon component of the above-roof free-stream wind speed U, the constant of proportionality being a function of the approach flow azimuth.[19] The same authors they found that $v = U \cos$

q, at least to first order, where q is the angle of incidence and U the horizontal wind speed out of the canyon. Nakamura and Oke reported that, for wind speeds up to 5 m/s, the general relationship between the two wind speeds appears to be linear, $v = pU$.[16] For wind speeds parallel to the canyon axis, and for a symmetric canyon with $H/W = 1$, Nakamura and Oke found that p varies between 0.37 and 0.68 under the condition that winds in and winds out are measured at about $0.06H$ and $1.2H$ respectively (H is the height of the buildings). Low p values are obtained as a result of the deflection of the flow by a side canyon. Measurements performed in a deep canyon with $H/W = 2.5$ have not shown any clear threshold value where coupling is lost.[25] Moreover, the correlation between the ambient speed parallel to the canyon and the wind speed along the canyon inside the canyon was not clear. This was mainly because most of the data corresponded to ambient wind speeds lower than 4 m/s, where the relationship between the two wind speeds is not clear. However, a statistical analysis of the data has shows that there is a statistical correlation between them. Figure 6.8 shows the variation of the median, upper and lower quartiles, as well as the outliers, of the along-canyon wind speed inside the canyon for four classes of increasing ambient wind speed. As shown in the figure, both the median and the quartiles increase as the ambient wind speed increases. The existence of outliers with a high wind speed inside the canyon but low ambient wind speeds does not allow any firm conclusions to be drawn from Figure 6.8.

The mean vertical velocity at the canyon top resulting from mass convergence or divergence in the along-canyon component of flow w can be expressed as:[20]

$$w = -H\, \partial v/\partial x, \qquad\qquad\qquad (6.3)$$

where H is the height of the lower canyon wall, x is the along-canyon coordinate and v is the along-canyon (x) component of motion within the canyon averaged over time and over the canyon cross section. Arnfield and Mills,[20] as well as Nunez and Oke,[41] found a linear relationship between the wind gradient $\partial v/\partial x$ in the canyon and the along-canyon wind speed. According to Arnfield and Mills, the value of $\partial v/\partial x$ varies

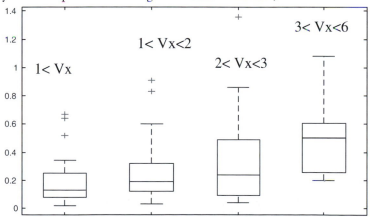

Figure 6.8. Box plot of wind speed inside the canyon for various values of the ambient wind speed for wind directions parallel to the canyon:[25] (a) $0 < V_x < 1$; (b) $1 < V_x < 2$; (c) $2 < V_x < 3$; (d) $3 < V_x < 6$ m/s

between -6.8×10^{-2} and 1.7×10^{-2} s^{-1}, while, according to Nunez and Oke, $\partial v/\partial x$ varies between -7.1×10^{-2} and 0 s^{-1}.

As far as the relationship between the vertical wind speeds at the top of the canyon and the along-canyon free-stream wind speed, Arnfield and Mills reported that they found that the vertical wind speed increases with the along-canyon free-stream velocity.[20] When the free-stream wind travels down a short section of the canyon and is still actively decelerating in response to the sudden imposition of the canyon facets, the relationship between the two wind speeds is almost linear. A positive, but much more scattered, relationship between the vertical wind speed and the along-canyon free-stream wind has been found for winds that have penetrated much further into the canyon and attained a partial equilibrium with the frictional effect of the walls and floor. In this case, the deceleration is reduced, as is the vertical outflow at the top of the canyon. In the results reported by Santamouris *et al.*,[25] an important downward flow is reported. The downward air movement close to the canyon walls could be the result of finite-length canyon effects associated with intermittent vortices created at the building corners, which are also responsible for the mechanism of downward advection flow from the building corners to the mid-block position in the canyon.[19,22]

Flow at an angle to the canyon axis

The most common case is where the air flows at a particular angle relative to the long axis of the canyon. Unfortunately, the volume of research to date on this topic is considerably smaller than for perpendicular and along-canyon flows. While some of the results that are available come from limited field experiments, they are mainly the result of wind-tunnel experiments and numerical calculations.

The main result of research so far has been that, when the flow above the roof is at some angle to the canyon axis, a spiral or corkscrew vortex is induced along the length of the canyon (Figure 6.9).[16]

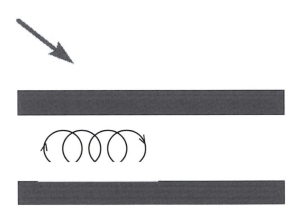

Figure 6.9. Corkscrew type of flow

Wind-tunnel research has also shown that a helical flow pattern develops in the canyon.[25,28,36] According to Yamartino and Wiegand,[19] for intermediate angles of incidence to the canyon long axis, the canyon air flow is the product of both the transverse and parallel components of the ambient wind, the former driving the canyon vortex and the latter determining the along-canyon stretching of the vortex.

Regarding the direction of the helical flow, Nakamura and Oke reported that, to a first approximation, the angle of incidence at the windward wall is the same as the angle of reflection from the wall that causes the return flow of the spiral vortex across the canyon floor.[16] However, Nakamura and Oke found some evidence that the angle of incidence is greater than the angle of reflection and that this could be caused by along-wind entrainment in the canyon.

Regarding the wind speed inside the canyon, Lee et al., reported the results of numerical studies in a canyon with $H/W = 1$ and a free-stream wind speed equal to 5 m/s, flowing at 45° to the long axis of the canyon.[24] Lee et al. reported that a vortex was developed inside the canyon with a strength that was less than the wind speed above the roof level by about an order of magnitude. The maximum cross-canyon air speed inside the canyon was 0.6 m/s and this occurred in the highest part of the canyon. The vortex was centred at the upper middle part of the cavity and specifically at about 0.65 of the height of the buildings. The maximum along-canyon wind speed was close to 0.8 m/s. Much higher along-canyon wind speeds (0.6–0.8 m/s)were reported for the downward facade than for the upward facade (0.2 m/s). The maximum vertical wind speed inside the canyon was close to 1.0 m/s. Much higher vertical velocities (0.8–1.0 m/s) were reported for the downward facade than for the upward one (0.6 m/s). The results of Santamouris et al. show that an increase in the ambient wind speed almost always corresponds to an increase of the along-canyon wind speed, for the median, lower and upper quartiles of the speed.[25]

Hoydysh and Daberdt reported the results of wind tunnel studies on the distribution of pollutant concentration in symmetric, even, step-down and step-up canyons, when the wind flows at a certain angle to the canyon axis.[21] They calculated the wind angle for which the minimum concentration occurs. They reported that, for the step-down configuration, the minimum concentration occurs for along-canyon winds (angle of incidence equal to 90°). For a symmetric, even, configuration the minimum on the leeward facade occurs for an angle of incidence of 30°, while on the windward facade the minimum occurs for angles between 20° and 70°. Finally, for step-up canyon configurations, the minimum on the leeward facade occurs at incidence angles between 0° and 40°, while for the windward facade the minimum is found for incidence angles between 0° and 60°.

PREDICTION OF THE AIR SPEED IN CANYONS

Knowledge of the air speed inside urban canyons is of great importance for many reasons, but mainly for passive cooling applications and, in particular, for naturally ventilated buildings. Various methods, both simple and detailed, have been proposed for calculating the wind speed inside a canyon.

Nakamura and Oke suggested the following simple linear equation for calculating the mean horizontal wind speed u_b inside a canyon:[16]

$$u_b = pu_{\text{roof}}, \tag{6.4}$$

where p is a reduction factor that depends on H/W and on the measurement levels. For wind speeds up to 5 m/s, with $H/W = 1$ and the canyon centre and above-roof measurements at heights of about $0.06H$ and $1.2H$ respectively, $p = 2/3$. At smaller H/W, p approaches unity and there is no longer any shelter.

Nicholson has proposed a simple model for calculating the vortex circulation produced in street canyons when the wind blows perpendicular to the street.[42] By applying mass-conservation techniques to the layer of air between the height of the building h_b and the height h' at which the effects of the canyon on the overall flow become negligible, Nicholson proposed the following expression for calculating a representative speed of the upward current of the vortex w_m (Figure 6.10):

$$w_m = 2(h' - h_b)([u]_B - [u]_A)/W, \tag{6.5}$$

where [ ] implies an average value across the layer from h_b to h', W is the canyon width and subscripts A and B refer to the locations in Fig 6.10.

According to Mills,[43] h' may be calculated from:

$$h' = (0.15d + z_0 h_b)/(0.15 + z_0), \tag{6.6}$$

where d is the zero-plane displacement, which is calculated with equations (3.10) and (3.11). Values of z_0 can be obtained from Table 3.1.

For winds parallel to the street, the exponential profile given in equation (3.1), using z_0 given by equation (3.4), can be used to predict the wind velocity inside the canyon.

Yamartino and Wiegand proposed a more representative canyon velocity calculated from the following expression:[19]

$$V_c = (w_m^2 + u_m^2)^{0.5}, \tag{6.7}$$

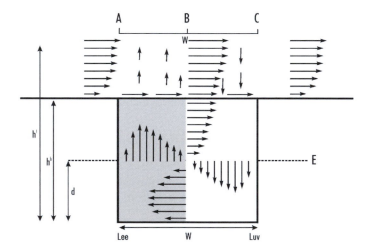

Figure 6.10. Theoretical wind profiles above city buildings and in street canyons for the case of wind blowing perpendicular to the street[42]

where w_m is the wind speed due to a vortex and u_m represents the along-canyon flow at mid-canyon height. Figure 6.11 (from Mills[43]) shows the relationship between V_c, w_m and u_m as a function of ambient wind azimuth.

Paciuk used wind tunnel experiments to try to identify the effects of building height and of distances between buildings on the wind speed in the open spaces between the buildings, when the buildings are perpendicular to the wind direction.[44] He developed a formula predicting the relative wind speed:

$$V_{r(u,b)} = 10 + (66(q - e^{-0.08b}))e^{-0.18D/W}, \qquad (6.8)$$

where $V_{r(u,b)}$ is the relative wind speed expressed as a percentage of the wind speed at the same height well away from the first line of buildings, D is the distance travelled by the wind in metres ($D = n(b + W) - 0.5W$), b is the depth of the buildings in metres, n is the serial number of the space (downwind), h is the height of the buildings in metres and W is the width of the space between the buildings in metres.

The formula shows that, as the wind approaches an urban area of long buildings with uniform height perpendicular to the wind direction, the initial turbulence 'agitation' above the first rows of buildings gradually declines until a uniform wind speed is reached in the spaces between the buildings.

Hotchkiss and Harlow proposed a model for calculating the canyon transverse air speed, $(w^2 + u^2)^{0.5}$, where u and w are the cross-canyon and vertical air speeds.[22] The model assumes incompressible flow, an absence of vortex sources and sinks within the canyon and appropriate boundary conditions for the simple 2D rectangular notch of depth H and width B. The following expressions for the u and w air speed components are proposed:

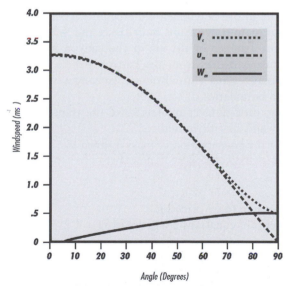

Figure 6.11. The relationship between V_c, w_m and u_m as a function of ambient wind azimuth (expressed as the angle of approach to the canyon axis).[43] The ambient wind speed is constant at 5 m/s

$$u = u_0(1-\beta)^{-1}[\gamma(1+k\gamma)-\beta(1-k\gamma)/\gamma]\sin(kx) \qquad (6.9)$$

and

$$w = -u_0(1-\beta)^{-1}ky[\gamma-\beta/\gamma]\cos(kx), \qquad (6.10)$$

where $k = \pi/B$, $\beta = \exp(-2kH)$, $\gamma = \exp(ky)$, $y = z - H$ and u_0 is the wind speed above the canyon at the point $x = B/2$, $z = H$.

This model was tested by Yamartino and Wiegard with very great success.[19] The same authors proposed the following expression to calculate the along-canyon component $v(z)$:

$$v(z) = v_r \log[(z+z_0)/z_0]/\log[(z_r+z_0)/z_0], \qquad (6.11)$$

where v_r is the value at reference height and z_r and z_0 are the surface roughnesses.

URBAN CANYON MODELS

As already discussed, the urban canopy layer is concerned with an unlimited number of microclimates generated by the various urban configurations. Canopy-layer models aim, in general, to simulate the thermal and flow characteristics in similar and common urban structural forms, such as the urban canyons. Canopy-layer models are, in general, energy-balance models for calculating the temperature distribution of the air and of the surfaces in the canyon.

Various complete urban-canyon models or models that predict specific parts of canyon models, have been proposed.[41,36,43,45-51]

Urban-canyon models are used to calculate the energy budget of each surface or panel in the canyon, as well as the energy balance of the air in the canyon. The energy balance of each panel in the canyon should take into account short- and long-wave radiation exchanges, convective gains and losses and subsurface conductive heat transfer. The energy balance of the air in the canyon should take into account convective heat gains or losses from the canyon panels, possible latent-heat exchanges, anthropogenic gains and advective exchanges with the surrounding air through the canyon boundaries.

In a general way, canopy-layer models solve the transport equations for mass, momentum, energy and specific humidity.[52]

The equation for the conservation of mass is given by:

$$\partial u_j / \partial x_j = 0, \qquad (6.12)$$

where u_j is the velocity in the x_j direction.

The three momentum equations are given by ($I = 1, 2, 3$):

$$\partial u_i / \partial t + u_j \partial u_j / \partial x_j = 1/\rho \, \partial p / \partial x_i - g_i + \partial[K_M \partial u_j / \partial x_j]/\partial x_j, \qquad (6.13)$$

where p is the pressure, g_i is the gravitational acceleration, ($g_i = 0$ for $i = 1$ or 2), ρ is the density, and K_M is the turbulent exchange coefficient for momentum. The Bousinesq approximation may be applied to the vertical momentum equation.

The energy equation is given by:

$$\partial T / \partial t + u_j \partial T / \partial x_j = Q_T / (\rho_0 C_p) + \partial [K_H \partial T / \partial x_j] / \partial x_j, \qquad (6.14)$$

where T is temperature, Q_T is the energy source term, C_p is constant-pressure specific heat and K_H is the thermal conductivity.

Although estimation of the radiation and conductive heat transfer is a routine calculation procedure, exact knowledge of the convective and advective exchanges requires an estimation of the air field both inside and at the boundaries of the canyon. Where the air speed is low, advective exchanges can be neglected and convective heat transfer can be calculated easily using standard knowledge of natural convection.[53]

Urban-canyon models are very useful tools for calculating the main climatic characteristics of the urban canopy layer. However, their main drawback is that, since their vertical extent terminates at several different building heights, accurate ambient boundary conditions should be known at these heights. Another problem is related to the exact knowledge of the surface properties, and in particular of the surface roughness, required by these models. As stated by Sailor,[52] surface roughness is intimately involved in determining atmospheric turbulence and local wind profiles in the vertical domain of interest. Thus, the use of a canyon model requires detailed specification of the urban geometry, a task than can be very tedious, especially in studies of large urban areas.

Earlier in this chapter, some models for predicting the air flow inside urban canyons were discussed. Some of these models have been used together with thermal-balance models for the canyon panels and air in order to develop complete heat- and mass-transfer energy canyon models. In the following sections, two of the more representative canyon models are presented. The Mills model[43] is based on the principle of control volumes and uses the canyon air flow model proposed by Nicholson.[42] This model is more appropriate for predicting canyon temperature distribution for air flows vertical to the canyon axis. The Nunez and Oke canyon model[41] is more appropriate for air flows parallel to the canyon and considers the air in the canyon as a single volume.

The Mills canyon model

Mills[43] has proposed an energy-canyon model using the air flow model of Nicholson,[42] as described above. The model is more appropriate for air flows vertical to the canyon axis. Mills proposes using expression (6.5) to calculate the vertical velocity of the air that enters or leaves the canyon at its top. The model divides the canyon into control volumes (Figure 6.12). Then, the energy budgets of the canyon surfaces and of the whole canyon unit are solved by considering the energy budgets for each control volume and its associated panels.

For each panel the energy balance is given by the expression given in equation (6.1), while the radiation balance is given by equation (6.2). The model proposes using measured values of the solar radiation or estimating it with atmospheric models like that of Davies and Hay.[54] It is proposed that the incident long-wave radiation is calculated with the empirical model of Idso and Jackson.[55] Multiple reflection of the short- and long-wave radiation is taken into account using the technique proposed by Arnfield.[47]

All the remaining terms in the energy budget are functions of the unknown surface temperature T_s and are adjusted continuously as there is convergence towards an

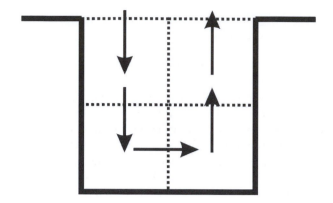

Figure 6.12. Control volumes used by Mills[44] to model the canyon energy budget

equilibrium surface temperature.

The model calculates the emitted long-wave radiation using $L\uparrow_e = \varepsilon\sigma T_s^4$, where ε is the surface emissivity and σ is the Stefan–Boltzmann constant. Conductive heat exchanges are modelled using the classical expression:

$$Q_G = k_s\{(T_s - T_i) / d_s\},\qquad\qquad(6.15)$$

in which T_i represents the temperature at a distance d_s within the substrate and k_s represents the conductivity of the material. Convective heat exchanges are calculated using the classical formula $Q_H = h_c(T_s - T_a)$, where T_a represents the temperature of the air in the adjacent control volume. It is proposed that the heat transfer coefficient h_c be evaluated using the formula proposed by Rowley and Eckley:[56,57]

$$h_c = (11.8 + 4.2V_c)\qquad\qquad(6.16)$$

where V_c is a representative wind speed for the canyon air volume.

Once an equilibrium surface temperature that satisfies the energy balance is found, the temperature of the air contained in the control volume is calculated. The model begins calculating at the panel adjacent to the control volume, where the ambient air of known temperature is drawn in. Then, the panel energy budgets are solved in a sequence determined by the direction of rotation of the canyon circulation.

By applying the energy budget to the air of each control volume, the exit temperature from the control volume is calculated:

$$T_{a(\text{exit})} = T_{a(\text{enter})} + \{[Q_H L_p t] / A_c C_a\},\qquad\qquad(6.17)$$

where L_p refers to the length of the panel and A_c to the cross-sectional area of the control volume, C_a is the heat capacity of air and t is the length of time taken by the air to traverse L_p.

This process is completed for each control volume in the sequence of the air circulation until the air leaves the canyon at a new temperature. If it is taken into account that the air leaves the canyon top with a vertical velocity w_m, (equation 6.5) the heat exchange at this point can be calculated from:

$$Q_{Ht} = (1-R_c)w_m\{C_a(T_{out}-T_{in})_t\} = (1-R_c)w_m\{C_a\Delta T_t\}, \qquad (6.18)$$

where R_c represents the proportion of the circulating canyon air that does not exit, but instead enters the adjacent control volume at the top of the canyon.

Once the calculations for all control volumes have been completed, each surface in the canyon has a new temperature that alters the incident long-wave radiation at every other panel in the canyon. The process is repeated until the individual panel energy budget remain stable; the model is considered to have reached a stable solution when each new value for Q_{Ht} differs from the previous value by less than 1 W/m².

The Mills model has been compared with experimental data and found to give satisfactory predictions.[58] Further comparisons with experimental data have been reported by Koronakis et al.[59] Predictions of the surface and air temperatures have been compared with experimental data collected during a long monitoring period under conditions of very low wind speed. Figures 6.13 and 6.14 show the calculated and measured values of the surface and air temperatures. As can be seen, the agreement achieved is very satisfactory for both the air and surface temperatures and the difference does not exceed 1.0 K.

The Nunez and Oke canyon model

According to this model, the energy balance of the *i*th canyon surface is given by the following:

$$Q_i^* = Q_{Hi} + Q_{Ei} + Q_{Gi}, \qquad (6.19)$$

where Q_{Hi} are the sensible heat exchanges, Q_{Ei} are the latent heat exchanges and Q_{Gi} is the subsurface heat flux.

The energy balance of the canyon air volume for a south–north oriented canyon, assuming that there are no water phase changes in the volume, is given by:

$$Q_A + \{H(Q_{He} + Q_{Hw}) + W(Q_{Hf} - Q_{Ht})\}L = \rho c_p \int_V (\partial\Theta/\partial t + \mathrm{div}\,Q^*/\rho c_p)\,dV, \qquad (6.20)$$

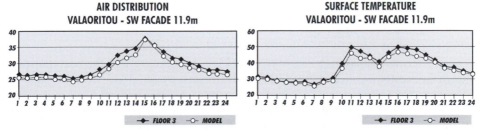

Figure 6.13. Comparison of measured and calculated air temperatures over a 24-hour period at the Valaoritou urban canyon[60]

Figure 6.14. Comparison of measured and calculated surface temperatures over a 24-hour period at the Valaoritou urban canyon[60]

where Q_A is the net advective heat due to horizontal transport, H is the canyon height, W is the canyon width, L the nominal canyon length, ρ is the air density, c_p is the specific heat of air, Θ is the time-averaged air temperature, t is time, Q^* is the net all-wave radiation flux and V the canyon air volume. The subscripts e and w indicate the canyon east and west walls respectively, while t and f indicate the canyon top and floor.

For a flow parallel to the canyon sides and when there is no water phase change, such that the energy storage change and the radiative flux divergence are negligible, then equation (6.20) can be rewritten as:

$$\rho c_p \{\int_{A_1} u_1 \theta_1 \, dA_1 - \int_{A_2} u_2 \theta_2 \, dA_2 - \int_{A_3} u_3 \theta_3 \, dA_3 \}_{\text{time average}} + HL(Q_{He} + Q_{Hw}) + Q_{Hf} A_3, \tag{6.21}$$

where $A_1 = WH$, (upwind face), and $A_3 = LW$. Also, u_1 and θ_1 are the speed and temperature of the air entering the canyon parallel to its axis (upwind face), u_2 and θ_2 the corresponding values of the air when it exits the canyon from the downwind face and u_3 and θ_3 the velocity and temperature of the air when it exits from the top of the canyon.

Nunez and Oke proposed expanding the integrals of equation (6.21) using the relations:

$$u_i = u_{mi} + \Delta u_{mi}, \quad \theta_i = \theta_{mi} + \Delta\theta_{mi}, \tag{6.22}$$

where subscript m denotes a space average over the area of a canyon air volume side. If it assumed that the integral of all fluctuating variables is negligible, then equations (6.22) can be rewritten in terms of spatial mean quantities:

$$\rho c_p \{u_{m1}\theta_{m1} A_1 - u_{m2}\theta_{m2} A_2 - u_{m3}\theta_{m3} A_3\} + (HL(Q_{He} + Q_{Hw}) + Q_{Hf} A_3 = 0 \tag{6.23}$$

and the following substitutions can be made:

$$u_{mi} = U_{mi} + u'_{mi}, \quad \theta_{mi} = \Theta_{mi} + \theta'_{mi}, \quad U_{m2} = U_{m1} - \Delta U_{m1}, \quad \Theta_{m2} = \Theta_{m1} \pm \Delta\Theta_{m1}, \tag{6.24}$$

where the primes indicate instantaneous departures from the mean, U is the time-averaged wind speed and Θ the time-averaged air temperature.

Substituting equation (6.24) into (6.23) and neglecting divergence in the horizontal turbulent transport term, i.e. $u'_{m1}\theta_{m1} - u'_{m2}\theta_{m2} \cong 0$, gives:

$$\rho c_p \{\pm\Delta\Theta_{m1}(U_{m1} - \Delta U_{m1}) + \Theta_{m1}\Delta U_{m1})A_1 - (U_{m3}\Theta_{m3} + u'_{m3}\theta'_{m3})A_3\} + HL(Q_{He} + Q_{Hw}) + Q_{Hf} A_3 = 0 \tag{6.25}$$

The first term in equation (6.25), $\rho c_p(\pm\Delta\Theta_{m1}(U_{m1} - \Delta U_{m1})$ represents the heat transport due to the mean wind, in combination with the mean horizontal temperature gradient. This term has an upper limit determined by the fact that increased wind speeds will destroy the temperature gradient $\Delta\Theta_{m1}$. The second and third terms in equation (6.25), i.e. $\rho c_p(\Theta_{m1}\Delta U_{m1}A_1 - U_{m3}\Theta_{m3}A_3)$, are due to the frictional retardation of the air flow along the canyon surfaces. As stated by Nunez and Oke, deceleration of the horizontal velocity must result in uplift, since both resulted from the equation of continuity for incompressible flow:

$$U_{m3} = \Delta U_{m1} A_1 / A_3. \tag{6.26}$$

The vertical flow U_{m3} will transport energy out of the canyon volume. By substituting U_{m3} in the previous expression, the advective term due to vertical transport can be calculated as:

$$\rho c_p \Delta U_{m1} A_1 (\Theta_{m1} - \Theta_{m3}), \tag{6.5.2.9}$$

such that if $\Theta_{m1} > \Theta_{m3}$ the air volume will be warmed, while if $\Theta_{m1} < \Theta_{m3}$ it will be cooled by the vertical transport.

REFERENCES

1. Oke, T.R. (1987). *Boundary Layer Climates*. Cambridge University Press.
2. Oke, T.R. (1976). 'The Distance between Canopy and Boundary Layer Urban Heat Island'. *Atmosphere*, Vol. 14, No. 4, pp. 268–277.
3. Park, H.S. (1987). 'City Size and Urban Heat Island Intensity for Japanese and Korean cities'. *Geographcial Review of Japan*, Sep. A, Vol. 60, pp. 238–250.
4. Yamashita, S, Sekine, K., Shoda, M., Yamashita, K. and Hara, Y. (1986). 'On the Relationships between Heat Island And Sky View Factor in the Cities of Tama River Basin, Japan'. *Atmospheric Environment*, Vol. 20, No. 4, pp. 681–686.
5. Barring, L., Mattsson, J.O. and Lindovist, S. (1985). 'Canyon Geometry, Street Temperatures and Urban Heat Island in Malmo, Sweden'. *Journal of Climatology*, Vol. 5, pp. 433–444.
6. Eliason, I. (1990/91). 'Urban Geometry, Surface Temperature and Air Temperature'. *Energy and Buildings*, Vol. 15–16, pp. 141–145.
7. Eliasson, I. (1996). 'Urban Nocturnal Temperatures, Street Geometry and Land Use'. *Atmospheric Environment*, Vol. 30, No. 3, pp. 379–392.
8. Arnfield, A.J. (1990). 'Canyon Geometry, the Urban Fabric and Nocturnal Cooling: a Simulation Approach'. *Physical Geography*, Vol. 11, pp. 209–239.
9. Yap, D. (1975). 'Seasonal Excess Urban Energy and the Nocturnal Heat Island – Toronto', *Archives for Meteorology, Geophysics, and Bioclimatology*, Ser. B23, pp. 68–80.
10. Santamouris, M., Papanikolaou, N. and Koronakis, I. (1997). 'Urban Canyon Experiments in Athens – Part A. Temperature Distribution'. Internal Report to the POLIS Research Project, European Commission, Directorate General for Science, Research and Technology. Available from the authors.
11. Roth, M., Oke, T.R. and Emery, W.J. (1989). 'Satellite-derived Urban Heat Islands From Three Coastal Cities and the Utility of such Data in Urban Climatology'. *International Journal of Remote Sensing*, Vol. 10, pp. 1699–1720.
12. Stoll, M.J. and Brazel, A.J. (1992). 'Surface Air–Temperature Relationships in the Urban Environments of Phoenix, Arizona'. *Physical Geography*, Vol. 13, pp. 160–179.
13. Oke, T.R (1988). 'Street Design and Urban Canopy Layer Climate'. *Energy and Buildings*, Vol. 11, pp. 103–113.
14. Hussain, M. and Lee, B.E (1980). 'An Investigation of Wind Forces on Three-dimensional Roughness Elements in a Simulated Atmospheric Boundary Layer Flow – Part II. Flow over Large Arrays of Identical Roughness Elements and the Effect of Frontal and Side Aspect Ratio Variations'. Report No BS 56, Department of Building Sciences, University of Sheffield.
15. Albrecht, F. (1933). 'Untersuchungen der verticalen Luftzirkulation in der Grossstadt'. *Meteorologische Zeitschrift*, Vol. 50, pp. 93–98.
16. Nakamura, Y and Oke, T.R (1988). 'Wind, Temperature and Stability Conditions in an E–W Oriented Urban Canyon'. *Atmospheric Environment*, Vol. 22, No. 12, pp. 2691–2700.
17. De Paul, F.T. and Shieh, C.M. (1986). 'Measurements of Wind Velocities in a Street Canyon'. *Atmospheric Environment*, Vol. 20, pp. 455–459.

18. McCormick, R.A. (1971). 'Air Pollution in the Locality of Buildings'. *Philosophical Transactions of the Royal Society of London*, Ser. A., Vol. 269, pp. 515–526.
19. Yamartino, R.J. and Wiegand, G. (1986). 'Development and Evaluation of Simple Models for the Flow, Turbulence and Pollution Concentration Fields within an Urban Street Canyon'. *Atmospheric Environment*, Vol. 20, pp. 2137–2156.
20. Arnfield, A.J. and Mills, G. (1994). 'An Analysis of the Circulation Characteristics and Energy Budget of a Dry, Asymmetric, East, West Urban Canyon. I. Circulation Characteristics'. *International Journal of Climatology*, Vol. 14, pp. 119–134.
21. Hoydysh, W. and Dabbert, W.F. (1988). 'Kinematics and Dispersion Characteristics of Flows in Asymmetric Street Canyons'. *Atmospheric Environment*, Vol. 22, No. 12, pp. 2677–2689.
22. Hotchkiss, R.S. and Harlow, F.H. (1973). 'Air Pollution Transport in Street Canyons'. Report by Los Alamos Scientific Laboratory for US Environmental Protection Agency, EPA-R4-73-029, NTIS PB-233 252.
23. Albrecht, F. and Grunow, J. (1935). 'Ein Beitrag zur Frage der vertikalen Luftzirkulation in der Grossstandt'. *Meteoroligische Zeitschrift*, Vol. 52, pp. 103–108.
24. Lee, I.Y., Shannon, J.D. and Park, H.M. (1994). 'Evaluation of Parameterizations for Pollutant Transport and Dispersion in an Urban Street Canyon using a Three-dimensional Dynamic Flow Model'. *Proceedings of the 87th Annual Meeting and Exhibition of the American Meteorological Society*, Cincinnati, Ohio, 19–24 June.
25. Santamouris, M., Papanikolaou, N., Koronakis, I., Livada, I. and Assimakopoulos, D.N. (1999). 'Thermal and Air Flow Characteristics in a Deep Pedestrian Canyon and Hot Weather Conditions', *Atmospheric Environment*. In the press.
26. Georgii, H.W., Busch, E. and Weber, E. (1967). 'Investigation of the Temporal and Spatial Distribution of the Concentration of Carbon Monoxide in Frankfurt Main'. Report No 11, Institute of Meteorology and Geophysics, University of Frankfurt/Main.
27. Johnson, W.B., Dabberdt, W., Ludwig, F. and Allen, R. (1971). 'Field Study for Initial Evaluation of an Urban Diffusion Model for Carbon Monoxide'. Contract CAPA-3-68(1-69), SRI Project 8563, Stanford Research Institute, Menlo Park, CA 94025.
28. Dabberdt, W.F., Ludwig, F.L. and Johnson, W.B. (1973). 'Validation and Applications of an Urban Diffusion Model for Vehicular Emissions'. *Atmospheric Environment*, pp. 603–618.
29. Waldeyer, H., Leisen, P. and Mueller, W.R. (1981). 'Die Abhangigkeit der Immissionsbelastung in Strassensluchten von meteorologischen und verkehrsbedingten Einflussgrossen'. *Proceedings Abgasbelastungen durch den Kraftfahrzengverkehr*, pp. 85–114, TUV Rheinland, Koln, Germany.
30. Jumard, R. (1981). 'Ausbreitungsmodelle fur Verkehrs – immissionen in Strassenschucten und Vergleich zu franzosischenMessungen'. *Proceedings Abgasbelastungen durch den Kraftfahrzengverkehr*, pp. 187–206, TUV Rheinland, Koln, Germany.
31. Hoydysh, W.G (1971). 'An Experimental Investigation of the Dispersion of Carbon Monoxide in the Urban Complex'. AIAA Paper No. 71-523, Urban Technology Conference, New York City, 24–26 May.
32. Jacko, R.B. (1972). 'A Wind Tunnel Parametric Study to Determine the Air Pollution Dispersion Characteristics in an Urban Center'. PhD Thesis, Purdue University.
33. Odaira, T., Asakuno, K., Fukuosa, S. Udagawa, M., Izumikawa, S., Funeshida, M., Ho, M., Yokota, H., Nishida, K. and Yamamoto, T. (1972. 'Using Wind Tunnel Experiments to Forecast Automobile Exhaust Gas Concentrations along Streets'. Department of Sanitary Engineering, Kyoto University, Japan.
34. Hoydysh, W.G. and Ogawa, Y. (1972). 'Characteristics of Wind, Turbulence and Pollutant Concentration in and above a Model City'. Report No. ERL TR 110, NASA Grant No. N6R 33-016-149, New York University, Center for Interdisciplinary Studies, New York.
35. Hoydysh, W.G., Ogaw.a Y. and Griffiths, R.A (1974). 'A Scale Model Study of Dispersion of Pollution in Street Canyons'. APCA Paper No 74-1157, 67th Annual Meeting of the Air Pollution Control Association, Denver, Colorado, 9–13 June.

36. Wedding, J.B., Lombardi, D.J. and Cermak, J.E. (1977). 'A Wind Tunnel Study of Gaseous Pollutants in City Street Canyons'. *Journal of the Air Pollution Control Association*, Vol. 27, 557–566.

37. Dabberdt, W.F., Shelar, E., Marimont, D. and Skinas, W. (1981). 'Analyses, Experimental Studies, and Evaluation of Control Measures for Air Flow and Air Quality on Highways'. Federal Highway Administartion, FHWA 81/051, Final Report, SRI International, Menlo Park, CA 94025.

38. Leisen, P., Jost, P. and Sonnborn, K.S (1981). 'Modellierung der Schadstoffausbreitung in Strassenschluchten Vergleich von Aussenmessungen mit rechnerischer und Windkanalsimulation'. *Proceedings Abgasbelastungen durch den Kraftfahrzengverkehr*, pp. 207–234, TUV Rheinland, Cologne, Germany.

39. Zhang, Y.Q. and Huber, A.H. (1995). 'The Influence of Upstream Wind Shear and Turbulence on the Wind Pattern and Pollutant Concentrations within Street Canyons. A Numerical Simulation Study'. *Proceedings of the 11th Symposium on Boundary Layers and Turbulence*, pp. 196–199. American Meteorological Society, Washington, DC.

40. Chang, P.C., Wang, P.N. and Lin, A. (1971). 'Turbulent Diffusion in a City Street'. *Proceedings of the Symposium on Air Pollution and Turbulent Diffusion*, 7–10 December, Las Cruces, New Mexico, pp. 137–144.

41. Nunez, M. and Oke, T.R (1976). 'Long Wave Radiative Flux Divergence and Nocturnal Cooling of the Urban Atmosphere. II. Within an Urban Canyon'. *Boundary Layer Meteorology*, Vol. 10, pp. 121–135.

42. Nicholson, S.E. (1975). 'A pollution Model for Street-level Air'. *Atmospheric Environment*, Vol. 9, pp. 19–31.

43. Mills, G.M. (1993). 'Simulation of the Energy Budget of an Urban Canyon – 1. Model Structure and Sensitivity Test'. *Atmospheric Environment*, Vol. 27B, No. 2, pp. 157–170.

44. Paciuk, M. (1975). 'Urban Wind Fields – An Experimental Study on the Effects of High Rise Buildings on Air Flow around them'. MSc Thesis, Technion, Haifa, Israel.

45. Aida, M. (1982). 'On the Urban Surface Albedo'. *Japanese Progress in Climatology*, pp. 4–5.

46. Aida, M. and Gotoh, K. (1982). 'Urban Albedo as a Function of the Urban Structure – a Two Dimensional Numerical Solution'. *Boundary Layer Meteorology*, Vol. 23, pp. 415–424.

47. Arnfield, A.J. (1976). 'Numerical Modelling of Urban Surface Radiative Parameters'. In J.A. Davies (ed.), *Papers in Climatology – The Cam Allen Memorial Volume*, pp. 1–28. McMaster University, Hamilton, Ontario, Canada.

48. Fuggle, R.F. and Oke, T.R. (1976). 'Long Wave Radiative Flux Divergence And Nocturnal Cooling of the Urban Atmosphere. I. Above Roof Level'. *Boundary Layer Meteorology*, Vol. 10, pp. 113–120.

49. Johnson, G.T. and Watson, I.T. (1984). 'The Determination of View Factors in Urban Canyons'. *Journal of Applied Meteorology*, Vol. 23, pp. 329–335.

50. Oke, T.R. (1981). 'Canyon Geometry and the Nocturnal Urban Heat Island: Comparison of Scale Model and Field Observations'. *Journal of Climatology*, Vol. 1, pp. 237–254.

51. Swaid, H. and Hoffman, M.E. (1990). 'Prediction of Urban Air Temperature Variations using the Analytical CTTC model', *Energy and Buildings*, Vol. 14, pp. 313–324.

52. Sailor, D.J. (1993). 'Role of Surface Characteristics in Urban Meteorology and Air Quality'. PhD Thesis, Mechanical Engineering Department, University of California.

53. Oke, T.R, Johnson, G.T., Steyn, D.G. and Watson, I.D. (1991). 'Simulation of Surface Urban Heat Islands under 'Ideal' Conditions at Night – Part 2: Diagnosis and Causation', *Boundary Layer Meteorology*, Vol. 56, pp. 339–358.

54. Davies, J.A and Hay, J.E (1980). 'Calculation of the Solar Radiation Incident on a Horizontal Surface'. In J.E. Hay and T.K. Won (eds), *Proceedings of the First Canadian Solar Radiation Workshop*, pp. 32–58.

55. Idso, S.B. and Jackson, R.D. (1969). 'Thermal Radiation from the Atmosphere', *Journal of Geophysical Research*, Vol. 74, pp. 5397–5403.

56. Rowley, F.B, Algren, A.B. and Blackshaw, J.W. (1930).'Surface Conductances as Affected by Air Velocity, Temperature and Character of Surface'. *ASHRAE Transactions*, Vol. 36, pp. 429–446.
57. Rowley, F.B. and Eckley, W.A. (1932). 'Surface Coefficients as Affected by Wind Direction'. *ASHRAE Transactions*, Vol. 38, pp. 33–46.
58. Mills, G. and Arnfield, A.J. (1993). 'Simulation of the Energy Budget of an Urban Canyon – II. Comparison of Model Results with Measurements'. *Atmospheric Environment*, Vol. 27B, No. 2, pp. 171–181.
59. Koronakis, I, Santamouris, M. and Papanikolaou, N. (1997). 'Thermal Modelling of an Urban Canyon'. Internal Report to the POLIS program of the European Commission, Directorate General for Science Research and Development, University of Athens, Physics Department. Available from the authors.

7

———

The energy impact of the urban environment

M. Santamouris,
Department of Applied Physics, University of Athens

ঙ০

URBAN ENVIRONMENT AND THE ENERGY CONSUMPTION OF BUILDINGS

Today, it is well accepted that urbanization leads to a very high increase of energy use. Data on the energy and specific electricity consumption of major European cities are given by Eurostat.[1] Data on the electricity consumption in European cities range from 60 GWh/year for Valetta to 26,452 GWh/year for London. The average electricity consumption calculated on the basis of available data for cities with more than a million inhabitants is around 4500 GWh per year. However, these data cannot be used to draw any conclusions.

Other, statistical, data show that the amount of energy consumed by cities for heating and cooling of offices and residential buildings in western and southern Europe has increased significantly in the last two decades.[2] A recent analysis showed that a 1% increase in the per capita GNP leads to an almost equal (1.03%) increase in energy consumption.[3] However, as reported, an increase of the urban population by 1% increases the energy consumption by 2.2%, i.e. the rate of change in energy use is twice the rate of change in urbanization. These data show clearly the impact that urbanization may have on energy use.

Increased urban temperatures have a direct effect on the energy consumption of buildings during the summer and winter periods. In fact, it is found that during summer, higher urban temperatures increase the electricity demand for cooling and the production of carbon dioxide and other pollutants, while higher temperatures may reduce the heating load of buildings during the winter period.

In parallel, the wind and the temperature regime in canyons dramatically affect the potential for natural ventilation of urban buildings and thus the possibility of using passive cooling techniques instead of air conditioning.

ANALYSIS OF EXISTING DATA CORRELATING URBAN TEMPERATURES AND ENERGY CONSUMPTION

Unfortunately, very few studies have been carried out on the impact of the urban climate on the energy consumption of urban buildings for heating and cooling purposes. Existing studies either correlate increased urban temperatures and the corresponding electricity demand for selected utility districts or use sets of local temperature data to calculate the breakdown of the cooling and heating load in a city suffering from increased temperatures.

Both methodologies and techniques have important advantages. When correlations between temperatures and energy use are established by relating utility-wide electricity loads to temperatures at the same time of the day, a very clear picture of the real impact of high urban temperatures is established. However, to achieve this, it is necessary to minimize the effects on the electricity demand that are not climate related, which is not always possible; and, when it is possible, it may not always be accurate. Although this technique gives an estimation of the increase in the energy consumption in an integrated way, it does not permit the investigation of local effects and the impact of the specific urban layout and characteristics on the energy consumption of the buildings.

When temporally extended data for the temperature breakdown in a city are used to calculate either the energy load of a reference building located in or around a city or the distribution of the energy consumption in a city, very useful information on the relative energy consumption of the various urban subregions, having different layouts and climatic characteristics, is established. However, it is either not possible or very difficult to evaluate the overall impact of high urban temperatures on the global energy consumption of the city. It is obvious that a combination of the two techniques may offer a more complete view of the impact of urban temperatures on the energy load of buildings.

In the following, two specific studies that make use of the two techniques are presented. First, the increased summer electricity load of some American cities, calculated mainly from utility data, is presented, followed by the spatial variation of the monthly cooling and heating load of a reference building in Athens, Greece, calculated using hourly temperature data measured at almost 30 stations installed in and around the city. Finally, some data from Japan and Singapore, reporting a possible increase in the energy consumption for air conditioning due to a heat-island effect, are presented.

Urban energy studies in the USA

The heat-island effect in warm to hot climates exacerbates energy use for cooling in summer. As reported for US cities with populations larger than 100,000, the peak electricity load will increase by 1.5% to 2% for every 1°F rise in temperature.[4] Since urban temperatures during summer afternoons in the USA have increased by 2 to 4°F during the last 40 years, it can be assumed that 3% to 8% of the current urban electricity demand is used to compensate for the heat-island effect alone.

Comparisons of high ambient temperatures with utility loads for the Los Angeles area have shown that an important correlation exists. It is found that the net rate of increase in electricity demand is almost 300 MW per °F. As there has been a 5°F increase in the peak temperature in Los Angeles since 1940, this can be translated into an added electricity demand of 1.5 GW due to the heat-island effect.

Similar correlations between temperatures and electricity demand have been established for selected utility districts in the USA. Table 7.1 gives the rate of increase in the electricity demand in MW/°F, as well as the increase in the electricity demand, as a percentage, for each utility.[4]

Table 7.1. Correlation between temperature and electricity demand in selected utility districts, based on measured data for 1986[4]

Utility district	Increase (MW/°F)	Increase (%/°F)
Los Angeles (LADWP)	75	2.0
Los Angeles (SCE)	225	1.6
Washington DC	100	2.0
Salt River Project, (Phoenix)	56	2.0
Dallas – Fort Worth (TX)	250	1.7
Tucson (AZ)	12	1.0
Colorado Springs (CO)	4	1.0

Based on the above rates of increase, it has been calculated that for the USA the electricity cost for the summer heat island alone could be as much as $1 million per hour, or over $1 billion per year.[4] Computer studies for the whole country have shown the possible increase in the peak cooling electricity load due to the heat-island effect could range from 0.5% to 3% for each 1°F rise in temperature.

Urban energy studies in Athens

As described in previous chapters, extended urban climate measurements have been carried out in Athens, Greece during the period 1996–1999. Almost 30 temperature and humidity stations have been installed in and around Athens measuring ambient temperature and humidity on an hourly basis since June 1996. The collected data have been used to calculate the distribution of the cooling and heating needs of a representative office building for all locations where climatic data were available.[5]

The building considered is constructed on seven different levels and has a total surface area of 500 m². It is used by 25 people and is a low-energy building involving many energy conservation features to decrease its heating and cooling needs. The building has been monitored for about two years and extended simulations have been carried out using the TRNSYS software to check the agreement of the experimental results with the theoretical predictions.[6] A theoretical model has been created in TRNSYS predicting in an accurate way the overall thermal performance of the building.

Using hourly data for the ambient temperature collected at all urban climate stations, simulations of the cooling and heating load of the reference building were performed for August 1996. The same solar radiation data were used for all the stations considered, since a non-significant spatial variation of solar radiation has been observed in Athens.[7] All other operational data, such as internal gains, were selected to correspond exactly to the measured conditions.

The calculated spatial variation of the cooling load of the reference building for a set point of 27°C is given in Figure 7.1. Very local phenomena and conditions are not shown on the maps unless a measuring station was placed locally. Figure 7.1 gives the iso-cooling load lines (in kWh/m² per month), indicating the spatial variation of the cooling load for the whole of Athens. As shown, the cooling load at the centre is approximately double that in the surrounding Athens region. The maximum cooling load always corresponds to the very central area of Athens and especially to a measuring station very close to a road with a high traffic density. Minimum values

Figure 7.1. Distribution of the cooling load in the city of Athens for a set-point temperature of 27°C in August 1996

were calculated in the south-east Athens region, a mean-density residential area close to the Hemetus forest. Much higher cooling loads were calculated for the western Athens region, which is characterized by high-density plots, lack of green spaces, important industrial activity and higher traffic density than the eastern Athens region.

As well as increasing the energy loads for cooling of buildings, high ambient temperatures increase peak electricity loads and cause a serious strain on the local utilities. Thus, knowledge of the possible increase of the peak electricity load due to higher urban ambient temperatures may be very important.

The instant peak cooling load of the reference building was calculated with TRNSYS for August 1996 and for various set-point temperatures ranging between 26°C and 28°C. The spatial variation of the peak cooling load obtained for the whole Athens region, for a set point of 26°C, is given in Figure 7.2; values are in kW. Note that the reported peak cooling loads refer to the whole month and do not occur at the same time at all stations. As expected, much higher peak cooling loads were calculated for the central Athens area. For a set-point temperature of 26°C, the highest peak load of the reference building was calculated as close to 27.5 kW, while the minimum peak load was close to 13.7 kW. Thus, the impact of higher urban temperatures is extremely important and almost doubles the peak cooling load of the reference building.

When the set point temperature was 28°C, much higher differences were found. In this case, the maximum cooling load was close to 23.5 kW while the minimum was close to 7.3 kW. Thus, while the maximum peak cooling load was reduced to about 4.3 kW, the minimum peak load was reduced by 6.4 kW. The results show that, for the central Athens area, the peak cooling load is mainly due to persisting very high ambient temperatures and is not sensitive to the change of the set-point temperature. In contrast, the calculated minimum peak cooling load changes dramatically as the set-point temperature increases. This is mainly due to the minimization of the load induced by the indoor–outdoor temperature differences as the set point increases. In

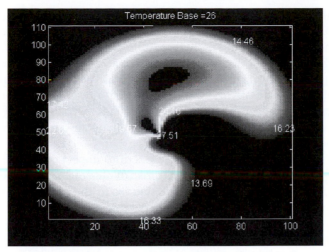

*Figure 7.2. Spatial variation of the peak cooling load of a
reference building in Athens during August 1996 for a set point
temperature of 26°C; values are in kW*

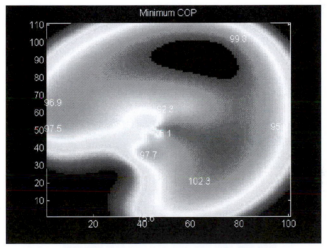

*Figure 7.3. Spatial distribution of the minimum COP of a
conventional air conditioner in Athens (values are percentages)*

this case the calculated peak cooling load is mainly due to solar radiation and the other
sources.

High ambient temperatures have a very serious impact on the efficiency of
conventional air conditioning (A/C) systems. The coefficient of performance (COP)
is directly affected by relative humidity and ambient temperature; thus it is of great
interest to investigate a possible decrease of the COP due to the heat-island effect.

The distribution of the COP value of a conventional A/C system was calculated, for
the whole summer of 1996, on the basis of hourly temperature and humidity data from
all stations in and around Athens. The minimum value for each station was calculated.
Figure 7. 3 gives the spatial distribution of the minimum COP values in Athens. As

shown, the absolute minimum COP values are for the very central area of Athens (close to 75%), because of the high ambient temperatures, and for the coastal area, because of the high humidity. The highest minimum COP value was calculated as close to 102% and this was for the south-eastern area of Athens. The results show clearly that, quite apart from the high cooling loads and peak electricity problems, the heat-island effect significantly reduces (by about 25%), the efficiency of air-conditioning systems and thus may oblige designers to increase the size of the installed A/C systems and thus increase peak electricity problems and energy consumption for cooling purposes.

Increased urban temperatures may have a serious impact on the heating load of urban buildings. Calculations of the heating load of the same reference building were performed for a number of representative stations in the major Athens area using hourly temperatures for January 1998. The stations were grouped into three clusters: stations located in the very central Athens area, suburban stations and stations in urban parks and green areas. The corresponding heating load for each type of station is given in Figure 7.4. The heating load in the central Athens region was estimated close to 3.7 kWh/m²/month, while the corresponding load for the suburban stations was close to 5.1 kWh/m²/month and the mean load for the stations located in green areas was close to 7.3 kWh/m²/month.

As shown in the figure, increased urban temperatures decrease the heating load of urban temperatures by about 30%, while the maximum difference between suburban and urban stations was close to 55%.

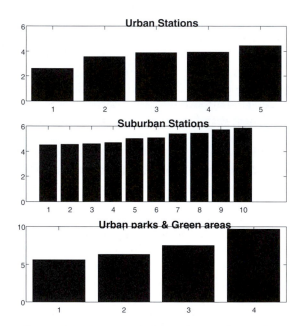

Figure 7.4. Heating load of the reference office building calculated for urban measuring stations, suburban stations and stations located in urban parks and green areas

Table 7.2. Sensible and latent loads and the compressor electrical consumption of the residential buildings in the Aegaleo munici- pality, calculated with locally recorded climatic data (Aegaleo), as well as with data from an almost rural area (Penteli) and a suburban measuring station (Ilioupolis), much less affected by the heat island; values are in GWh

Area	Sensible load	Latent load	Electrical energy
Penteli	87	24	50
Ilioupolis	95	35	48
Aegaleo	145	54	90

Based on the distribution of the urban temperature in the Athens metropolitan area, Hassid has calculated the additional energy spend for cooling purposes to counterbalance the heat-island effect in the major western Athens area (Aegaleo).[8] The area studied is 5,354,685 m^2 in size and has a total of 25,155 households, or 30,648 residences, with 75,629 people living in them, of which 74,605 people are living in households. Calculations were performed with a Geographic Information System and detailed building simulation techniques, in order to report on the condition of each main building block. Locally recorded temperatures were used to calculate the global sensible and latent loads in the area. In parallel, calculations were performed using climatic data from reference stations much less affected by the heat island. The total cooling load and the compressor electrical consumption of the residential buildings calculated with data for the western Athens area (Aegaleo), as well as with data for an almost rural area (Penteli) and a suburban reference station (Ilioupolis), are given in Table 7.2.

From Table 7.2 it can be seen that the difference between the energy consumption of the Aegaleo buildings and the energy consumption, had the climate been similar to that of Ilioupolis, is equal to 41.3 GWh. The cost to compensate for the heat island is calculated to be close to €4.13 million or €164 per household.

Hassid has also calculated the additional total peak and peak cooling electrical loads to compensate for the heat island in the same area.[8] The calculated values are given in Table 7.3.

From Table 7.3 it can be seen that satisfying the peak electric demand is 82.4 MW, which is 25 MW more than the figure for the area almost unaffected by the heat island. According to Hassid, if one assumes a diversity factor of 0.8 and a marginal price for

Table 7.3. Total peak and peak cooling electrical loads of the residential buildings in the Aegaleo municipality, calculated with locally recorded climatic data (Aegaleo), as well as with data from an almost rural area (Penteli) and a suburban measuring station (Ilioupolis), much less affected by the heat island; values are in MW.

Area	Total peak electrical load	Peak cooling electrical load
Penteli	68.9	51.6
Ilioupolis	57.5	30.5
Aegaleo	82.4	61.6

an additional kW from the grid of €1000, this means an additional investment of approximately €20 million in generation and distribution equipment.

Other studies

Heat-island studies in Singapore show a possible increase of the urban temperature close to 1 K.[9] If there were to be similar changes in temperatures 50 years from now, the anticipated increase in building energy consumption, mainly for air conditioning, would be of the order of 33 GWh per annum for the whole island.[9]

Watanabe et al., using LANDSAT-5 data analysed the land temperature distribution and the thermal environment of the Tokyo metropolitan area for some days of September 1973.[10] On the basis of these data, as well as the results of a field survey by the metropolitan government, energy consumption distribution maps were prepared. Although no further information is provided, the overall analysis shows clearly that much higher energy consumption was calculated for the central Tokyo area.

Another study of the Tokyo area also shows clearly that during the ten years from 1965 to 1975, as a result of the heat-island phenomenon, the cooling load of existing buildings increased by 10–20% on average.[11] If the cooling load continues to increase at the same rate, it will have increased by more than 50% by 2000.

NATURAL VENTILATION IN URBAN AREAS

As shown in Chapter 3, the wind speed in an urban canyon is seriously reduced compared to the undisturbed wind velocity. In parallel, the wind direction inside canyons is almost completely different from the one measured by regular meteorological stations.

Natural ventilation is one of the most effective passive cooling techniques.[12] Natural ventilation applied in northern buildings can provide effective cooling during both day and night, while night ventilation is a very effective strategy in hot climates.

The appropriate design of openings to make use of natural ventilation techniques in urban environments requires knowledge of appropriate wind speed and direction data. Methods to calculate these parameters have been already discussed in Chapter 3.

Estimations of the real potential of natural ventilation techniques when applied to buildings located in an urban canyon have been reported by Santamouris *et al.*[13] In order to evaluate the natural ventilation potential of urban buildings, as well as a possible decrease in natural ventilation due to canyon-related phenomena, simulations of the air flow processes were carried and reported for ten different canyons in which wind speed and temperature data have been collected. Single-sided and cross-ventilation configurations were considered. A typical building zone of 36 m^2 and 144 m^3, having a window of 1.5 × 1.5 m in each canyon facade was also considered (Figure 7.5).

Two types of simulation have been performed for each configuration. The first was based on the wind and temperature data measured inside the canyon, while the second one was based on the undisturbed temperature and the wind speed measured above the buildings. Comparison of the two simulation results permitted the decrease in the natural ventilation potential in urban canyons to be assessed.

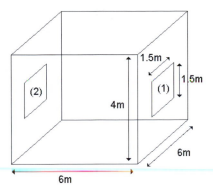

Figure 7.5. Dimensions of the considered zone

Simulations were performed using the AIOLOS natural ventilation simulation program.[12] The software used has been well validated in a high number of experiments within the framework of the PASCOOL research project.[14]

Figures 7.6 and 7.7 give the air flow rate for the ten canyons and for the single-sided and cross-ventilation configurations respectively. The two flow rates, one when the undisturbed ambient temperature and wind speed are used and the second corresponding to the data measured inside the canyon, are given. Analysis of the results has made it possible to extract the following conclusions:

- During the daytime, when the ambient wind speed is considerably higher than wind speed inside the canyon and inertia phenomena dominate the gravitational forces, the natural ventilation potential in single-sided and cross-ventilation configurations is seriously decreased inside the canyon. In practice, this happens when the ambient wind speed is higher than 4 m/s. For single-sided ventilation configurations the air flow is reduced up to five times, while in cross-ventilation configurations the flow is sometimes reduced up to ten times.
- During the daytime when the ambient wind speed is lower than 3–4 m/s, gravitational forces dominate the air flow processes. In this case, the difference between the wind speed inside and outside the canyon does not play an important role, especially in single-sided configurations.
- During the night the ambient wind speed is seriously decreased and is comparable to the wind speed inside the canyon. In this case the air flow calculated for inside and outside the canyon is almost the same.
- The calculated reduction of the air flow inside the canyon is mainly a function of the wind direction inside the canyon. When the ambient flow is almost normal to the canyon axis, the flow inside the canyon is almost vertical and parallel to the window. In this case the much higher pressure coefficient corresponds to the conditions outside the canyon and thus a much higher flow is calculated when the ambient conditions are considered and inertia forces dominate. When the ambient flow is parallel to the canyon axis, a similar flow is observed inside the canyon and thus the pressure coefficients are almost similar.

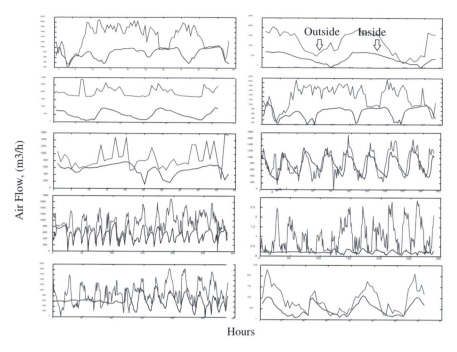

Figure 7.6. Air flow rates calculated for ten different canyons and for single-side building configurations

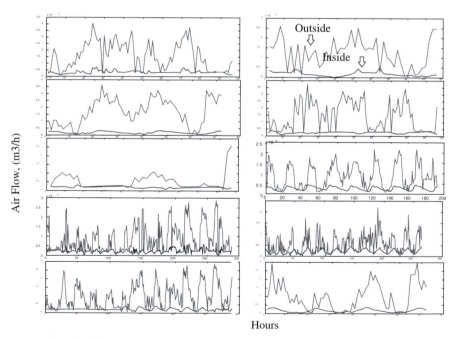

Figure 7.7. Air flow rates calculated for ten different canyons and for cross-ventilation configurations

NIGHT VENTILATION IN URBAN AREAS

Based on the potential for natural ventilation in urban canyons, defined previously, Geros has calculated the reduction of the performance of night-ventilation techniques when applied to naturally ventilated buildings located in urban canyons.[15] The study was performed for the same zone as the above and for all the ten canyons in Athens for which experimental data were available.

The cases of an air-conditioned and a free-floating building were considered. The air flows calculated for single-sided and cross-ventilated zones in and out the canyon, as given above, as well as the corresponding climatic data, were used as inputs to the TRNSYS simulation program. The cooling load of the air-conditioned building and the temperature decrease of the free-floating building were calculated for all cases.

Table 7.4 gives the calculated cooling load for all the cases studied. As shown for both single-sided and cross-ventilation configurations, the cooling load on the buildings inside the canyons is much higher than that calculated for buildings where the wind is not obstructed. In particular, in single-sided configurations the cooling load is between 6% and 89% higher, while in cross-ventilation configurations the cooling load increases by 18% to 72%. Thus, the canyon effect has a very considerable effect on the performance of night ventilation techniques in air-conditioned buildings.

Table 7.5 gives the results obtained for the free-floating building. Simulations were used to calculate the mean and maximum temperature differences between buildings located outside and inside the canyon. Results are given for the whole 24-hour period, as well as for the operational period, 9:00–18:00. As shown, for single-sided configurations, the mean temperature difference varies between 0.1 and 2.8 K for the whole day and between 0.0 and 1.1 K for the operational period of the buildings. The maximum difference during the day varies between 0.3 and 4.1 C, while during the night it varies between 1.0 and 5.6 K.

For cross-ventilation configurations, the mean temperature difference varies between 0.4 and 4.0 K for the whole day and between 0.2 to 3.5 for the operational

Table 7.4. Cooling load when single-sided or natural cross-ventilation techniques are applied to an air-conditioned building during the night period

Canyon code	Cooling load			
	Single-sided ventilation		Cross ventilation	
	Wind data inside the canyon (Wh/m^2)	Undisturbed wind speed (Wh/m^2)	Wind data inside the canyon (Wh/m^2)	Undisturbed wind speed (Wh/m^2)
Ippokratous	4240	3850	5330	4320
Solonos	4660	4120	5190	4160
Kavalas	1390	650	1660	470
Papastratos	50	46	42	48
Valaoritou	1070	1130	1350	1280
Mavromihali	1500	1200	1570	1190
Kodrou	870	830	940	820
Giannitson	1030	780	1080	670
Omirou	22	2	84	48
Evrota	180	140	110	121

Table 7.5 Mean and maximum temperature difference during the whole day and the operational period of two buildings located inside and outside the canyon.

	Temperature difference							
	Single-sided ventilation				Cross ventilation			
	For the whole period (K)		For the operating period of the zone (K)		For the whole period (K)		For the operating period of the zone (K)	
Canyon code	Mean	Max	Mean	Max	Mean	Max	Mean	Max
Ippokratous	0.7	2.2	0.6	1.3	1.6	4.1	1.3	2.2
Solonos	1.0	2.7	1.0	1.8	2.0	4.6	1.7	3.3
Kavalas	2.9	5.0	2.6	3.1	4.0	7.9	3.5	4.4
Papastratos	0.7	2.2	0.3	1.1	0.7	2.3	0.2	1.2
Valaoritou	0.1	1.0	0.02	0.3	0.8	3.6	0.6	1.4
Mavromihali	2.8	5.6	2.3	4.1	2.4	5.0	2.0	3.5
Kodrou	1.0	3.6	0.7	1.6	2.4	7.0	1.6	3.9
Giannitson	0.7	1.6	0.6	1.0	1.1	2.7	0.9	1.6
Omirou	1.2	2.3	1.1	1.7	1.5	3.2	1.3	2.2
Evrota	1.0	2.5	0.9	1.7	1.2	3.0	1.0	1.7

period of the buildings. The maximum difference during the day varies between 1.2 to 4.4 C, while during the night period it varies between 2.3 to 7.9 C.

REFERENCES

1. Eurostat (1995). *Europe's Environment. Statistical Compendium for the Dobris Assessment.* Statistical Office of the European Communities, Luxembourg.
2. Stanners, D. and Bourdeau, P. (eds) (1995). *Europe's Environment: The Dobris Assessment.* European Environmental Agency, Copenhagen.
3. Jones, B.G. (1992). 'Population Growth, Urbanization and Disaster Risk and Vulnerability in Metropolitan Areas : A conceptual Framework'. In Kreimer, Alcira and Mohan Munasinghe, *Environmental Management and Urban Vulnerability*, World Bank Discussion Paper No. 168.
4. Akbari, H., Davis, S., Dorsano, S., Huang, J. and Winett, S. (1992). *Cooling our Communities – A Guidebook on Tree Planting and Light Colored Surfacing.* US Environmental Protection Agency. Office of Policy Analysis, Climate Change Division.
5. Santamouris, M., Papanikolaou, N., Livada, I., Koronakis, I., Georgakis, C., Argiriou, A. and Assimakopoulos, D.N. (1999). 'On the Impact of Urban Climate on the Energy Consumption of Buildings'. *Solar Energy*, in the press.
6. Geros, V., Santamouris, M., Tombazis, A. N. and Guarraccino, G. (1996). 'On the cooling Efficiency of Night Ventilation Techniques', PLEA International Conference, Louvain La Neuve, Belgium.
7. Psiloglou, V. 1997) 'Development of an Atmospheric Solar Radiation Model', PhD Dissertation, Physics Department, University of Athens.
8. Hassid, S. (1998). 'The Effect of the Athens Heat Island on Air Conditioning Load', Internal Report, University of Athens, Physics Department, Group Building Environmental Studies, Athens.
9. Tso, C.P. (1994). 'The Impact of Urban Development on the Thermal Environment of Singapore'. In *Report of the Technical Conference on Tropical Urban Climates*. WMO, Dhaka.
10. Watanabe, H., Yoda, H. and Ojima T. (1990/91). 'Urban Environmental Design of Land Use in Tokyo Metropolitan Area'. *Energy and Buildings*, Vol. 15–16, 133–137.

11. Ojima, T. (1990/91).'Changing Tokyo Metropolitan Area and its Heat Island Model'. *Energy and Buildings*, Vol. 15–16, 191–203.
12. Allard, F. (ed.) (1998). *Natural Ventilation of Buildings*. James and James Science Publishers, London.
13. Santamouris, M., Papanikolaou, N., Koronakis, I., Georgakis, C. and Assimakopoulos, D. N. (1998). 'Natural Ventilation in Urban Environments'. *Proceedigs of the 1998 AIVC Conference, Oslo*. AIVC, Oslo.
14. Limam, K., Allard, F. and Dascalaki, E. (1997). 'Natural Ventilation Research Activities Undertaken in the Framework of Pascool', *International Journal of Solar Energy*, Vol. 19, 81–119.
15. Geros, V. (1998). 'Contribution to the Energy Potential of Night Ventilation Techniques'. PhD Thesis, INSA Lyon.

8

Short-wave radiation

A. Tsangrassoulis
Department of Applied Physics, University of Athens

℘

The Sun is practically the only source of energy that influences our climate. The rate of total solar energy at all wavelengths incident on a unit area exposed normally to rays of the Sun at one astronomical unit is called the solar constant. The solar constant has been studied extensively since the beginning of the century. The value of this constant according to NASA measurements[1] is 1353 Wm^{-2}. More recently, Frohlich *et al.* using the new World Radiometric Reference recommended the value of 1367 Wm^{-2} for solar constant.[2] Table 8.1 shows the distribution of the extraterrestrial energy within different wavelength bands.

Extraterrestrial solar radiation is modified when passing through the atmosphere. The portion of the incoming radiation that is reflected and scattered, together with the portion that is multiply reflected between the surface and the atmosphere (back-scattered), gives the diffuse solar radiation. Finally, the portion of the incoming solar radiation that arrives at the Earth's surface, without being absorbed or diffused, is called direct-beam short-wave radiation.

The study of the losses in terms of energy received at surfaces below the urban 'dust' shield was initiated by various researchers years ego. Chandler reported that the London inner city has 16% fewer hours of sunshine, and the suburbs of the city 5% fewer than the countryside.[3] As mentioned above, the urban reduction in energy received at the ground is greatest at low solar elevations when the relative thickness of the turbid layers is greatest (Figure 8.1).

The considerable decrease in the incoming light at small wavelength intervals is due to the absorption of the gases H_2O, CO_2, O_3 and O_2. These gases, together with CH_4, N_2O and halocarbons, exhibit strong light absorption in the far infrared, which

Table 8.1. *Distribution of extraterrestrial energy within different wavelength bands*

	Wavelength band (μm)	Irradiance (Wm^{-2})	Percentage of the solar constant
Ultraviolet	< 0.4	109.81	8.03
Visible	0.39–0.77	634.4	46.41
Infrared	> 0.77	634.4	46.41

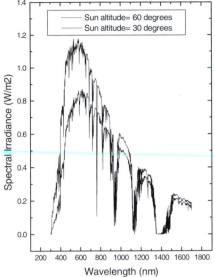

Figure 8.1. Solar spectral irradiance for two values of solar altitude

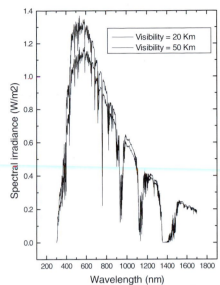

Figure 8.2. Solar spectral irradiance for two values of visibility

is important for the greenhouse effect. Consequently, the attenuation is not evenly distributed over the solar spectrum. The shorter wavelengths are particularly affected. Figure 8.2 shows the influence of visibility – which is related to Ångstrom's turbidity coefficient – in the spectral distribution of irradiance.

The ultraviolet wavelengths suffer far more loss than the infrared. In many industrial cities in winter, nearly all energy in wavelengths below 400 nm is completely absorbed. Table 8.2 is a representative example for the spectral distribution of solar radiation in and near Paris, as reported by Maurain in 1947.[4]

Table 8.2. Spectral distribution of solar radiation in and near Paris (percentage of total intensity)

Spectral region	Centre of Paris (%)	Outskirts (%)
Ultraviolet	0.3	3
Extreme violet	2.5	5
Visible	43	40
Infrared	54	52

Two different mechanisms cause the attenuation of light, or any radiation, in the atmosphere:

- Scattering of the light in the atmosphere occurs as a result of the presence of aerosol particles and air molecules. Scattering by the molecules of trace gases can be neglected. Part of the incident radiation is diverted by scattering to directions that are not parallel to that of the incident radiation. No light is lost and all the energy is converted as light.

- As a result of the absorption of the light, part of the energy of the incident light is transferred to the absorbing molecule or particle, the internal energy of which is therefore increased. Finally, however, the absorbed energy is transferred to the surrounding gas as heat.

Both absorption and scattering of the light in the atmosphere decrease the flux of the sunlight reaching the ground. As a result of scattering, part of the light can be reflected back into space and the energy input, both to the atmosphere and to the surface of the Earth, is reduced, causing a decrease in temperature. In contrast, absorption of light increases the energy input into the atmosphere, thus increasing its temperature. Therefore any change to the particles and gases in the atmosphere will change the light absorption and scattering properties of the atmosphere, which will have an influence on climate change. As well as climatic effects, light absorption in the visible spectrum has optical effects, ranging from the dark and dirty appearance of industrial hazes to coloration of cloud plumes or the discoloration of the blue sky.

In more detail, the direct spectral irradiance at any wavelength on a surface normal to the Sun's rays and at mean Sun–Earth distance is as follows:

$$I_\lambda = I_{0\lambda} T_\lambda \tag{8.1}$$

where T_λ is the monochromatic transmittance due to the combined effects of continuum attenuation and molecular absorption and $I_{0\lambda}$ is the extraterrestrial normal spectral irradiance. This transmittance is defined as:

$$T_\lambda = T_{r\lambda} T_{a\lambda} T_{o\lambda} T_{g\lambda} T_{wa\lambda} \tag{8.2}$$

where $T_{r\lambda}$ is the transmittance of the direct beam due to molecular scattering, $T_{a\lambda}$ is the transmittance due to scattering and absorption by aerosols, $T_{o\lambda}$ is the transmittance due to absorption by the ozone, $T_{g\lambda}$ is the transmittance due to absorption by uniformly mixed gases such as CO_2 and O_2, and $T_{wa\lambda}$ is the transmittance due to absorption by water vapour.

These transmittances are calculated as follows:

$$T_{r\lambda} = \exp(-0.008753\lambda^{-4.08}m), \tag{8.3}$$

where m is the relative optical air mass at the actual pressure;

$$T_{a\lambda} = \exp(-0.008635\lambda^{-2}wm_r) \times \exp[-0.08128\lambda^{-0.75}(d/800)m], \tag{8.4}$$

where m_r is the air mass at standard pressure, w is the thickness of perceptible water (cm) and d is the number of dust particles per cubic centimetre;

$$T_{o\lambda} = \exp(-k_{0\lambda} l m_r), \tag{8.5}$$

where $k_{0\lambda}$ is the attenuation coefficient for ozone absorption, l is the amount of ozone in cm (NTP);

$$T_{g\lambda} = \exp[-1.41k_{g\lambda}m_a / (1+118.93k_{g\lambda}m_a)^{0.45}], \qquad (8.6)$$

where $k_{g\lambda}$ is the coefficient of attenuation due to absorption by mixed gases (dimensionless);

$$T_{wa\lambda} = \exp[-0.2385k_{wa\lambda}wm / (1+20.07k_{wa\lambda}wm)^{0.45}], \qquad (8.7)$$

where $k_{wa\lambda}$ is the coefficient of attenuation due to absorption by water vapour (cm^{-1}).

The diffuse spectral irradiance ($I_{\text{diffuse}, \lambda}$) arriving at the ground can be separated into three parts: the diffuse spectral irradiance produced by Rayleigh scattering that arrives at the ground after the first pass through the atmosphere ($I_{\text{diffuseR}, \lambda}$); then the diffuse spectral irradiance produced by aerosols after the first pass through the atmosphere ($I_{\text{diffuseA}, \lambda}$); and, finally, the diffuse spectral irradiance produced by multiple reflections ($I_{\text{diffuseM}, \lambda}$).

Consequently,

$$I_{\text{diffuse},\lambda} = I_{\text{diffuseR},\lambda} + I_{\text{diffuseA},\lambda} + I_{\text{diffuseM},\lambda}. \qquad (8.8)$$

$I_{\text{diffuseR}, \lambda}$ can be calculated on the basis that half of the scattered radiation is directed downwards as follows:

$$I_{\text{diffuseR},\lambda} = I_{0\lambda} \cos(\text{Sun's zenith angle}) T_{ma\lambda}[0.5(1-T_{r\lambda})T_{a\lambda}], \qquad (8.9)$$

where $T_{ma\lambda}$ is the spectral transmittance due to absorption by ozone, uniformly mixed gases and water vapour, and $I_{0\lambda}$ is the extraterrestrial spectral irradiance. $I_{\text{diffuseA},\lambda}$ is estimated as follows:

$$I_{\text{diffuseA},\lambda} = I_{0\lambda} \cos(\text{Sun's zenith angle}) T_{ma\lambda}[F_c\omega_0(1-T_{r\lambda})T_{a\lambda}], \qquad (8.10)$$

where F_c is the ratio between the energy scattered in the forward direction and the total scattered energy. This variable depends on particle size, shape and wavelength and, of course, on the zenith angle as well. The single scattering albedo ω_0 is defined as the ratio of energy scattered by aerosols to total attenuation under the first impingement of direct radiation. In practice, the determination of ω_0 is almost impossible since it depends on the material, shape, size and optical properties of the aerosol particles. The variables F_c and ω_0 are assumed to be invariant with wavelength.

For the calculation of the multiple-reflected spectral irradiance, the albedo of the atmosphere is needed. In an approximate form this is as follows:

$$R = T_{ma\lambda}[0.5(1-T_{r\lambda})T_{a\lambda} +(1-F_c)\omega_0(1-T_{a\lambda})T_{r\lambda}]. \qquad (8.11)$$

Thus, $I_{\text{diffuseM},\lambda}$ can be estimated as follows:

$$I_{\text{diffuseM},\lambda} = Q_\lambda (R_g R / (1 - R_g R), \tag{8.12}$$

where Q_λ is defined as:

$$Q_\lambda = (I_{\text{diffuseM},\lambda} + I_{\text{diffuseA},\lambda}) + I_{n\lambda} \cos(\text{Sun's zenith angle}). \tag{8.13}$$

In practice, what is needed for most applications is the determination of the radiation over the complete solar spectrum. Such a procedure can be accomplished by integration of the above equations, but has the disadvantage of being time consuming. Consequently, simple broadband equations have been developed to account for the attenuation by each of the atmospheric constituents.

A schematic description of the annual energy balance of the Earth–Atmosphere system is given in Figure 8.3. The Earth–Atmosphere system is a closed one, which means that mass cannot be imported or exported, but it does allow exchange of energy with outer space. As mentioned above, the input to the system is called the solar constant and is equal to 1367 W m^{-2}. When this value is averaged over a year at the top of the atmosphere, the mean annual value is exactly 25% of the solar constant, 342 W m^{-2}. In Figure 3 all fluxes are represented as percentages of this value. Over the period of a year exactly the same amount of energy must be lost to space from the Earth–Atmosphere system. If this were not so, the system would experience a net energy gain or loss, resulting in a net storage change and a rise or fall in the average temperature of the Earth–Atmosphere system. Equally, if the subsystems were not in balance, the system as a whole would not be in equilibrium.

In the atmosphere the clouds reflect about 19% of the mean annual energy back to space and absorb about 5%. Atmospheric constituents scatter and reflect about 6% to space and absorb about 20%. The remainder of the original beam is transmitted to the Earth's surface, where approximately 3% is reflected back into space and the

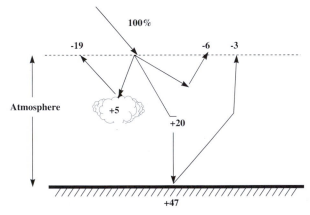

Figure 8.3. Schematic diagram for the distribution of annual solar energy in the atmosphere

remaining 47% is absorbed. Thus, in summary, the solar radiation input is disposed of in the following manner:

$$100 = \underset{\substack{\text{Atmospheric}\\\text{reflection}}}{19+6} + \underset{\substack{\text{Atmospheric}\\\text{absorption}}}{5+20} + \underset{\substack{\text{Earth}\\\text{reflection}}}{3} + \underset{\substack{\text{Earth}\\\text{absorption}}}{47} .$$

(8.14)

SOLAR RADIATION IN URBAN AREAS

The city has a marked impact upon the short-wave components of the net radiation budget as result of both the presence of radiatively active pollutants in the air and the changes in the surface radiative properties. This situation is presented in Figure 8.4.[5] In its path through the urban boundary layer, the incoming short-wave radiation (Flux 1) and the short-wave radiation reflected from the surface of the city (Flux 3) are subject to greater attenuation than the equivalent fluxes in the rural case. The amount received at the surface (the direct beam and the diffuse contribution – Flux 2 – plus the net back-scattered radiation – Flux 4) is typically 2–10 % lower in the city.[6] On the other hand, urban albedo values in the mid latitudes are typically 0.05 to 0.10 lower than for the countryside.[7] As a result urban/rural net short-wave radiation differences are considered to be rather small. The sign of any differences will depend upon the relative strengths of the pollution attenuation and the albedo.

The total albedo of an urban system and therefore its ability to absorb solar radiation depend upon the albedo of the component materials and their geometrical arrangement.

The importance of geometry has been demonstrated by using observations and both numerical and scale models of canyon radiative exchange.[8,9,10] All approaches show that for buildings of equal height the albedo of a crenellated surface is lower than that of a flat plane composed of the same materials. It also appears that the effect increases with latitude and is more pronounced in the low-sun season.[9,10] Further, it increases with height-to-width ratio (H/W) and is greater in E–W canyons than in those oriented N–S. Using different scale-model arrays, Aida found absorption increments of 13–27% for H/W in the range 0.5–2.0 when the flat plane albedo was 0.4.[10] He also suggested that this increment is mainly dependent on the plan density. In a numerical simulation Aida and Gotoh studied the albedo of different canyon

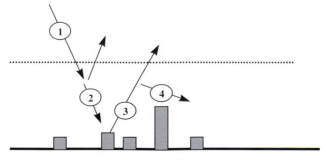

Figure 8.4. Schematic description of radiative exchanges in a polluted urban boundary layer

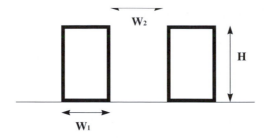

Figure 8.5. Geometrical configuration of an urban canyon

arrangements as a function of the ratio W_1/W_2, where W_1 is the width of the block elements and W_2 is the inter-block or street width (Figure 8.5).[11]

The interesting result of Aida and Gotoh is that, when H/W is held constant at unity, there is a minimum albedo for the system at $\log(W_1/W_2) = -0.3$ to -0.6. This means that in cases where we wish to maximize solar absorption, such as mid-latitude cities in the cold season, the ideal W_1/W_2 is approximately 0.5.

Although minimizing the albedo increases the absorptance of the total system for solar radiation, it does not necessarily benefit positions near the canyon floor, which may be in the shade. Indeed, certain geometrical arrangements can create absorptance that is less than that of a flat plane.[9]

The relevant limit for solar access to street canyons depends upon the degree of radiation penetration at the winter solstice. Penetration is needed to facilitate solar-energy gain by equator-facing walls, to provide sufficient daylight for building interiors and to help improve the comfort of pedestrians and assist their psychological attitude.

Since many mid-latitude cities are frequently cloudy in winter, the amount of diffuse radiation for daylighting is also important. The conventional criterion of daylighting specialists is that an H/W of 0.58 is appropriate at a latitude of 45°. The corresponding ratios at 40° and 50° are 0.7 and 0.46.[12] Figure 8.6 shows the radiation loss caused by buildings on the opposite sides of the street

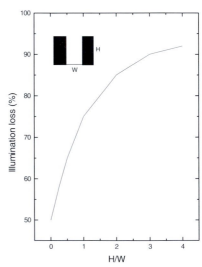

Figure 8.6. Radiation loss caused by the urban canyon on the vertical facade.

ENERGY BALANCE IN THE URBAN CANYON

In order to study the radiation distribution at the surfaces of an urban canyon, one should identify the basic units of structures that are repeated in the urban area. There are probably no truly representative urban surfaces, such as those that can be identified as characteristic of rural and other horizontal natural terrains. Such units that are formed consist of the more or less geometric combinations of horizontal and vertical surfaces arising from the block-like arrangement of buildings and streets. These units reflect the essentially three-dimensional nature of the urban canopy. Here, we identify a basic urban surface unit as the urban canyon. The canyon (Figure 8.7) consists of the walls and the ground between two adjacent buildings. The air volume of the canyon is the air contained within the canyon structure and bounded at the top by an imaginary lid, approximately at roof level. The top of this air volume, together with the rooftops, forms the lower boundary for most urban boundary layers, at least at the centre of most cities.[13]

The energy balance of the ith canyon surface is given by the equation:

$$Q_i^* = Q_{Hi} + Q_{Ei} + Q_{Gi,}$$

(8.15)

where Q^* is the net all-wave radiation flux (W m^{-2}), Q_E is the latent heat flux (W m^{-2}), Q_G is the subsurface heat flux (W m^{-2}) and Q_H is the flux of the sensible heat to the air (W m^{-2}); see Figure 8.8.

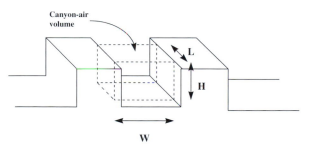

Figure 8.7. Schematic description of the urban/atmosphere interface, showing how the air volume of the urban canyon is defined

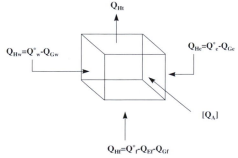

Figure 8.8. Sensible heat exchanges into and out of the canyon air volume. Subscripts have the following meaning: f floor; t top; e east; w west

The energy balance of the air canyon volume (in the case of a N–S oriented canyon if there are no phase changes of the water in the air volume) is given by the equation:

$$[Q_A] + \{H(Q_{He} + Q_{Hw}) + W(Q_{Hf} - Q_{Ht})\}L = \rho c_p \int_v \left(\frac{\partial \Theta}{\partial t} + \frac{\mathrm{div}Q^*}{\rho c_p}\right) \mathrm{d}V, \quad (8.16)$$

where Q_A is the net advected heat due to horizontal transport (Joule s^{-1}), ρ is the air density (kg m^{-3}), Θ is the time-averaged air temperature (K) and V is the air volume of the canyon.

The left-hand side of the above equation is the net sensible heat input (or output) of the six sides of the air volume of the canyon as a result of advection and turbulent transport. The right-hand side indicates that changes in the net energy status of the volume will be manifested as changes in sensible energy storage (changes in air temperature) and/or energy change due to volume radiative flux divergence.

Yoshida *et al.* have presented measurements on the energy balance in a urban canyon in Kyoto with E–W orientation.[14] Figure 8.9 shows the components of the mean energy balance at the roof and canyon-top levels in the day (10:00–16:00) and in the night (00:00–05:00) during the measurement period. The values of the components are normalized in terms of the net radiative heat transfer Q^* at roof level in the daytime.

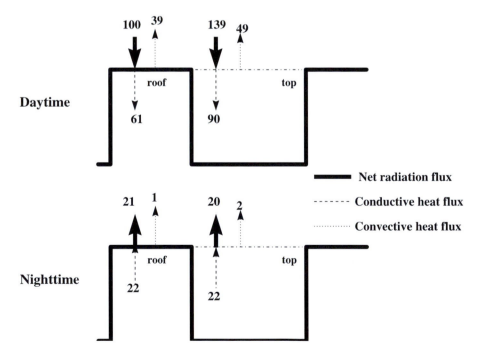

Figure 8.9. The components of the mean energy balance at the roof level and in the urban canyon system during the measurement period

METHODS FOR THE CALCULATION OF SHORT-WAVE RADIATION

The energy balance calculation for the surfaces of an urban canyon can be performed by estimating the exchange of energy, mass and momentum and, of course, reradiation phenomena should be taken into account. The resolution of such a theoretical model should be sufficiently detailed for cities to be examined on a building-by-building basis. Each street canyon represents a set of buildings or surface areas.

Such an approach can be accomplished by subdividing surfaces into finite strips that commence at the top of a building, continue down the side of that building, cross the street, ascend the side of the building across the street and finally terminate at the top of the second building.[15] Each area on the strip is subdivided into sunlit and shaded portions. For every subdivision the amounts of short- and long-wave radiation, and the associated view-factors due to reflection and reradiation, are computed.

A brief summary of the main equations for calculating the short-wave radiation in this procedure (URBAN 3) is given below.[16]

Horizontal surface

Direct beam from the sky:

$$Q_b = SV^{-2}t^m \cos Z, \tag{8.17}$$

where

$$t^m = \exp[-0.089(P_z m / 1013)^{0.75} - 0.174(wm / 20)^{0.6} - 0.083(\Theta m)^{0.9}]. \tag{8.18}$$

Diffuse from the sky:

$$q_b = 0.08203 + 0.253Q_s - 0.1421Q_b - 0.01742m + 0.332\log(w/10). \tag{8.19}$$

Reflected from the environment:

$$q_r = \sum (Q+q)_i a_i F_{i-A}. \tag{8.20}$$

Total absorbed:

$$(Q_b + q_b + q_r)(1-a). \tag{8.21}$$

Vertical surface

Direct beam from the sky:

$$Q_u = SV^{-2}t^m \cos A_s \cos(a_s - a'). \tag{8.22}$$

Diffuse from the sky:

$$q_u = q_b F_{sky-A} + q_b [f(\cos A_s \cos \theta) - 0.45].$$ (8.23)

Total absorbed:

$$(Q_u + q_u + q_r)(1 - a).$$ (8.24)

where S is the solar constant; V is the radius vector of the distance from Earth to Sun; t is the transmissivity of air; m is the optical air mass; Z is the solar zenith angle; P_z is the air pressure at elevation z; w is the precipitable water vapour; Q_s is the solar radiation on top of the atmosphere; F_{i-A} is the view factor of each surrounding block for block A; a is the albedo; A_s is the solar altitude; a_s is the solar azimuth; a' is the block orientation; F_{sky-A} is the view-factor of the sky for block A; and θ is the wall–solar azimuth.

In general, models that are used to calculate the short-wave radiation incident on building surfaces are invariably difficult to write and expensive to run. Researchers are therefore often forced to neglect some of the factors (such as the calculation of multiple reflections) in order to simplify the procedure. The following is a more detailed model based on the radiosity method for the calculation of short-wave radiation falling on different facades of an urban canyon.[17]

The short-wave radiosity $\mathcal{J}_{k,i}$ of surface i is a function of the incoming short-wave flux density $G_{k,i}$ and the surface reflectance $\mathcal{J}_{k,i}$:

$$\mathcal{J}_{k,i} = \rho_{k,i} G_{k,i},$$ (8.25)

where $G_{k,i}$ consists of radiation received from the sky and from the N wall surfaces viewed:

$$G_{k,i} = \sum_{j=1}^{N} F_{i,j} \mathcal{J}_{k,j} + G_{k,is},$$ (8.26)

where $F_{i,j}$ represents the view factor of wall i for wall j, $\mathcal{J}_{k,i}$ is the short-wave radiosity of wall j and $G_{k,is}$ is the short-wave radiation received by the wall i from the sky. When this is substituted for $G_{k,i}$ equation (8.25) becomes:

$$\mathcal{J}_{k,i} - \rho_{k,i} \sum_{j=1}^{N} F_{i,j} \mathcal{J}_{k,j} = \rho_{k,i} G_{k,is}.$$ (8.27)

Thus, for an enclosure consisting of N walls, we have a system of N linear equations in N unknown radiosities, where the ith equation is of the form:

$$\sum_{j=1}^{N} a_{i,j} \mathcal{J}_{k,j} = c_i,$$ (8.28)

where

$$a_{i,j} = \delta_{i,j} - \rho_{k,i} F_{i,j} \qquad (8.29)$$

and

$$c_i = \rho_{k,i} G_{k,is}. \qquad (8.30)$$

If $\rho_{k,s}$, $F_{i,j}$ and $G_{k,is}$ are known for each wall i, the vector of radiosities can be evaluated using the Gauss–Jordan method. Finally, the incoming short-wave flux density $G_{k,i}$ can be obtained.

With this complicated model the influence of different sky types on the energy balance of the urban canyon can be examined. The effects of neglecting horizon obstructions under overcast skies vary little among walls of different orientations of the urban canyon at a given level, because of the azimuthal isotropy of the overcast sky. Furthermore, the neglect of horizon obstructions leads to overestimation of the incoming radiation as a result of the replacement of shady walls with the comparatively bright sky. Under clear skies, diffuse radiation reaches its greatest intensity in the circumsolar region. Thus the assumption of sky isotropy leads to an underestimation of the incoming radiation on sunlit vertical surfaces and a smaller overestimation for the shaded walls. The neglect of multiple reflections will obviously cause an underestimation of the incoming short-wave radiation, the magnitude of which increases with increasing wall view-factors and wall reflectivities.

DAYLIGHT IN URBAN AREAS

Daylight is the part of the energy spectrum of electromagnetic radiation emitted by the sun within the visible waveband that is received at the surface of the earth after absorption and scattering.

Sky luminance varies according to a series of meteorological and seasonal parameters. In order to simplify calculations concerning daylight, some models for standard skies have been developed.

- *Uniform luminance sky distribution*. This model represents a sky with constant value of luminance.
- *CIE overcast-sky distribution*. In this model the luminance is varied with elevation and corresponds to a situation where the sky is covered with light clouds and the sun is not visible. The luminance at a point (L) of the sky with elevation q is given by:

$$L = L_z \frac{1 + 2\sin\theta}{3}, \qquad (8.31)$$

where L_z is the zenith luminance (Figure 8.10). The transfer process of the radiation through clouds, in the absence of severe pollution, produces white

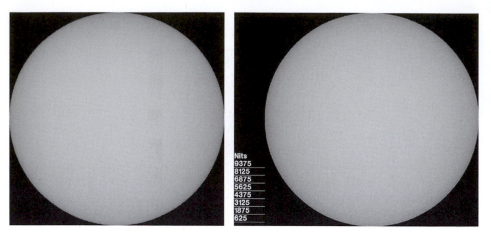

Figure 8.10. Hemispherical view and luminance distribution of an overcast sky (simulation with RADIANCE)

light by colour mixing. If the atmosphere is heavily polluted, the overcast sky colour appears yellow.

- *Clear-sky luminance distribution.* This type of sky has a strongly non-uniform luminance distribution. The CIE has recommended standard luminance distributions for a clear sky that are simplifications of a complex process of atmospheric scattering in the presence of different pollutants. Most clear-sky luminance models are variants of the same basic form. The following equation relates the luminance L at one point of the sky vault to the zenith luminance L_z:

$$\frac{L}{L_z} = \frac{(a + be^{-3\theta} + c\cos^2\theta)(1 - e^{-d\sec z})}{(a + be^{-3z_s} + c\cos^2 z_s)(1 - e^{-d})}, \tag{8.32}$$

where a, b, c and d are adjustable coefficients, θ is the angle (in radians) between the point with luminance L and the Sun, z is the zenith angle of the point and z_s is the zenith angle of the Sun (Figures 8.11 and 8.12).

The CIE clear-sky model has $a = 0.91$, $b = 10$, $c = 0.45$ and $d = 0.32$.[18] Additionally, CIE has standardized a clear sky for polluted atmospheres, using the coefficients $a = 0.856$, $b = 16$, $c = 0.3$ and $d = 0.32$.

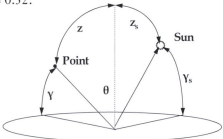

Figure 8.11. Definition of parameters used in sky luminance distribution functions

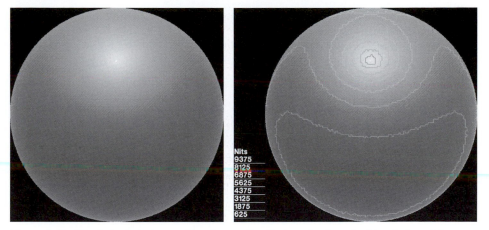

Figure 8.12. Hemispherical view and luminance distribution of a clear sky (simulation with RADIANCE)

The blue colour of the clear sky is strongly dependent on the height of the site above sea level and on the amount of the atmospheric pollution. Nitrogen dioxide makes the colour of the atmosphere brown and this can be seen when looking towards an urban area from the surrounding countryside. High water-vapour levels, in the absence of pollution, tend to give the sky a white appearance.

Since sky luminance distribution data are available only for a handful of locations, models describing actual luminance distributions have been developed. Perez *et al.* presented an all-sky model that is a generalization of the CIE standard clear sky formula.[19] This expression includes five coefficients, which can be adjusted to account for luminance distributions that range from totally overcast to clear skies.

Studies at De Montfort University have evaluated the performance of four sky luminance distribution models by comparing estimates with vertical illuminance measurements.[20] The models evaluated were : CIE overcast sky,[21] CIE clear sky,[18] intermediate sky[22] and Perez *et al.* all sky.[23] The reported results may be enumerated as follows:

- The CIE overcast-sky model shows an overall negative bias in the predictions of illuminances due to north- and east-facing windows.
- The CIE clear-sky model is the worst of all four models, with MBEs being three to four times those associated with the intermediate sky model and almost ten times higher than those from the Perez *et al.* all-sky model.
- For all but one aspect the Perez *et al.* model generates single-digit mean bias errors (MBEs).

Mean bias errors is defined as:

$$MBE = \Sigma(Y_c - Y_o) / n$$

where Y_c is the calculated value, Y_o the observed value and n the number of measurements.

Urban planning for daylight is essential in establishing the maximum benefit to be derived from natural light. The inclusion of daylight in the urban planning process can have a tremendous impact on the density of buildings (i.e. the exploitation factor), the use of land and finally the cost of the land.

Urban planning efforts to take into consideration daylight in the UK and the USA have led to the development of some design targets in order to ensure that daylight will be available within new buildings or to protect the building's site from future development. These targets are as follows:

- Daylight should be present on the facades of the buildings in order to provide good transmission of light into the interior.
- Daylight should be present either at particular locations on the exterior or within the interior.
- Daylight should be adequate at all sites likely to be developed or redeveloped.
- Daylight must be safeguarded for buildings adjacent to the proposed project.

Daylight usage in buildings differs according to their functional usage, the distribution of spaces and the building's lighting needs. It is obvious that, because of the different lighting needs of commercial and residential buildings, different design approaches should be adopted.

The first step in planning for daylight in a building is to ensure that daylight can be provided both in the interior and on the exterior of the building as well as in and around adjacent buildings. This task can be achieved by examining the position of the proposed building on the site, by determining the impact of natural exterior obstacles on the distribution of daylight and finally by determining the impact of man-made exterior obstacles.

The main factors that can affect the daylight (as skylight or sunlight) on buildings are:

- the distance between buildings;
- the height of the facing buildings;
- the size of openings;
- the size of shading devices;
- the orientation of and reflectance from the facing buildings.

In hot climates high-density housing with narrow streets is observed. This causes a large part of the radiation to be reflected away, minimizing the build-up of the air temperature near the ground. In colder climates, open spaces are used more frequently in order to collect solar radiation and to allow adequate ventilation.

Furthermore, the orientation of the urban canyon plays an important role in shading and daylighting. In hot regions the best orientation is N–S in order to achieve the best shading conditions in summer and the best lighting conditions in winter.

There are some design aids that can be used as help during the building design process. These are transparent overlays, which are used to ensure that some portion of the building has access to enough of the sky dome and their design (for either overcast or clear sky) is based on the calculation of horizontal and vertical angle of acceptance. These two angles together form the cone of viewable sky, which in turn determines the amount of daylight that is falling on the façades in question (Figure 8.13). For clear-sky conditions (without sun),[24] about 5% of the sky dome needs to be

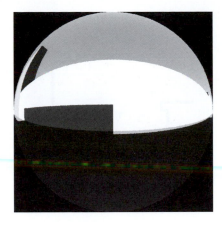

Figure 8.13. Estimate of the percentage of the sky dome as seen from a point on a vertical facade (simulation with RADIANCE)

visible from the reference point in order to illuminate the interior spaces. For overcast sky conditions 4% is satisfactory.

All the overlays are used in order to check a series of reference points 2 m above the ground on the facades of the building. If enough daylight can reach the reference 2 m height on the exterior walls, more daylight will be available at higher elevations. The reference height corresponds to the top of the ground-floor windows for buildings with basement windows.

There are different overlays for:

- clear skies;
- overcast skies;
- newly developed buildings;
- existing buildings.

Under overcast sky conditions, instead of the above-mentioned vertical and horizontal angles of acceptance, values of the vertical sky component can be used. In the UK, a vertical sky component equal to 27% at 2 m above the ground is sufficient to illuminate the interior (Figure 8.14).[25] Any reduction below this level, due to presence of a facing building, should be kept to a minimum. If the vertical sky component, with the new development in place, is both less than 27% and less than 0.8 times its former value, then the occupants of the existing building will notice a reduction in the amount of skylight. The impact on the daylighting distribution in the existing building can be found by plotting the no-sky line in each of the main rooms. For houses this would include living rooms, dining rooms and kitchens. In non-domestic buildings each main room where daylight is expected should be investigated. The no-sky line divides the points on the working plane that can and cannot see the sky. Areas beyond the no-sky line, since they receive no direct daylight, usually seem dark and gloomy compared with the rest of the room, however bright it is outside.

SUNLIGHTING

Sunlight offers a psychological benefit in many daylighting systems. In some cases, it may also provide additional heat and thermal comfort. Sunlight should not be ignored

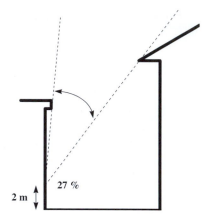

Figure 8.14. Vertical sky component of 27% at 2 m above the ground is sufficient to illuminate the interior

during the design process because it can significantly affect thermal and visual comfort, as well as the size of the cooling system of the building. Where prolonged access to sunlight is available, shading devices will also be needed so as to avoid overheating and glare from the sun. In site planning for sunlighting the key issues are:

- Understanding the shadow patterns of the proposed design, to assess their impact on the sunlighting of areas surrounding the buildings (Figure 8.15).
- The protection of future buildings by not placing them in the shade.

Access to sunlight can be improved by:

- having taller buildings to the north of the site, taking into account any overshadowing problems that may occur;
- having low-density housing in the area south of the building;

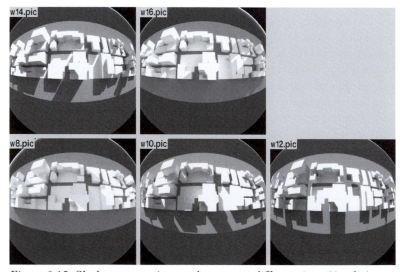

Figure 8.15. Shadow patterns in an urban area at different times (simulation with RADIANCE)

- creating courtyards in the southern part of the building area.

For daylighting systems, direct light may be needed all year round, while the opposite is the case for thermal applications, in which sunlight is desirable during winter but not in summer. Furthermore, the interior surface materials when sunlight is used for lighting are quite different from those for a solar thermal application where energy is typically stored for later use.

For estimating the potential for having sunlight on the facades, overlays similar to building-to-building overlays can be used. For the determination of the impact of the proposed building on surrounding buildings, the solar envelope technique can be used.

During the design phase of the building, transparent overlays can be used to study the potential for allowing sunlight on the facades. These overlays are:

- the sunlight availability overlay (Figure 8.16), which is used to find the probable sunlight hours received at any point in a building.
- the sunpath indicator, which is used to find the times of day and year for which sunlight is available on a window;
- the solar gain overlay, which is used to find the incident solar radiation on a south vertical window wall.

These overlays are constructed on the basis of the solar location at different latitudes and the heights of obstructions that might block the light. The major parts of these overlays are:

- one reference point, which is the point at which the sunlight potential will be evaluated;
- the solar azimuth lines;
- the lines of any obstruction's height.

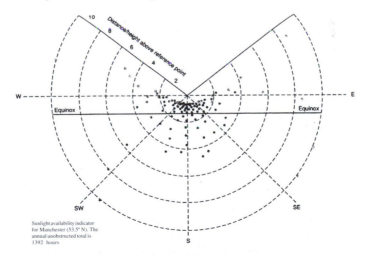

Figure 8.16. Sunlight availability indicator for Manchester. Each of the 100 existing spots represents 1% of annual probable sunlight hours

In addition to these, the azimuth corresponding to a 10° solar altitude is an important design issue, since for summer, spring or fall this line does not cover a major portion of the working day; furthermore, the sun's light stabilizes and its efficacy remains constant as soon as the 10° elevation is reached.

The British Standard recommends that interiors where the occupants expect sunlight should receive at least one quarter of annual probable sunlight hours, including at least 5% of annual probable sunlight hours during winter months, between 21 September and 21 March. The probable sunlight hours means the total number of hours in the year that the sun is expected to shine on unobstructed ground, allowing for average levels of cloudiness for the location in question. At the site-layout stage the positions of windows may not have been decided. It is suggested that sunlight availability be checked at points 2 m above the lowest storey level on each main window wall that faces within 90° of due south. The building facade as a whole should have good sunlighting potential if every point on the 2 m high reference line is within 4 m – measured sideways – of a point that meets the British Standard criterion for probable sunlight hours. This criterion is met provided either of the following is true:

- The window walls face within 90° of due south and no obstruction, measured in the section perpendicular to the window wall, subtends an angle of more than 25° with the horizontal. Obstruction within 90° of due north of the reference point need not count here.
- The window wall faces within 20° of due south and the reference point has a vertical sky component of 27% or more.

The recommendation that has been included in the German DIN 5034 on daylight provision is that at least four hours possible sunlight should fall on the middle of the window on 20 March.

REFLECTED SUNLIGHT AND DAYLIGHT

Reflected light from another buildings or from the ground can be used to illuminate interiors, influencing not only the quantity of light, but the quality as well. The colour of daylight under a clear blue sky derives from the additive mixing of coloured light from four sources: the blue sky, the rather more yellow-coloured sunlight, the ground which, if covered with growing vegetation, is green and, finally, other reflecting external surfaces.

Specific conditions, such as south-oriented obstructions reflecting sunlight, can cause glare.

Under clear-sky conditions, the impact of ground-reflected light on the interior illuminance is minor. Under overcast-sky conditions the light from the ground plane may increase the illuminance on a vertical surface by as much as 50%.

Conversely, on clear days vertical surfaces can significantly increase light levels on facing vertical surfaces, while their impact is minimal under overcast conditions. For these reasons, ground-reflected light is more useful under overcast conditions, while reflected light from vertical surfaces is more useful in clear-sky climates, as shown in Figure 8.17.

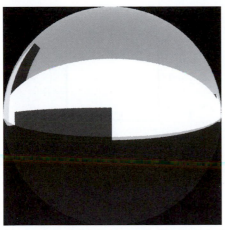
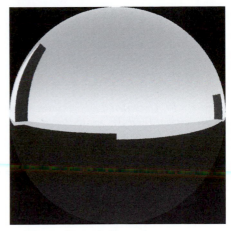

Figure 8.17. With facing sunlit facades light levels can increase dramatically on the facing wall and can create glare problems (simulation with RADIANCE)

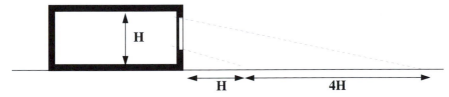

Figure 8.18. Impact of the ground plane on a one-storey building

Figure 8.18 shows the area of the ground, according to the dimensions of the window, that is critical in increasing or decreasing the amount of light that enters the space under consideration.[24]

The light that is reflected from vertical surfaces (such as facing building walls) is affected by:

- their dimensions;
- their reflectivity;
- their orientation.

One advantage of these surfaces is that they can be most effective during winter months, when it may be difficult to maintain the design interior illuminance in the building. For the geometrical configuration presented in Figure 8.19, the monthly average values of the Obstruction Illuminance Multiplier (OIM) have been calculated. The OIM is defined as the ratio of the illuminance received on a vertical surface, due to light received from the sky, ground and the obstruction, to the illuminance on the same surface without the presence of the obstruction. The results are presented in Figure 8.20.

Walls opposite north-facing windows have the best opportunity to reflect sunlight, while east- or west-oriented walls can be useful only in the morning or in the evening.

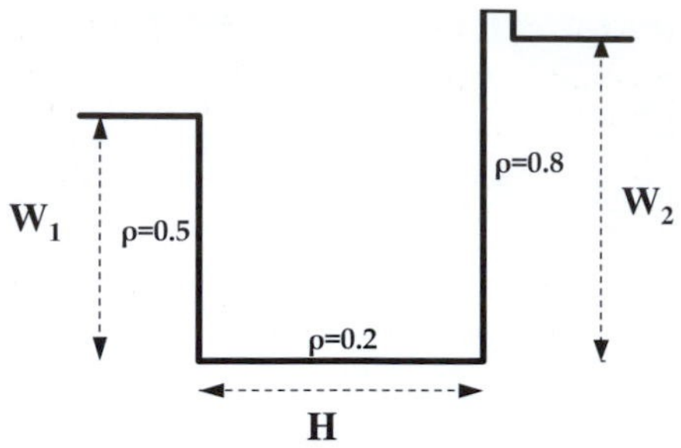

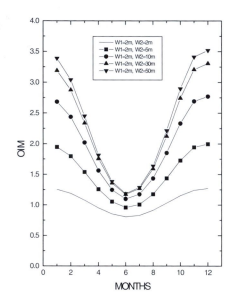

Figure 8.19. Geometrical configuration of the urban canyon. Wall (W₁) is the target surface, while wall (W₂) faces south

Figure 8.20. Obstruction illuminance multiplier for different geometric configurations (H = 10 m)

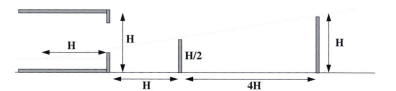

Figure 8.21. Typical vertical surfaces that are to be used as reflectors

The use of reflected light to increase the light penetration into a space can be accomplished by placing reflective surfaces in the field of view of the window. Since the lower parts of the window reflect light more deeply into the space, a good design target is to maintain a view of the sky from the upper part of the window, leaving the lower part for reflected light. The heights of different vertical surfaces that are to be used as reflectors for north-facing apertures are presented in Figure 8.21.

With the increasing interest in passive solar design, many architects have come to realize that the design of the building facade is critical if radiant energy from the sky is to be used. During recent years coated glazings have found widespread application in order to – among other reasons – reflect heat or reduce glare. This may be desirable for the occupants of the building in question, but people opposite or on the road may have problems due to glare. Vertical glazed surfaces may only cause problems when the Sun is low in the sky (during sunset or sunrise). If the facade is inclined with respect to the vertical, problems may occur even when the Sun is high in the sky.

According to Littlefair,[26] if the sun's elevation is h_2 and its azimuth with respect to the normal of a vertical facade is φ_2, the reflected ray will have an altitude h_1 and an azimuth φ_1 that are given by the following equations:

$$h_1 = -h_2 \tag{8.33}$$

$$\varphi_1 = -\varphi_2. \tag{8.34}$$

If the facade slopes back from the vertical, with sloping angle equal to s, then the above equations are transformed to:

$$\sin h_2 = (\sin 2s \cos h_1 \cos \varphi_1) - (\cos 2s \sin h_1) \tag{8.35}$$

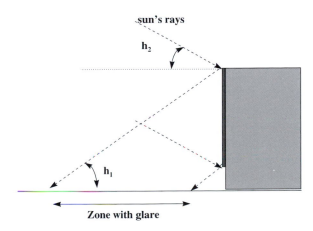

sun's rays

h_2

h_1

Zone with glare

Figure 8.22. Definition of the zone with glare due to reflected sunlight

$$\tan \varphi_2 = -\varphi_1 / (\cos 2s \cos \varphi_1 + (\cos 2s \cos \varphi_1 + \sin 2s \tan h_1). \tag{8.36}$$

If these equations apply for a set of points of known altitude and azimuth angles h_1 and φ_1, which are located at the corners of the façade, then a set of points (h_2, φ_2) can be calculated that bound the sun positions at which reflection can occur These points can be plotted on a sunpath diagram to give times of day and year at which reflected sunlight is likely to be a problem (Figure 8.22).

METHODS FOR CALCULATING DAYLIGHT IN URBAN AREAS

Illuminance at a given reference point in a particular geometry varies with time, orientation and sky condition. To make predictions of illuminance at a given reference point, mathematical methods offers several advantages:

- They allow quick analyses of different geometrical configurations.
- They can be used to determine daylighting over an extended period of time (a month or a year).
- They allow analyses of environments having non-diffuse surfaces.

The following section describes the different approaches to solving the problems of the quantitative estimation of daylighting in a canyon.

Simplified calculation of illuminance at various points on canyon facades

These methods can be mathematical, graphical or tabular and give illuminance at given points in a single-stage procedure. This often results in reduced accuracy, but saves time, which may be particularly useful in the first stages of the design process. These methods include:

- *A skylight indicator to find the vertical sky component.* This is independent of latitude and may be used anywhere.
- *Waldram diagram.* With this method the vertical sky component can be estimated more accurately.
- *Sunlight availability indicators.* These are used to find the probable sunlight hours received by a window wall, or at any other point in a building layout.
- *Sunpath indicators.* These are used to find the times of day and year for which sunlight is available on a window wall or at a point in a layout.
- *Solar gain indicators.* These are used to find the incident solar radiation on a south-facing vertical window wall.
- *Use of the solar envelope.* This is defined as the greatest volume that a building can fill on a site without causing significant overshadowing of adjacent sites.
- *Tregenza's method.*[27] With this procedure the illuminance on urban canyon facades can be estimated. Direct sun illuminance is taken into account.
- *Building daylight analysis.* This is parametric analysis under a uniform sky luminance distribution for four building typologies and gives sky components at any point on the facades.[28]
- *Daylight and urban form.* This method can be applied under overcast or clear sky without sun and gives daylight characteristics for four different building typologies, allowing for surface reflectivity, street width and orientation.
- *Permissible height indicators.* This tool is used to ensure that all or part of the building has a view of enough sky dome to ensure that sufficient daylighting can reach the facades.[29] Uniform, overcast and clear skies can be used.
- *Sky-dome projection.* This is a photographic tool using full-field camera techniques or sky-dome projection applied to the urban layers.

Radiosity

In this numerical method all surfaces are assumed to be diffuse reflectors.[30] This means that these surfaces have constant luminance independent of the viewing direction. Each surface is divided into a mesh of patches. Simple geometrical rules allow the determination of patches that are visible from light sources and therefore the determination of their illuminance. Each illuminated patch in turn reflects some of the incident flux back into the room. From Lambert's cosine law the illuminances of all patches that can receive the reflected flux can be determined. The whole process is iterative and proceeds until all the reflected flux has been absorbed. The process involves the following stages:

1. Definition of the geometry.
2. Determination of form factors.
3. Solution of simultaneous equations for patch illuminances.
4. Calculation of point illuminances and luminances from patch illuminances.
5. Production of rendered images.

The disadvantages of this method are:

- The 3D mesh requires more memory than the original surfaces.
- The surface sampling algorithm is more susceptible to imaging artefacts than ray tracing.
- The method does not account for specular reflections or transparency effects.

The advantages of the method are:

- Diffuse interreflections between surfaces can be calculated.
- It is possible to obtain view-independent solutions for fast display of arbitrary views.
- There are immediate visual results, which progressively improve in accuracy and quality.

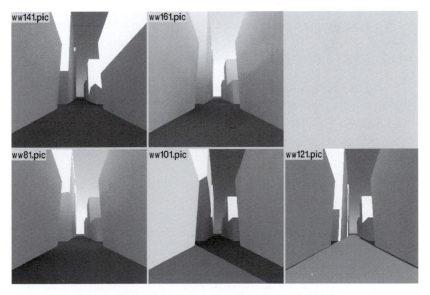

Figure 8.23. Simulation of an urban canyon with the ray-tracing method.

Ray-tracing methods

Ray tracing is a method for computing luminance by following light backwards (or forwards) from the point of measurement to the sources of light. With this approach, production of synthetic images can be performed quite simply; see, for example, Figure 8.23.

At a point in the image plane, the surface is visible and hence colour and intensity at that point can be obtained by tracing a ray backwards from the eye through the point into the scene. If this ray intersects an object, then local calculations will determine the colour and the intensity that are the result of direct illumination at that point. This is the light from a light source directly reflected from the surface. If the object is partially reflective or partially transparent or both, the colour and the intensity of the point in the image plane will include a contribution from reflected or transmitted rays. These rays must be traced backwards to discover their contribution. The trace of a particular ray terminates when no more objects are intersected or if so many surface intersections separate it from the observer that its colour and intensity contribution to the image is likely to be negligible. In principle, a ray is traced through the centre of each pixel in the image plane using zero spatial coherence (all rays are traced independently) and the sampling produces aliasing.

In more detail, each ray can be thought of as the luminance value that results either directly from an emitting surface or indirectly from a reflecting surface. Computing luminance from a reflecting surface requires convolution of the surface reflectance distribution function with the luminous flux arriving at the surface. New rays are traced to determine the necessary values of luminance, making the process recursive. Because ray tracing does not impose any restrictions on the distribution of the surface reflectance, specularity and other phenomena can be readily incorporated in the model. The difference between a simple reflection model and a ray-traced model[31] is the depth to which interaction between light rays and objects in the scene is examined.

Computer simulation of illuminated spaces involves two main areas of modelling, surface geometry and surface reflectance. In addition, synthetic imaging requires the modelling of a camera.

The ray-tracing algorithm has the following advantages:

- Direct illumination, shadows, specular reflections and transparency effects are accurately rendered.
- It is memory-efficient.

The disadvantages are:

- The method is computationally expensive. The time required to produce an image is greatly affected by the number of light sources.
- The process must be repeated for each view, i.e. it is view dependent.
- The method does not account for diffuse interreflections.

REFERENCES

1.   NASA (1971). *Solar Electromagnetic Radiation*. Report SP-8005, NASA.

2. Frohlich, C. and Brusa, R.W. (1981). 'Solar Radiation and its Variation in Time'. *Solar Physics*, Vol. 74, pp. 209–215.
3. Chandler, T. (1965). *The Climate in London*. Hutchinson, London.
4. Maurain, C. (1957). *Le Climat Parisien*. Presses Universitaires, Paris.
5. Oke, T. (1982). 'The Energetic Basis of the Urban Heat Island'. *Quarterly Journal of the Royal Meteorological Society*, Vol. 108, pp. 1–24.
6. Peterson, J. and Stoffel, T. (1980). Analysis of Urban–Rural Solar Radiation Data from St. Louis'. *Journal of Applied Meteorology*, Vol. 19, pp. 275–283.
7. Oke, T.R. (1974). *Review of Urban Climatology*. WMO Technical Note No. 134, WMO, Geneva.
8. Craig, C.D. and Lowry, W.P. (1972). 'Reflections on the Urban Albedo'. Preprint, Conference on Urban Environment and 2nd Conference on Biometeorology, American Meteorological Society, Boston, pp. 159–164.
9. Terjung, W. and Louie, S.F. (1973). 'Solar Radiation and Urban Heat Island'. *Annals of the Association of American Geographers*, Vol. 63, pp. 181–207.
10. Aida, M. (1982). 'Urban Albedo as a Function of the Urban Structure – a Model Experiment'. *Boundary Layer Meteorology*, Vol. 23, pp. 405–414.
11. Aida, M. and Gotoh, K. (1982). 'Urban Albedo as a Function of the Urban Structure – a Two Dimensional Numerical Simulation'. *Boundary Layer Meteorology*, Vol. 23, pp. 415–424.
12. Evans, M. (1980). *Housing, Climate and Comfort*. The Architectural Press, London.
13. Nunez, M. and Oke, T. (1977). 'The Energy Balance of an Urban Canyon'. *Journal of Applied Meteorology*, Vol. , pp. 11–19.
14. Yoshida, A., Tominaga, K. and Watatani, S. (1990/91). 'Field Measurements on Energy Balance of an Urban Canyon in the Summer Season'. *Energy and Buildings*, Vol. 15–16, pp. 417–423.
15. Terjung, W. and O'Rourke, P. (1980). 'Simulating the Causal Elements of Urban Heat Islands'. *Boundary Layer Meteorology*, Vol. 19, No.1, pp. 93–118.
16. Terjung, W. and O'Rourke, P. (1979). *A Climatic Model of Urban Energy Budgets*. LA, Academic Publishing Services, Los Angeles, CA.
17. Werseghy, D. and Munro, D. (1989). 'Sensitivity Studies on the Calculation of the Radiation Balance of Urban Surfaces. I. Shortwave Radiation'. *Boundary Layer Meteorology*, Vol. 46, pp. 309–331.
18. CIE (1973). 'Standardisation of Luminous Distribution on Clear Skies'. International Conference on Illumination, CIE, Paris.
19. Perez, R., Seals, R. and Michalsky, J. (1993). 'All Weather Model for Sky Luminance Distribution. Preliminary Configuration and Validation'. *Solar Energy*, Vol. 50, No. 3, pp. 235–245.
20. Mardaljevic, J. (1995). 'Validation of a Lighting Simulation Program under Real Sky Conditions'. *Lighting Research and Technology*, Vol. 27, No. 4, pp 118–188.
21. Moon, P. and Spencer, P. (1942). 'Illumination from a Non-uniform Sky'. *Transactions of the Illumination Engineering Society*, Vol. NY37, pp. 707.
22. Matsuura, K. and Iwata, T. (1990). 'A Model of Daylight Source for the Daylight Illuminance Calculation on All Weather Conditions', A. Spiridonor (ed). *Proceedings of the Third International Daylighting Conference, Moscow*.
23. Perez, R., Seals, R. and Michalsky, J. (1991). 'Modelling Sky luminance Angular Distribution from Real Sky Conditions'. *Proceedings of the ISES World Congress*, Denver, CO. See *Journal of the Illumination Engineering Society*, (1992) Vol. 21, No. 2, pp. 84–92.
24. Robbins, C.L. (1986). *Daylighting Design and Analysis*. Van Nostrand Reinhold, New York.
25. Littlefair, P. (1991). *Site Layout Planning for Daylight and Sunlight. A Guide to Good Practice*. BR209, BRE, Garston.
26. Littlefair, P. (1987). *Solar Dazzle Reflected from Sloping Glazed Facades*. BRE information paper IP3/87.

27. Tregenza, P. (1995). 'Mean Daylight Illuminance in Rooms Facing Sunlit Rooms'. *Building and Environment*, Vol. 30, No. 1, pp. 83–89.
28. Vio, M. (1988). 'Per una Valutazione Sintentica delle Prestazioni Iluminotecniche ed Energetiche di Alcune Tipologie Edilizie'. *Energie Alternative THE*, No. 51.
29. Bryan, H. and Stuebing, S. (1987). 'Daylight:The Third Dimension of the City'. *Sun World*, Vol. 11, No.2.
30. Goral, C., Torrance, K., Greenbery, D. and Battaile, B. (1984). 'Modelling the Interaction of Light between Diffuse Surfaces', *Computer Graphics*, Vol. 18, No. 3, pp. 212–222.
31. Ward, G. (1994). 'RADIANCE: Lighting Simulation System'. SIGGRAPH. (ACM Special Interest Group on Computer Graphics). Software downloadable from URL http://radsite.lbl.gov.

FURTHER READING

Baker, N., Fanchiotti, A. and Steemers, K. (eds) (1993). *Daylighting in Architecture*. James & James Science Publishers, London.
Goulding, J., Owen-Lewis, J. and Steemers, T. (eds) (1983). *Energy in Architecture*. Batsford, London.
Hopkinson, R., Longmore, J. and Petherbridge, P. (1966). *Daylighting*. Heinemann, London.
Iqbal, M. (1983). *An Introduction to Solar Radiation*. Academic Press, New York.
Landsberg, H. (1981). *The Urban Climate*. Academic Press, New York.
Muneer, T. (1997). *Solar Radiation and Daylight Models*. Architectural Press, London.

9

——

Urban pollution

V.D. Asimakopoulos,
Centre of Environmental Technology, Imperial College, London

෨

Despite the fact that, all over the world, many people already live in large metropolitan areas, there is a clear tendency for urbanization to continue for many more years in most geographical regions. It is estimated that within the next decade more than half of the world's population will be living in big cities. Europe is the most urbanized continent in the world, with some 80% of the population permanently residing in densely populated areas. This is associated with the increase in all kind of human activities, which in turn results in the degradation of the local natural atmospheric environment.

Under the same urbanization pressure, the local climate in large metropolitan areas is also altered. This is especially apparent when certain climatic characteristics are considered, e.g. temperature, humidity and wind. In fact, all the main meteorological parameters are severely affected, resulting in the development of a local climatic regime, which is characterized by increases in temperature (the heat-island effect) and reduction of humidity and wind. Furthermore, in central areas particularly, the continuous replacement of vegetation with buildings and roads severely affects the radiation balance and this further influences the temperature regime of the environment. Under these circumstances the comfort index for those living in big cities is quite different from that for those living in suburban and rural areas.

URBAN POLLUTION

Environmental quality is the most important issue for the urban environment. Of major concern are air pollution, noise and traffic. In this section urban air pollution, which is a significant problem in most metropolitan areas, will be discussed briefly.

The relative importance of different air pollutants and sources in different geographical areas has changed with time and culture. In certain European cities the dominant sources of atmospheric pollution nowadays are motor vehicles and the combustion of gaseous fuels. The receptors of all pollutants are human beings, buildings and animals. In this respect the World Health Organization has set certain air quality guidelines,[1] a summary of which is presented in Table 9.1.

The values in Table 1 are the guideline values based on health or environmental effects and on other considerations, such as abatement strategies and economic and social conditions. In addition to the above, it should be noted that the actual exposure of the urban inhabitants to air pollutants is difficult to estimate, because concentrations

Table 9.1. WHO air quality guidelines and effect levels

Pollution type	Guide value	Averaging time (mg/m³)	Effect level	Classification
Short term O_3	120	8 h	200 µg/m³	Mild
SO_2	500	10 min	400 µg/m³	Moderate
SO_2	125	24 h	400 µg/m³	Moderate
NO_2	200	1 h	–	–
CO	100,000	15 min	–	–
CO	60,000	30 min	–	–
CO	30,000	1 h	–	–
NO_2 (long term)	40	1 year	–	–
Lead	0.5	1 year	–	Moderate
SO_2	50	1 year	–	mild

Source of information: WHO[2]

vary over time and space and the receptors' intake amounts are related to location, level of physical activity and distance from the local source. An example is a main road; a pedestrian takes in a completely different amount of a pollutant from an inhabitant living on the sixth floor of a building. The problem gets more complicated if local airflow patterns are also considered. Recent publications indicate that the air mass can either circulate or move with the synoptic wind if the wind direction is transverse or parallel to the street canyon orientation respectively.[3] The various pollutants follow the same pattern.

Urban areas in general are subject to a wide range of pollutants. Most air pollutants stem from three sources: industry, motor vehicles and the burning of fossil fuels for heating or electricity generation. The contribution of industrial sources to air pollution varies considerably from one town to another, depending on the density and type of industry in an area, its precise location and the extent to which it has adopted restricting measures to control emissions or disperse them over long distances. In many cases, industrial pollution is exclusively an urban problem. On the other hand, air pollution problems related to city transport and buildings are more closely linked to the internal functioning of the city. The contribution of these energy-using activities to the levels of particular pollutants is set out in Table 9.2.[4]

Depending on the energy source, space heating can be one of the most important sources of air pollution. In Dublin, for example, domestic heating is a major source of SO_2 and particulate. A gradual shift away from coal has removed some of the worst effects of particulate and SO_2 pollution on a local scale. However, a shift to electricity does not solve the problem at the global level, owing to the pollution resulting from forms of electricity generation. In this wider global perspective, the use of all forms of fossil fuel contributes to problems of acid rain and, indirectly, to the greenhouse effect.

Table 9.2. Sources of air pollution

Sector	CO_2	SO_2	NO_x
Energy generation	37.5	71.3	28.1
Industry	18.6	15.4	7.9
Transport	22.0	4.0	57.7
Others	21.9	9.3	6.3

While the worst problems of local air pollution caused by heating have been solved, they have been replaced by increased levels of transport pollution. Automobile engines are major sources of NO_x, CO, particulates and lead. As far as CO_2 is concerned, it is worth noting that almost half of transport combustion is estimated to be due to urban traffic, while in many cities, the transport sector is responsible for almost 90% of carbon monoxide emissions.

In a number of southern European cities, such as Athens and Naples, there has been significant increase in most air pollutants and a corresponding reduction in air quality. This can be traced to the major growth in population of these cities, resulting from an exodus from the rural areas of southern Europe. The growth of urban industry and vehicle emissions in Athens, for example, has been so great that the authorities have had to introduce a system whereby vehicles are only permitted to enter the city centre an alternate days – and not at all when pollution levels exceed safety limits. These measures have only succeeded in preventing the situation from worsening. They cannot be considered as long-term solutions.

While cases such as Athens are widely known, it would be wrong to assume that cities in northern Europe do not have critical problems as well. While data obtained by monitoring may indicate an overall improvement, the use of average estimates from a restricted number of recording stations may well be obscuring the presence of harmful concentrations at certain times and locations. In Brussels, for example, rush-hour traffic in road tunnels results in pollution levels many times the limit recommended by the World Health Organization.

Air pollutants, such as sulphur and nitrogen oxides, can be transported over long distances downwind from the released position. These pollutants either reach the surface in dry form (dry deposition) or are removed from the air and carried to the ground by means of rain or snow (wet deposition or acid rain). Acid deposition contains both dry and wet deposition. Sulphur dioxide and oxides of nitrogen that are on the ground are transformed into acids interacting with the water. In addition, the air pollutants remaining aloft may be transformed into drops of sulphuric acid (H_2SO_4) and nitric acid (HNO_3) and fall to the earth. As a result of this toxic precipitation, different areas of the world become acidic, causing severe effects on the natural environment. It is well known that Sweden and other places in western European suffer from acid precipitation that comes from factories in the eastern part of England. Acid deposition may affect plants and water resources, as is the case in Germany where more than 200,000 acres of woodland have been seriously damaged. The same applies to thousands of lakes all over the world. The foundations of structures, building surfaces, monuments and other structures in many cities have been seriously affected. This is a major environmental problem that will become much more serious if adequate precautions are not taken.

AIR POLLUTION IN STREET CANYONS

The street-canyon domain is a region where large amounts of pollutants are released near the ground from motor vehicle exhausts and near the roof level from domestic or industrial chimneys. In urban and industrial areas the sources of pollutants vary significantly. The analysis of a single source contribution to the total pollution requires a fairly precise knowledge of dispersion processes in complex flows. In the highly complex geometry of urban canopies, the wind flows are characteristic, incorporating recirculation regions overlapping with complex structures, and acceleration systems as well as blocking effects. The fluid motion governing the dispersion process is highly turbulent and influenced by both large-scale gravitational and thermal effects. To analyse these complex turbulent buoyant flows, it is necessary to develop specific techniques and tools.

Systematic studies that allow the derivation of general rules and relationships between the flow characteristics and the urban geometric pattern are rare.[5] Most studies are based on measurements in wind-simulation tunnels and investigate the flow in the immediate vicinity of, and the pressure distribution on, buildings or blocks of structures. Their aim is to aid to urban-comfort planning or to estimate either the effects of wind on constructions or air quality. Unfortunately, only few studies have been generic enough to derive laws and formulation. However, there are good examples of which a special mention should be made.[6] Generally speaking, the authors principally measure pressure distributions, which form the input data for further analysis of the flow regime.[3]

Although many wind-tunnel simulations of flow around isolated buildings have been validated by on-site measurements, the extremely small number of in-street detailed flow-structure measurements allows only very partial validation of simulations.

The climate of urban canyons is primarily controlled by micrometeorological effects of canyon geometry, rather than the mesoscale forces controlling boundary-layer climatic systems.[7] Significant components of the street-canyon microclimate are the characteristic wind-induced flow patterns. These are significant because air flows often directly affect the comfort of the city inhabitants and the quality of the air. Furthermore, the dynamic characteristics of the air circulation in a street canyon reflect the near-field geometry, whereas, depending on the canyon aspect ratio and the local roof circulation, different canyon vortex patterns can develop.

The urban environment shows well-known atmospheric chemical characteristics, which, since the first 1943 Los Angeles smog episode, have been thoroughly investigated by many research groups. An important finding of the early investigations was that a mixture of the products of pollutant emissions from petroleum combustion is not capable of producing the effects of smog unless exposed to sunlight. This is because ozone (O_3) and other pollutants are not produced directly, but are the result of secondary cycles through photochemical reactions involving solar radiation. Nitrogen dioxide (NO_2) is an efficient absorber of ultraviolet portions of the solar spectrum that cause the NO_2 molecule to photolyse into nitric oxide (NO) and atomic oxygen (O). The O can react with molecular oxygen (O_2) to form O_3. The O_3 in turn can react with NO to form NO_2 and O_2 again, completing the atmospheric NO photolytic cycle. Naturally, at night, the chemical cycle is different. A typical daily variation of NO_2, NO, O_3 and hydrocarbons (HC) is shown in Figure 9.1.

To illustrate a real situation, a typical example air-quality monitoring station time series in an urban environment is presented in Figures 9.2 to 9.6. In particular, a very central station of the Athens city-monitoring network (Patision) is used to show recent annual concentrations of the most important pollutants.[8]

Figure 9.2 shows the hourly mean concentrations of CO from the Patision station in Athens for 1996. The trace especially indicates the local traffic since the sampling unit is located close to the road. As expected for this particular location, the concentrations remain high since the traffic load is nearly constant throughout the day. The two peaks indicate the periods with higher local traffic loads. Similar behaviour is shown in Figure

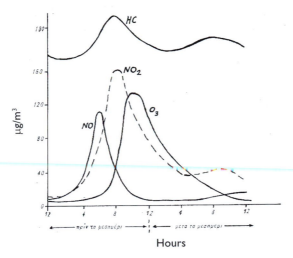

Figure 9.1. Typical variations of NO, NO₂, O₃ and
HC from an urban monitoring station

9.3 but for sulphur dioxide (SO$_2$). Here again the same logic applies as for the CO
concentrations. Figures 9.4, 9.5 and 9.6 show the concentrations of O$_3$, NO$_2$ and NO
respectively for the same station and time period as for Figures 9.2 and 9.3. Naturally,
here we have the photochemical cycle (briefly described previously), which changes the
logic of the concentration variations. However, the photochemical interactions and
chemical reactions are apparent for the three pollutants. Thus, as expected, the maximum
values of ozone (14:00 hours) are different from the nitrogen oxides peaks around 11:00
hours. Many researchers have reported similar results for various urban monitoring
stations.

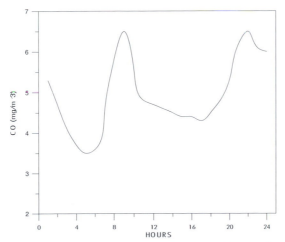

Figure 9.2. Hourly mean concentrations of CO for the
year 1996 at the Patision (Athens) monitoring station

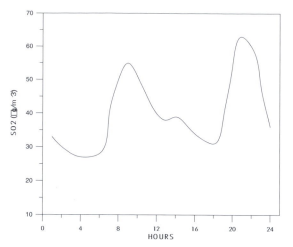

*Figure 9.3. Hourly mean concentrations of SO₂ for the year
1996 at the Patision (Athens) monitoring station*

In recent years, emphasis has been put on the canyon atmospheric chemistry problems, which, in association with the problems of flow circulation that are very difficult to describe mathematically, form the leading edge of research in this field.

URBAN NOISE

Noise is one urban form of pollution that seriously affects the quality of life of the inhabitants. The most important sources of noise pollution are road traffic, aircraft and railways, as well as the building and development sites. The workplaces of many city inhabitants can also be of importance with respect to the noise problem.

Many communities have proposed noise-level standards for motor vehicles as part of the norm-setting process of the market. Naturally these standards have been modified to cope with new technological achievements.

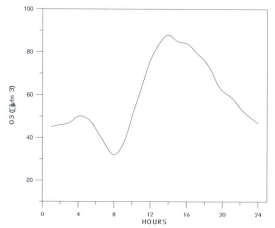

*Figure 9.4. Hourly mean concentrations of O₃ for the year
1996 at the Patision (Athens) monitoring station*

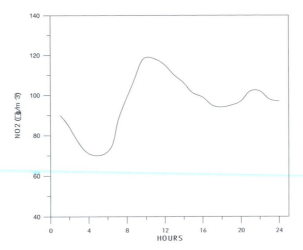

Figure 9.5. Hourly mean concentrations of NO₂ for the year 1996 at the Patision (Athens) monitoring station

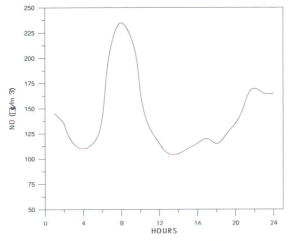

Figure 9.6. Hourly mean concentrations of NO for the year 1996 at the Patision (Athens) monitoring station

Many studies on the effects of noise on human health suggest that the outdoor level of noise should not exceed a daytime equivalent sound pressure level (despl) of 65 dB, the level at which serious impact of noise is noticeable. In fact, areas where the noise level exceeds 75 dB are considered dangerous, since the noise can cause hearing loss.

According to the European Environment Agency recent records, 113 million people in Europe (17% of the population) are exposed to despl over 65 dB and 450 million people (65% of the population) to above despl 24h 55 dB. About 9.7 million people are exposed to unacceptable noise levels above despl 24h 75 db. In large cities, the percentage of people exposed to unacceptable levels is two to three times higher than the national average.

The existing data do not allow trends in noise exposure to be plotted for many major European cities. However, levels higher than the maximum acceptable level of 65 dB occur in most cities (Figure 9.7).

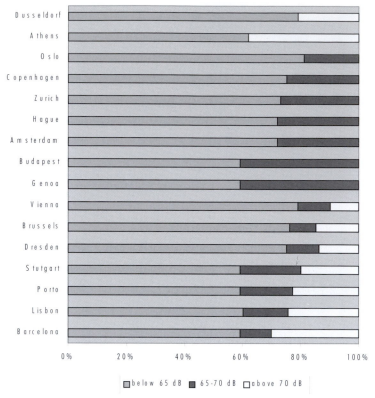

Figure 9.7. Excessive noise levels in some European cities

REFERENCES

1. WHO (1987). *Air Quality Guide Lines for Europe*. WHO Regional Publications, European Series No 23. World Health Organization, Copenhagen.
2. WHO (1998). *Revised WHO Air Quality Guide Lines for Europe* (2nd edn). WHO European Center for Environment and Health, Bilthoven, The Netherlands.
3. Oke, T.R. (1988). *Boundary Layer Climates* (2nd edn). Routledge, London.
4. European Environment Agency (1998). *Europe's Environment: The Second Assessment*. European Environment Agency, Copenhagen.
5. Sini, J.F., Anquetin, S. and Mestayer, P.G. (1996). Pollutant Dispersion and Thermal Effects in Urban Street Canyons. *Atmospheric Environment*, Vol. 30, No. 15, pp 2659–2677.
6. Hussain, M. and, Lee B.E. (1980). An Investigation of Wind Forces on Three Dimensional Roughness Elements In A Simulated Atmospheric Boundary Layer Flow. Report BS 56, Department of Building Science, Faculty of Architectural Studies, University of Sheffield.
7. Hunter, L.J., Johnson, G.T. and Watson, I.D. (1992). 'An investigation of three-dimensional characteristics of flow regimes within the urban canyon'. *Atmospheric Environment*, Vol.26B, No.4, pp. 425–432.
8. *Atmospheric Pollution in Athens* (1997). Technical report published by the Ministry of Environment, Planing and Public Works.

10

The role of green spaces

M. Santamouris,
Department of Applied Physics, University of Athens

જી

GENERAL – PHYSICAL PRINCIPLES

Trees and green spaces contribute significantly to cooling our cities and saving energy. Trees can provide solar protection to individual houses during the summer period, while, in addition, evapotranspiration from trees can reduce urban temperatures. In parallel, trees absorb sound and block erosion-causing rainfall, filter dangerous pollutants, reduce wind speed and stabilize soil, and prevent erosion (Figure 10.1). The American Forestry Association estimated that the value of an urban tree is close to $57,000 for a 50-year old mature specimen.[1] As mentioned by Akbari and Taha,[2] the above estimate includes a mean annual value of $73 for air conditioning, $75 for soil benefits and erosion control, $50 for air pollution control and $75 for wildlife habitats.

Evapotranspiration, defined as the combined loss of water to the atmosphere by evaporation and transpiration, is the major mechanism through which trees contribute to decreasing urban temperatures. Evaporation is the process by which a liquid is

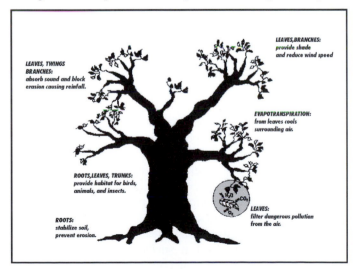

Figure 10.1 Ecological qualities of trees. Adapted from Akbari[2]

transformed into gas,[3] and in the atmosphere usually water changes to water vapour. Transpiration is the process by which water in plants is transferred as water vapour to the atmosphere. The energy transfer to the latent heat from plants is very high, 2324 kJ/kg of water evaporated.[4] Moffat and Schiller report that 1460 kg of water is evaporated from an average tree during a sunny summer day,[5] and this consumes about 860 MJ of energy, a cooling effect outside a home that is 'equal to five average air conditioners'; the exact methodologies for calculating the water losses due to transpiration are discussed below. The same authors report that the latent heat transfer from wet grass can result in temperatures 6–8 K cooler than over-exposed soil and that one acre of grass can transfer more than 50 GJ on a sunny day.

Evapotraspiration contributes to creating lower-temperature spaces in an urban environment; this is known as 'the oasis phenomenon'. The magnitude of the temperature reduction is related to the overall energy balance of the area, but in general, the oasis phenomenon is characterized by the Bowen ratio, which is the ratio of the sensible heat flux to the latent heat flux. Taha reports than under average oasis conditions, Bowen ratios in vegetative canopies are within 0.5–2.[6] A Bowen ratio of 2 was measured in a pine forest in England in July around midday;[7] this corresponds to sensible and latent heat fluxes of 400 and 200 W/m² respectively. In a Douglas fir stand in British Columbia, a Bowen ratio of 0.66 was found, corresponding to sensible and latent heat fluxes of 200 and 300 W/m² respectively.[8] As also reported by Taha,[6] the Bowen ratio is typically around 5 in urban areas, while in a desert it is close to 110 and over tropical oceans about 0.1.

The positive impact of trees on energy savings has gained an increasing acceptance during recent years. For example, the Sacramento Municipal Utility District is actually supporting and funding tree planting to offset the projected need for a new power plant in the next decade.[9] As reported by Akbari and Taha:[2]

> Field measurements have shown that through shading, trees and shrubs strategically planted next to buildings can reduce summer air conditioning costs typically by 15 to 35 per cent, and by as much as 50 per cent or more in certain specific situations. Simply shading the air conditioner – by using shrubs or a vine covered trellis – can save up to 10 per cent in annual cooling energy costs.

As a result of the above, several communities and non-profit organizations, mainly in the USA, have initiated and coordinated tree-planting efforts and encourage residents to plant trees.

Trees also help mitigate the greenhouse effect, filter pollutants, mask noise, prevent erosion and have a calming effect on people. As pointed out by Akbari et al.,[10] 'the effectiveness of vegetation depends on its intensity, shape, dimensions and placement. But in general, any tree, even one bereft of leaves, can have a noticeable impact on energy use'. Trees in paved urban areas intercept both the advected sensible heat and the long-wave radiation from high-temperature paved materials such as asphalt.[11,12]

Trees absorb gaseous pollutants through leaf stomata and can dissolve or bind water-soluble pollutants onto moist leaf surfaces, while tree canopies intercept particulates. In parallel, trees reduce ozone concentrations in the ambient air, either by absorbing the ozone (and other pollutants such as NO_2) directly or by reducing air temperatures, which reduces hydrocarbon emission and ozone formation rates.[13,14] Bernatsky reports that a street lined with healthy trees can reduce airborne dust particles by as much as 7000 particles per litre of air.[15] Trees also remove atmospheric CO_2 and store it as a woody biomass. However, biogenic hydrocarbon emissions from trees may play a role in ozone formation.[16,17]

In parallel, trees reduce and filter urban noise. As pointed out by Akbari and Taha,[2] leaves, twigs and branches absorb high-frequency sounds that are most bothersome to humans. The same authors report that a belt of trees 33 m wide and 15 m tall reduced highway noise by 6 to 10 dB – a sound reduction close to 50%. Data on the attenuation effect of vegetation as a function of the amount of vegetation are given by Broban,[18] who reports that in order to obtain significant reductions in the noise level, of the order of 110 dB, dense vegetation at least a 100 m deep is required.

Reduction of wind speeds by trees is important and may contribute to important energy savings. Heisler reports that an increase by 10% in the tree cover in a residential area can reduce wind speeds by 10–20%,[19] while an added 30% cover can reduce wind speed by 15 to 35%. Huang *et al.* simulated the effect of wind shielding by 30% tree cover on the heating and cooling energy use of older houses, for various US cities, and found that for all locations the heating load was significantly reduced.[20] Studies by DeWalle, dealing with the role of windbreaks of height H on infiltration and heating energy needs of a mobile house, have shown that, at distances of 1–4H from the wind break, the air speed was reduced by 40–50% of the undisturbed wind.[21] The reduction of the infiltration was from 55% at 1H to 30% at 4H and 8H. The corresponding heating energy was reduced by about 20% at 1H and by about 10% at 4H.

Trees have an important social impact. Anderson and Cordell found that the presence of trees increases property values,[22] while Getz *et al.* reported that city residents said that trees influence their decision about where to live.[23] As reported by Akbari and Taha,[2] studies have shown that trees can increase property values by between 3 and 20%. It is evident that spaces with vegetation make urban settings more attractive,[24] and have a positive impact on moods.[25] The Dobris assessment of the European Commission has suggested using as an indicator of the urban environmental quality the walking distance of 15 minutes or less from homes to green spaces.[26] The same authors have compiled data on the proportion of green spaces relative to the city area, as shown in Figure 10.2. In parallel, Ulrich has demonstrated that natural stimuli promote healing and recovery from stress.[27] This study proved that patients who saw trees, rather than a brick wall, from their windows had shorter post operative stays, fewer negative evaluative comments in nurses' notes and fewer post-surgical complications; they also needed fewer painkillers.

Ulrich also studied the reaction of humans to colour slides or rural and urban scenes.[28] He reported that humans were more interested in, and felt more positively about, rural rather than urban scenes. He recorded higher-amplitude alpha waves from subjects' brains when they saw rural scenes; brain alpha waves are correlated with feelings of relaxation. Ulrich *et al.* reported that they showed 120 humans a stressful movie and then showed them one of six different videotapes of urban and natural scenes.[29] Records of psychological and physiological parameters showed that humans recuperated from the stressful movie more rapidly, and more thoroughly, with exposure to natural settings.

EVALUATION OF THE TRANSPIRATION RATE FROM TREES

To evaluate the possible decrease in air temperature due to evapotranspiration, it is necessary to know the exact latent heat absorbed by the ambient air during the evaporation of water by a tree. As the latent heat of vaporization is known, calculation of the latent heat requires knowledge of the total transpiration rate E_{tot} of the tree. A method to calculate this has been proposed by Kjelgren and Montague,[30] based on an energy-balance model proposed by Green.[31]

The model assumes an isolated two-layer tree crown suspended over a surface unshaded by surrounding objects. The tree crown is classified into sunlit and shaded layers, each with a separate energy balance. No energy exchange is assumed between the two layers. The total tree transpiration, E_{tot}, in mm, is calculated as the fractional contribution of each layer:

$$E_{tot} = E_s(LAI_s/LAI_{tot}) + E_{sh}(LAI_{sh}/LAI_{tot}),   (10.1)$$

where E_s and E_{sh} are the transpiration rates for sunlit and shaded portions of the canopy respectively. LAI is the leaf area index, defined as the area under the crown dripline. According to Monteith and Unsworth,[32] the sunlit leaf area LAI_s can be calculated as a fraction of total LAI_{tot} such that:

$$LAI_s = (1 - e^{-kLAI})/k,   (10.2)$$

where k is the transmissivity or porousness of a tree crown to light and is a function of the leaf orientation relative to the ground surface and the solar elevation.[30] Thus, the amount of the shaded leaf area is:

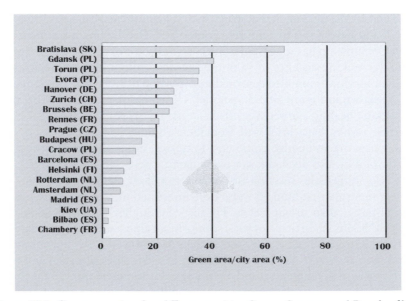

Figure 10.2. Green spaces in selected European cities. Source: Stanners and Bourdeau[26]

$$LAI_{sh} = LAI_{tot} - LAI_s. \tag{10.3}$$

Transpiration rates may be calculated for each layer as:[31]

$$\lambda E = (\nabla R_n^{s,sh} + \rho c_p e_a / r_a) / (\nabla + \gamma(2 + r_s^{s,sh}/r_a)), \tag{10.4}$$

where λ is the latent heat of vaporization (J/g), E is the transpiration rate, (g/sm² per unit leaf area), $R_s^{s,sh}$ is the net radiation flux density retained by sunlit and shaded layers (W/m²), respectively, e_a is the canopy-level vapour pressure deficit of air (Pa), r_a is the total tree leaf boundary layer resistance (s/m²) to vapour and heat movement, which are assumed to be equivalent,[33] $r_s^{s,sh}$ are the average leaf stomatal resistances (s/m) for the sunlit and shaded layers, ∇ is the slope of the saturation vapour pressure curve, (Pa/K), at T_a, γ is the psychrometric constant (66.2 Pa/K), r is the density of air (g/m³) and c_p is the specific heat capacity of air at constant pressure, (J/(gK)).

The variables r_s, $(1/g_s)$, and e_a can be measured directly. The net radiation R_n^s may be calculated from the global short-wave radiation for sunlit and shaded layers separately,[30] while r_a can be calculated using the following empirical formula proposed by Landsberg and Powell:[34]

$$r_a = 59 p^{0.56} (d/u)^{0.5}, \tag{10.5}$$

where d is the characteristic dimension of a leaf, u is the canopy-level wind speed and p is a dimensionless number derived from the ratio of total to crown silhouette area perpendicular to the horizontal wind flow.[30]

THE IMPACT OF TREES AND URBAN PARKS ON AMBIENT TEMPERATURE AND AIR QUALITY

Evapotranspiration from soil–vegetation systems can contribute significantly to reducing urban temperatures. Sailor considers that the low evaporative heat flux in cities is the most significant factor in the development of urban heat islands.[35] As already reported, evapotranspiration from plants at the National Park of Athens creates 'oases' of 1–5 K during the night. Duckworth and Sandberg found that temperatures in San Francisco's Golden Gate Park, which contains high levels of vegetation, average about 8 K cooler than nearby areas that have much less vegetation.[36] In Tokyo, zones with vegetation are 1.6 K cooler in summer than places without vegetation,[37,38] while in Montreal urban parks can be 2.5 K cooler than surrounding built-up areas.[39] Jauregui reported that the park in Mexico City was 2–3 K cooler than its boundaries.[40] Lindqvist has performed studies in Gothenburg, Sweden, and he has reported that on some occasions the air temperature increased by 6 K from 100 m inside a park to a point within the built-up area 150 m outside the park.[41] More frequently, the air temperature gradient in the transition zone was 0.3–0.4 K per 100 m outside the park. Similar results for Gothenburg have also been reported by Eliason.[42] Taha *et al.* have reported that evapotranspiration can create oases that are 2–8 K cooler than their surroundings,[43,44] while Bowen reports a 2–3 K temperature reduction due to evapotranspiration by plants.[45] Finally, Saito *et al.* studied the effect of green areas on the thermal environronment of Kumamoto city in Japan.[46] They reported that even small green areas of 60 m × 40 m showed the cooling effect. The maximum temperature difference between inside and outside the green area was 3 K.

Measurements of the ambient temperature in and around an urban park in Athens, Greece were performed for a period of ten days during August 1998. Measurements were taken simultaneously by five mobile stations moving from the centre of the park to the area around it. Experiments were performed throughout the day. The aim of the experiments was to investigate the temperature difference between the area with vegetation and the surrounding area, as well the gradient of the temperature increase as a function of the distance from the park. Although the analysis has not yet been completed, preliminary findings permits the following conclusions:

- The temperature inside the park varies principally as a function of shading and vegetation cover. The difference between the lowest and highest temperatures can by high as 1.5 K.
- The maximum temperature difference between the park and the surrounding urban area during the daytime is close to 3 K.
- A constant gradient of temperature increase with distance from the park is not found. At all four gates of the park, it has been found that leaving the park resulted in an immediate temperature increase of about 1 K. This change almost had the characteristics of a step function.
- The temperature around the park was mainly influenced by parameters other than the presence of the park, such as the density of the buildings, the rate of anthropogenic heat (mainly released by cars), the shading of the canyons, etc.
- The distribution of the temperature around the park (Figure 10.3) shows that the high-density and high-circulation area located in the northern part of the park has a temperature about 1.5 K higher than less dense west and southern parts, which have more vegetation.

Multiyear temperature measurements performed in two urban parks in Athens have shown that both parks present almost 40% fewer cooling degree hours than the surrounding urban stations. Furthermore, the parks also present the lowest recorded absolute minimum temperature of all stations, although, as shown in Table 5.9, during the daytime the absolute temperature inside the park is higher than in the suburban areas. During the heating period, urban parks present approximately 15% higher heating degree days than the surrounding urban stations. The distribution of the daily temperature in the parks, as well as for a reference suburban and urban stations, is given for a representative summer and winter day in Figures 10.4 and 10.5, respectively.

As shown in the figures, during both summer and winter periods, the park located closer to the central area presents, at about midday, a temperature 2–3 K higher than the park located in a less central area. The temperature of the two parks is similar during the night.

During the summer season and around noon, both parks have a temperature 4–7 K lower than that of the urban station, but 2–5 K higher than that of the suburban

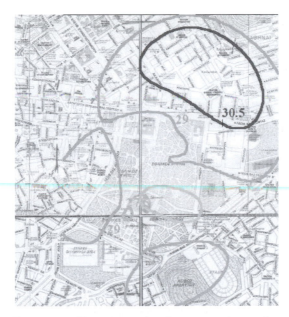

*Figure 10.3. Temperature distribution in and around
a park in Athens, Greece*

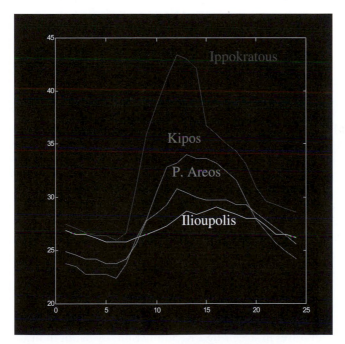

*Figure 10.4. Daily temperature distribution in two urban parks
(Kipos and P. Areos), at an urban station (Ipookratous) and at a
reference suburban station (Ilioupolis) during a typical summer day
in Athens, Greece*

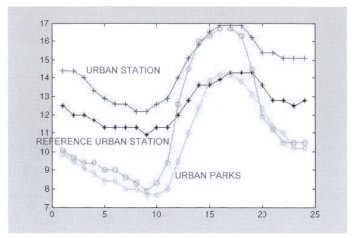

Figure 10.5. Daily temperature distribution in two urban parks, at an urban station and at a reference suburban stations during a typical winter day in Athens, Greece

reference station. During summer nights both parks have a temperature 5–6 K lower than the urban station and 2–3 C lower than the suburban reference station.

During the winter period, the centrally located park shows an almost similar temperature, during the peak hours, as that of the urban station and about 3 K higher than that of the reference station. During the same period, the second park shows a peak temperature 3 K lower than that of the urban station and almost the same temperature as the reference station. During the night both urban parks show a temperature 4 K lower than that of the reference station and 5–6 K lower than that of the urban station.

Numerical studies that try to simulate the effect of additional vegetation on the urban temperatures have been performed by various researchers and provide very useful information. Huang *et al.* have reported that computer simulations predict that increasing the tree cover by 25% in Sacramento and Phoenix, USA, would decrease air temperatures at 2:00 p.m. in July by 6 to 10.0°F, (Figures 10.6 and 10.7).[47,48] Taha has reported simulation results for Davis, California, using the URBMET PBL model.[49] He found that the vegetation canopy produced daytime temperature depressions and night-time excesses compared to the bare surroundings. The factors that affect temperature reduction are evaporative cooling and shading of the ground, whereas temperature increase during the night is the result of the reduced sky factor within the canopy. The results of the simulations show that a vegetation cover of 30% could produce a noontime oasis of up to 6 K in favorable conditions and a night-time heat island of 2 K.

Other numerical simulations reported by Gao,[50] show that green areas decrease maximum and average temperatures by 2 K, while the vegetation can reduce maximum air temperatures in streets by 2 K. Simulations by Sailor indicate a potential for reducing peak summertime temperatures in Los Angeles by more than 1.3 K, through the implementation of a 0.14 increase of fractional vegetation cover.[51] The same author reports results from simulations evaluating the impact of added vegetation on

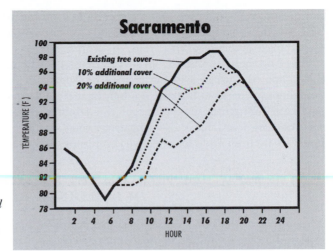

Figure 10.6. Temperature reductions in Sacramento due to added tree cover on a typical summer day in July, based on computer simulations. Source: Akbari et al.[10]

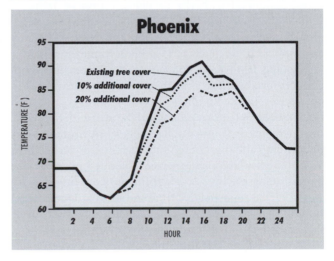

Figure 10.7. Temperature reductions in Phoenix due to added tree cover on a typical summer day in July, based on computer simulations. Source: Akbari et al.[10]

the heating, HDD, and cooling degree days, CDD, of cities located throughout the USA.[52] He found that increasing the vegetation cover by 0.15 over only the residential neighbourhoods, which is equivalent to 0.065 averaged over the city core, reduces the number of cooling degree days by 2–5% and increases the number of heating degree days by 0.5–3.5%. According to Sailor, one would expect a city-wide saving of up to 5% of summertime air-conditioning energy use.

Finally, simulations reported by Taha et al. indicate that the net effect of increased urban vegetation is a decrease in ozone concentration.[52] Hader has summarized the conclusions of different studies on the distribution of dust within and outside urban green areas.[54] He found that inside the green area, a diminution of dust is noticeable and he describes the processes of filtration of dust by vegetation. He concludes that green belts can be effective in reducing natural dust, as well as particles generated by motor vehicles on roads and particles generated by coal burning. Similar results are reported by HUD International, where it has been found that air pollution is significantly reduced only within the green belt itself and in the area directly behind it.[55] Such an effect also applies to trees planted along an avenue, thus helping, to a certain extent, to clean the street air.

Based on these results, Givoni advises that trees and public parks should be spaced throughout the urban area rather than concentrated in few spots.[56] He also gives guidelines for parks in hot dry, hot humid and cold regions.[57] Honjo and Takakura, based on numerical simulations of the cooling effects of green areas on their surrounding areas, have also suggested that smaller green areas with sufficient intervals are preferable for effective cooling of surrounding areas.[58]

THE IMPACT OF TREES ON THE ENERGY CONSUMPTION OF BUILDINGS

The impact of trees on the energy consumption of buildings is very important. As reported by the US National Academy of Sciences,[59] the plantation of 100 million trees, combined with the implementation of light surfacing programmes, could reduce electricity use by 50 billion kWh per year, which is equivalent to 2% of the annual electricity use in the USA, and reduce the amount of CO_2 dumped in the atmosphere by as much as 35 million tonnes per year.

Shading from trees contributes to decreasing significantly energy for cooling. Shaded surfaces have a much lower temperature and thus decrease the rate of heat convection to the building interior. However, the presence of vegetation may also decrease the radiation exchange of the wall with the sky and thus contribute, especially during night, to a higher wall temperature. Simulation studies using limited numbers of buildings and tree configurations for cities across the USA indicate that shade from a single well-placed, mature tree of about 8 m crown diameter, can reduce annual air conditioning use by 2–8% and the peak cooling demand by 2–10%.[2,20,47,48,60–64]

Parker has reported that trees and shrubs planted next to a South Florida residential building can reduce summer air conditioning costs by 40%.[65] Reductions in summer power demand of 59% during the morning and 58% in the afternoon were also measured. Parker has also reported that the average temperature of walls shaded by trees or by a combination of trees and shrubs was reduced by 13.5–15.5 K.[66–68] Climbing vines reduced the surface temperature by 10–12 K. Heisler has reported that shading from trees of a small mobile home can reduce air conditioning by up to 75%.[69] Akbari et al. have reported the results of a monitoring campaign in which the peak power and cooling energy savings from shading trees were monitored in two houses in Sacramento, USA.[70] They found that shading trees at the two monitored houses yielded seasonal cooling-energy savings of 30%, corresponding to average daily savings of 3.6 and 4.8 KWh. Peak-demand savings for the same houses were 0.6 and 0.8 kW, (about 27% savings in one house and 42% in the other). Simpson and McPherson have calculated the magnitude of tree shade at 254 residential properties in Sacramento, California.[71] They used 3.1 trees per property, which reduced annual and peak cooling energy use by 153 kWh (7.1%) and 0.08 kW (2.3%) per tree, respectively. The annual heating load increased by 0.85 GJ (1.9%) per tree. The authors found that changes in cooling load were smaller, but percentage changes were larger for newer buildings.

The effect of appropriate landscape treatments around a building may be very important for the temperature regime around the building. Measurements of different landscape treatments around a model building have been reported by McPherson et al.[72] The study involved grass turf around the building and no shade, rock mulch around the buildings and the walls shaded by shrubs, and rock mulch with neither grass nor shade. It is reported that the surface temperature of grass turf around noon was to about 15 K lower than that of the rocks. Furthermore, the air temperature at a height of 0.5 m above the turf was 2 K lower than above the rocks. Finally, the model building with the rock mulch consumed between 20–30% more cooling energy than the models with the turf and with the shrub shading. Givoni has reported that, during clear summer days, the air temperature close to a shrub fence is up to 3 K lower than the temperature over exposed pavements.[57]

Akbari et al. have presented the results of computer simulations aiming to study the combined effect of shading and evapotranspiration of vegetation on the energy use of typical one-storey buildings in various US cities.[10] Figure 10.8 summarizes these results. It was found that by adding one tree per house, the cooling energy savings ranged from 12% to 24%, while adding three trees per house could reduce the cooling load by between 17% and 57%. According to this study, the direct effects of shading account for only 10% to 35% of the total cooling-energy savings. The remaining savings result from temperatures lowered by evapotranspiration.

McPherson has reported the results of the Trees for Tucson/Global Releaf reforestation programme.[73] This programme proposes planting of 500,000 trees during a period of five years. A full economic analysis has been carried out covering all the major costs and benefits associated with the programmw. Figure 10.9 shows the main results. Projected net benefits are $236.5 million for the 40-year planning horizon.

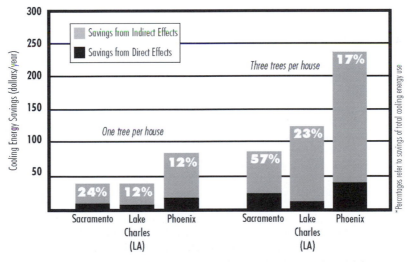

Figure 10.8. Estimated cooling-energy savings in a typical well-insulated new house from the combined direct and indirect effects of trees. Source: Akbari et al.[10]

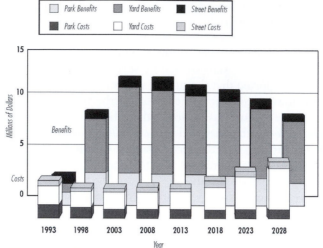

Figure 10.9. Projected annual costs and benefits of the Trees for Tucson/Global Releaf reforestation programme. Source: McPherson[72]

REFERENCES

1. American Forestry Association (1989). *Save Our Urban Trees: Citizens Action Guide*. Washington, DC.
2. Akbari, H. and Taha, H. (1992). 'The Impact of Trees and White Surfaces On Residential Heating and Cooling Energy use in Four Canadian Cities'. *Energy*, Vol. 17, No. 2, pp. 141–149.
3. Oke, T.R. (1987). *Boundary Layer Climates*, 2nd edn. Routledge, London.
4. Montgomery, D. (1987). 'Landscaping as a Passive Solar Strategy'. *Passive Solar Journal*, Vol. 4, No. 1, pp. 79–108.
5. Moffat, A. and Schiler, M. (1981). *Landscape Design that Saves Energy*. William Morrow and Company, New York.
6. Taha, H. (1997). 'Urban Climates and Heat Islands: Albedo, Evapotranspiration and Anthropogenic Heat'. *Energy and Buildings*, Vol. 25, pp. 99–103.
7. Gay, L.W. and Stewart, J.B. (1974). 'Energy Balance Studies in Coniferous Forests'. Report No. 23, Inst. Hydrology, Natural Environment Research Council, Wallingford, Berks, UK.
8. McNaughton, K. and Black, T.A. (1973). 'A Study of Evapotranspiration from a Douglas Forest using the Energy Balance Approach'. *Water Resources Research*, Vol. 9, p. 1579.
9. Summit, J. and Sommer, R. (1998). 'Urban Tree-planting Programs – A Model for Encouraging Environmentally Protective Behavior'. *Atmospheric Environment*, Vol. 32, pp. 1–5.
10. Akbari, H., Davis, S., Dorsano S., Huang J. and S. Winett (1992). *Cooling our Communities - A Guidebook on Tree Planting and Light Colored Surfacing*. US Environmental Protection Agency. Office of Policy Analysis, Climate Change Division.
11. Halvorson, H. and Potts, D. (1981). 'Water Requirements of Honeylocust (*Gleditsia Triacanthos f. inermis*) in the Urban Forest'. USDA Forest Service Research Paper, NE-487.
12. Heilman, J., Brittin, C. and Zajicek, J. (1989). 'Water Use by Shrubs as Affected by Energy Exchange with Building Walls'. *Agricultural and Forest Meteorology*, Vol. 48, pp. 345–357.
13. Cardelino, C.A. and Chameides, W.L. (1990). 'Natural Hydrocarbons, Urbanization and Urban Ozone'. *Journal of Geophysical Research*, Vol. 95, No. 13, pp. 971–979.
14. McPherson, E.G., Scott, K.I. and Simpson, J.R. (1998). 'Estimating Cost Effectiveness of Residential Yard Trees for Improving Air Quality in Sacramento, California, Using Existing Models'. *Atmospheric Environment*, Vol. 32, No. 1, pp. 75–84.

15. Bernatsky, A. (1978). *Tree Ecology and Preservation*. Elsevier Scientific Publishers, New York.
16. Winer, A. M., Fitz, D.R. and Miller, P.R. (1983). 'Investigation of the Role of Natural Hydrocarbons in Photochemical Smog Formation in California'. Final Report, Contract No. AO-056-32. Sacramento, CA. California Air Resources Board.
17. Chameides, W.L., Lindsay, R.W., Richardson, J. and Kiang, C.S. (1988). 'The Role of Biogenic Hydrocarbons in Urban Photochemical Smog: Atlanta as a Case Study'. *Science*, Vol. 241, pp. 1473–1475.
18. Broban, H.W. (1967). 'Stadebauliche Grundlagen des Schallschutzes'. *Deutsche Bauzeit*, Vol. 5.
19. Heisler, G.M. (1989). 'Site Design and Microclimate Research'. Final outcome to the Argonne National Laboratory, University Park, PA: US Department of Agriculture Forest Service, Northeast Forest Experimental Station.
20. Huang, Y.J., Akbari, H. and Taha, H.G. (1990). 'The Wind Shielding and Shading Effects of Trees on Residential Heating and Cooling Requirements'. 1990 *ASHRAE Transactions*, Atlanta GA, American Society of Heating, Refrigeration and Air-conditioning Engineers.
21. DeWalle, D.R. (1983). 'Windbreak Effects on Air Infiltration and Space Heating in a Mobile House'. *Energy and Buildings*, Vol. 5, pp. 279–288.
22. Anderson, L. and Cordell H. (1985). 'Residential Property Values Improve by Landscaping with Trees'. *Scandinavian Journal of Applied Forestry*, Vol. 9, pp. 162–166.
23. Getz, D., Karow, A. and Kielbaso, J.J. (1982). 'Inner City Preferences for Trees and Urban Forestry Programs'. *Journal of Arboriculture*, Vol. 8, pp. 258–263.
24. Sheets, V. and Manzer, C. (1991). 'Affect, Cognition, and Urban Vegetation: Some Effects of Adding Trees along City Streets'. *Environment and Behavior*, Vol. 23, pp. 285–304.
25. Hull, R.B. (1992). 'Brief Encounters with Urban Forests Produce Moods that Matter'. *Journal of Arboriculture*, Vol. 118, pp. 322–324.
26. Stanners, D. and Bourdeau, P. (eds) (1995). *Europe's Environment – The Dobris Assessment*. European Environmental Agency, Denmark.
27. Ulrich, R.S. (1984). 'View through a Window may Influence Recovery from Surgery'. *Science*, Vol. 224, pp. 420–421.
28. Ulrich R.S. (1981). 'Natural Versus Urban Scenes: Some Psychological Effects'. *Environment and Behavior*, Vol. 13, pp. 523–556.
29. Ulrich, R.S., Simons, R.F., Losito, B.D., Fiorito, E., Miles, M.A. and Zelson, M. (1993). 'Stress Recovery during Exposure to Natural and Urban Environments'. *Journal of Environmental Psychology*, Vol. 36, pp. 729–742.
30. Kjelgren, R. and Montague, T. (1998). 'Urban Tree Transpiration over Turf and Asphalt Surfaces'. *Atmospheric Environment*, Vol. 32, No. 1, 35–41.
31. Green, S. (1993). 'Radiation Balance, Transpiration and Photosynthesis of an Isolated Tree'. *Agricultural and Forest Meteorology*, Vol. 64, pp. 201–221.
32. Monteith, J. and Unsworth, M. (1990). *Principles of Environmental Physics*, 2nd edn. Edward Arnold, London, pp. 58–73.
33. Jones, H. (1992). *Plants and Microclimate: A Quantitative Apprioach to Environmental Plant Physiology*, 2nd edn. Cambridge University Press, pp. 9–44.
34. Landsberg, J. and Powell, D. (1973).'Surface Exchange Characteristics of Leaves Subject to Mutual Interference'. *Agricultural Meteorology*, Vol. 13, pp. 169–184.
35. Sailor, D.J. (1994). 'Sensitivity of Coastal Meteorology and Air Quality to Urban Surface Characteristics'. Preprints of the Eighth Joint Conference on the Applications of Air Pollution Meteorology, Vol. 8. American Meteorological Society, Boston, MA, pp. 286–293.
36. Duckworth, E. and Sandberg, J. (1954). 'The Effect of Cities upon Horizontal and Vertical Temperature Gradients'. *Bulletin of the American Meteorological Society*, Vol. 35, pp. 198–207.
37. Tatsou Oka (1980). 'Thermal Environment in Urban Areas'. Report D7: 1980, Swedish Council for Building Research, Stockholm, Sweden.
38. Gao W., Miura, S. and Ojima, T. (1994). 'Site Survey on Formation of Cool Island due to Park and Inner River in Kote -ku, Tokyo'. *Journal of Architectural Planning and Environmental Engineering AIJ*, No. 456.

39. Oke T.R. (1977). 'The Significance of the Atmosphere in Planning Human Settlements'. In E.B. Wilkin and Ironside, G. (eds), *Ecological Land Classification in Urban Areas, Ecological Land Classification Series*, No. 3, Canadian Government Publishing Center, Ottawa, Ontario.

40. Jauregui, E. (1990/91). 'Influence of a Large Urban Park on Temperature and Convective Precipitation in a Tropical City'. *Energy and Buildings*, Vol. 15–16, pp. 457–463.

41. Lindqvist, S. (1992). 'Local Climatological Modelling for Road Stretches and Urban Areas'. *Geografiska Annaler*, Vol. 74A, pp. 265–274.

42. Eliasson, I. (1996). 'Urban Nocturnal Temperatures, Street Geometry and Land Use'. *Atmospheric Environment*, Vol. 30, No. 3, pp. 379–392.

43. Taha, H., Akbari, H. and Rosenfeld, A. (1989). 'Vegetation Microclimate Measurements: the Davis Project'. Lawrence Berkeley Laboratory Report 24593, Berkeley, CA.

44. Taha, H., Akbari, H. and Rosenfeld, A. (1991). 'Heat Island and Oasis Effects of Vegetative Canopies: Microclimatological Field Measurements'. *Theoretical Applied Climatology*, Vol. 44, p. 123.

45. Bowen, A. (1980). 'Heating and Cooling of Buildings and Sites through Landscape Planning'. *Passive Cooling Handbook*. American Assiciation of the International Solar Energy Society, Newark, DE.

46. Saito, I., Ishihara, O. and Katayama, T. (1990/91). 'Study of the Effect of Green Areas on the Thermal Environment in an Urban Area'. *Energy and Building*, Vol. 15–16, pp. 493–498.

47. Huang, Y.J., Akbari, H., Taha, H.G. and A.H. Rosenfeld (1986). 'The Potential of Vegetation in Reducing Summer Cooling Loads in Residential Buildings'. Lawrence Berkeley Laboratory Report 21291, Berkeley, CA.

48. Huang, Y.J, Akbari, H., Taha. H. and Rosenfeld, A.H. (1987). 'The Potential of Vegetation in Reducing Cooling Loads in Residential Buildings'. *Journal of Climate and Applied Meteorology*, Vol. 26, pp. 1103–1116.

49. Taha, H. (1988). 'Site Specific Heat Island Simulations: Model Development and Application to Microclimate Conditions'. Lawrence Berkeley Laboratory Report 26105; M.Geogr. Thesis, University of California, Berkeley, CA.

50. Gao, W. (1993). 'Thermal Effects of Open Space with a Green Area on Urban Environment, Part I: A Theoretical Analysis and its Application'. *Journal of Architectural Planning and Environmental Engineering AIJ*, No. 488.

51. Sailor, D.J. (1995). 'Simulated Urban Climate Response to Modifications in Surface Albedo and Vegetative Cover'. *Journal of Applied Meteorology*, Vol. 34, pp. 1694–1704.

52. Sailor, D.J. (1998). 'Simulations of Annual Degree Day Impacts of Urban Vegetative Augmentation'. *Atmospheric Environment*, Vol. 32, No. 1, pp. 43–52.

53. Taha, H., Douglas, S. and Haney, J. (1997). 'Mesoscale Meteorological and Air Quality Impacts of Increased Urban Albedo and Vegetation', *Energy and Buildings*, Vol. 25, pp. 169–177.

54. Hader, F. (1970). 'The Climatic Influence of Green Areas, their Properties as Air Filters and Noise Abatement Agents'. *Climatology and Buildings*. Comití Internationale de Bâtiment, Vienna.

55. HUD International (1973). *Green Belts and Air Pollution*. Information Series 21, The Netherlands.

56. Givoni, B. (1989). *Urban Design in Different Climates*. WMO Technical Report 346.

57. Givoni, B. (1991). 'Impact of Planted Areas on Urban Environmental Quality. A Review'. *Atmospheric Environment*, Vol. 25B, pp. 289–299.

58. Honjo, T. and Takakura, T. (1990/91). 'Simulation of Thermal Effects of Urban Green Areas on their Surrounding Areas'. *Energy and Buildings*, Vol. 15–16, pp. 443–446.

59. National Academy of Sciences (1991). *Policy Implications of Greenhouse Warming*. Report of the Mitigation Panel. National Academy Press, Washington, DC.

60. Heisler, G.M. (1991). 'Computer Simulation for Optimizing Windbreak Placement to Save Energy for Heating and Cooling Buildings'. In *Trees and Sustainable Development: the*

3rd National Windbreaks and Agroforestry Symposium Proceedings, Ridgetown College, Ridgetown, pp. 100–104.

61. McPherson, E.G and Sacamano, P.L. (1992). 'Energy Savings with Trees in Southern California'. Technical Report US Department of Agriculture. Forest Service, Pacific Southwest Research Station, Western Center for Urban Forest Research, Davis, California.

62. Sand, M.A. and Huelman, P.H. (1993). 'Planting for Energy Conservation in Minnesota Communities'. Summary report for 1990-1993 LMCR research project, Department of Natural Resources, Forestry, St Paul, Minnesota.

63. McPherson, E.G. (1994). 'Benefits and Costs of Tree Planting and Care in Chicago'. In McPherson, E.G., Novak, D.J. and Rowntree, R. (eds), *Chicago Urban Forest Ecosystem. Final Report of the Chicago Urban Forest Climate Project*. USDA Forest Service, Northeastern Forest Experiment Station, Radnor, PA, pp. 117–135.

64. Simpson, J.R., McPherson, E.G. and Rowntree, R.A. (1994). 'Potential of Tree Shade for Reducing Building Energy Use in the PG and E Service Area'. Final Report to Pacific Gas and Electric Company, San Francisco, CA.

65. Parker, J.H. (1983). 'Landscaping to Reduce the Energy used in Cooling Buildings'. *Journal of Forestry*, Vol. 81, No. 2, 82-83.

66. Parker, J.H. (1983). 'The Effectiveness of Vegetation on Residential Cooling'. *Passive Solar Journal*, Vol. 2, pp. 123–132.

67. Parker, J.H. (1987). 'The Use of Shrubs in Energy Conservation Plantings'. *Landscape Journal*, Vol. 6, pp. 132–139.

68. Parker, J.H. (1989). 'The Impact of Vegetation on Air Conditioning Consumption'. *Proceedings of Conference on Controlling the Summer Heat Island*. Lawrence Berkeley Laboratory Report 27872, Berkeley, CA, pp. 46–52.

69. Heisler, G.M. (1986). 'Energy Savings with Trees'. *Journal of Arboriculture*, Vol. 12, No. 5, pp. 113–124.

70. Akbari, H., Kurn, D.M., Bretz, S.E. and Hanfold, J.W. (1977). 'Peak Power and Cooling Energy Energy Savings of Shade Trees'. *Energy and Buildings*, Vol. 25, pp. 139–148.

71. Simpson, J.R. and McPherson, E.G. (1998). 'Simulation of Tree Shade Impacts on Residential Energy Use for Space Conditioning in Sacramento'. *Atmospheric Environment*, Vol. 32, No. 1, pp. 69–74.

72. McPherson, E.G., Simpson, J.R. and Livingston, M. (1989).'Effect of Three Landscape Treatments on Residential Energy and Water Use in Tucson, AZ'. *Energy and Buildings*, Vol. 13, pp. 127–138.

73. McPherson, E.G. (1991). 'Economic Modeling for Large Scale Tree Plantings'. In Vine, E., Crawley, D. and Centoella, P. (eds), *Economic Efficiency and the Environment: Forging the Link*, Ch. 19. American Council for an Energy Efficiency Economy, Washington, DC.

11

Appropriate materials for the urban environment

M. Santamouris,
Department of Applied Physics, University of Athens

ℰↃ

GENERAL AND PHYSICAL PRINCIPLES

Traditionally, the materials used to build houses and cities tended to come from what lay around and what could be used. This is why the original American settlers built their houses from wood, mud or even straw and why Aberdeen was built of granite.[1] In contrast, the materials for modern houses and cities are international in origin and a much wider selection is available. The building sector is responsible for the 50% of the material resources taken from nature,[2] and thus appropriate selection of materials will have an important ecological and financial impact.

The technical characteristics of the materials used determine to a high degree the energy consumption and comfort conditions of individual houses, as well as of open spaces. In particular, the optical characteristics of materials used in urban environments, and especially the albedo to solar radiation and emissivity to long-wave radiation, have a very important impact on the urban energy balance.

The albedo of a surface is defined as its reflectivity, integrated hemispherically and over wavelength. The materials used in the external facades of buildings and in streets and pavements either absorb or reflect the incident solar radiation. Use of high-albedo materials reduces the amount of solar radiation absorbed through building envelopes and urban structures and thus keeps their surfaces cooler.

The albedo α is defined, with reference to a spherical coordinate system, as:

$$\alpha = \frac{\int_{\lambda_2}^{\lambda_2}\int_0^{2\pi} I\uparrow \cos\theta \, d\omega d\lambda}{\int_{\lambda_2}^{\lambda_2}\int_0^{2\pi} I\downarrow \cos\theta \, d\omega d\lambda} \tag{11.1}$$

where I is the radiant intensity (W/m²), θ is the zenith angle, defined as the angle between the normal to a surface and the incident beam, λ is the wavelength and ω is solid angle, defined as the ratio of a partial spherical area of interest to the square of the sphere radius. The up and down arrows indicate the reflected and incident radiation respectively. In order to find an average albedo, an additional integral over time has to be added to equation (11.1). The albedo of materials for solar radiation should take into account wavelengths in the band between 0.28 and 2.8 μm, which accounts for 98% of the solar irradiance at sea level.

At the same time, materials emit long-wave radiation. The emitted radiation is a function of the material's temperature and emissivity. Materials with high emissivities are good emitters of long-wave energy and readily release the energy that has been absorbed as short-wave radiation.

Although the impact of solar reflectivity and the material's emissivity are important, it is important to note that the temperature of a material is determined by its thermal balance, with conductive and convective phenomena taken into account. In particular, the role of convective flows is very important. Grigs *et al.*, using simulations, showed that the surface temperature of a black roofing membrane reached 82°C with no wind, but only 46°C with a wind of 15 m/s.[3]

As well as the albedo and emissivity, aspects related to the durability, cost, appearance and pollution emitted by the materials have to be considered. In particular, since many paints and coatings emit volatile organic compounds (VOCs), non-emitting materials have to be selected. Volatile organic compounds combine with nitrogen oxides and create ozone during the day.

Various studies have been performed in order to improve the understanding of the thermal and optical performance of materials, as well as their impact on the climate of cities. Yap reported that systematic urban–rural differences in surface emissivity are the potential cause of part of the heat-island effect.[4] Robinette reported relative temperatures of 38°C over grass, 61°C over asphalt, and 73°C over artificial turf,[5] while Santamouris *et al.* reported asphalt temperatures close to 63°C and white pavements close to 45°C.[6] Lower surface temperatures contribute to decreasing the temperature of the ambient air as heat-convection intensity from a cooler surface is lower. Such temperature reductions can have significant impacts on consumption of cooling energy in urban areas, a fact of particular importance in cities with a hot climate.

Oke *et al.* simulated the effect of the optical and thermal characteristics of the materials used to the heat island intensity during the night and found that the role of emissivity is minor.[7] As the emissivity increased from 0.85 to 1.0, there was a slight increase of 0.4 K in ΔT between the urban and rural environment for very tight canyons, while there was almost no change for higher view factors. In contrast, the effect of the thermal properties of the materials used appeared to be much more important. For flat country, Oke *et al.* found that, if the urban admittance is 2200 $J/m^2/K$ and the rural admittance 800 units lower, a heat island of about 2 K develops during the night, while, when the urban admittance is decreased to 600 $J/m^2/K$, a cool island of over 4 K may be formed during the night.

The use of appropriate materials to reduce the heat island and improve the urban environment has been of increasing interest during recent years. Much research has been carried to identify the possible energy and environmental gains when light-coloured surfaces are used. There have been investigations of the impact of the optical and thermal characteristics of the materials on the urban temperature, as well as of possible energy conservation during the summer period. A detailed guide to light-coloured surfaces has been published by the US EPA,[8] and research has shown that important energy gains are possible when light-coloured surfaces are used in combination with the plantation of new trees. For example, computer simulations by

Rosenfeld *et al.* showed that white roofs and shade trees in Los Angeles, USA, would lower the need for air conditioning by 18% or 1.04 billion kWh, equivalent to a financial gain close to $100 million per year.[9]

In the next section, information on the optical characteristics of materials used in the urban environment is given. The impact of light-coloured surfaces on the energy consumption of buildings is then discussed and the results of several research projects are presented.

OPTICAL CHARACTERISTICS OF MATERIALS

The optical characteristics of the materials used in the urban fabric largely define its thermal balance. As previously stated, reflectivity to solar radiation and the emissivity of the materials are the most important of the optical parameters. Study of the thermal performance and the energy impact of materials used in streets and pavements, as well as those used in the exterior facades and roofs of urban buildings, has been of increasing interest during recent years. Development of more appropriate materials is also a topic of great interest.

Simple uniform materials used in urban environments are characterized by various albedo values and these determine the complex reflectivity of a city. Table 11.1 gives the albedo of various typical urban materials and areas,[10–13] while Table 11.2 gives the emissivity and the reflectivity for selected materials.[10,12,14]

Regarding the thermal performance of various materials used in pavements and streets, Asaeda *et al.* reported that they had tested experimentally the impact of various

Table 11.1. Albedo of typical urban materials and areas[10–12,15]

Surface	Albedo	Surface	Albedo
Streets		*Paints*	
Asphalt (fresh 0.05, aged 0.2)	0.05-0.2	White, whitewash	0.50–0.90
		Red, brown, green	0.20–0.35
Walls		Black	0.02–0.15
Concrete	0.10–0.35		
Brick/Stone	0.20–0.40	*Urban areas*	
Whitewashed stone	0.80	Range	0.10–0.27
White marble chips	0.55	Average	0.15
Light-coloured brick	0.30–0.50		
Red brick	0.20–0.30	*Other*	
Dark brick and slate	0.20	Light-coloured sand	0.40–0.60
Limestone	0.30–0.45	Dry grass	0.30
		Average soil	0.30
Roofs		Dry sand	0.20–0.30
Smooth-surface asphalt (weathered)	0.07	Deciduous plants	0.20–0.30
Asphalt	0.10–0.15	Deciduous forests	0.15–0.20
Tar and gravel	0.08–0.18	Cultivated soil	0.20
Tile	0.10–0.35	Wet sand	0.10–0.20
Slate	0.10	Coniferous forests	0.10–0.15
Thatch	0.15–0.20	Wood (oak)	0.10
Corrugated iron	0.10–0.16	Dark cultivated soils	0.07–0.10
Highly reflective roof		Artificial turf	0.50–0.10
after weathering	0.6–0.7	Grass and leaf mulch	0.05

Table 11.2. *Albedo and emissivity for selected surfaces*[10,14]

Material	Albedo	Emissivity
Concrete	0.3	0.94
Red brick	0.3	0.90
Building brick	–	0.45
Concrete tiles	–	0.63
Wood (freshly planed)	0.4	0.90
White paper	0.75	0.95
Tar paper	0.05	0.93
White plaster	0.93	0.91
Bright galvanized iron	0.35	0.13
Bright aluminium foil	0.85	0.04
White pigment	0.85	0.96
Grey pigment	0.03	0.87
Green pigment	0.73	0.95
White paint on aluminium	0.80	0.91
Black paint on aluminium	0.04	0.88
Aluminium paint	0.80	0.27–0.67
Gravel	0.72	0.28
Sand	0.24	0.76

pavement materials commonly used in urban environments during the summer period.[16] They found that the surface temperature, heat storage and the subsequent emission of heat to the atmosphere were significantly greater for asphalt than for concrete and bare soil. At the maximum, asphalt pavement emitted an additional 150 W/m^2 in infrared radiation and 200 W/m^2 in sensible heat compared to a bare soil surface. They also found that the rate of infrared absorption by the lower atmosphere over an asphalt pavement was greater by 60 W/m^2 than over a soil surface or concrete pavement. Gustavson and Burgen also discussed the influence of road construction on road surface temperature.[17] On a test road, they had found a nocturnal maximum difference of 1.5 K between beds consisting of blast-furnace slag and those consisting of gravel.

Berg and Quinn reported that in midsummer white-painted roads with an albedo close to 0.55 have almost the same temperature as the ambient environment, while unpainted roads with an albedo close to 0.15 were approximately 11 K warmer than the air.[18]

Taha *et al.* measured the albedo and surface temperatures of a variety of materials used in urban structures. They found that white elastomeric coatings with an albedo of 0.72 were 45 K cooler than black coatings with an albedo of 0.08.[19] The experiments were performed in the early afternoon of a clear day in summer. They also reported that a white surface with an albedo of 0.61 was only 5 K warmer than ambient air, whereas conventional gravel with an albedo of 0.09 was 30 K warmer than the air.

Cantat estimated the albedo of various types of surfaces as well as their temperature in the major Paris area (Table 11.3).[20] As shown, urban areas have a much lower albedo. Also, there is no correlation between the albedo and the surface temperature.

Boutell and Salinas reported the results of an experiment in which the surface temperature and associated energy use were measured during the autumn period as a white elastomeric coating was applied over a built-up roof.[21] In parallel, the surface temperature of an almost identical building roofed with a standard black built-up

Table 11.3. Albedo and surface temperatures in selected areas around Paris[20]

Type of surface	Albedo	Surface temperature (°C)
Leaf bearing forest	0.168	13.5
Coniferous forest	0.137	14.5
Field of cereals	0.254	14.5
Kitchen garden	0.256	20.0
Grass	0.235	22.5
Dry soil	0.257	25.0
Industrial zone – Renault, (Boulogne – Billancourt)	0.140	28.5
Industrial zone – railway station at Ivry	0.180	27.0
Industrial zone – 5ndustry at Gennevilliers	0.265	25.0
Dense urban area (Paris)	0.159	23.0
Group of high buildings	0.194	21.5
Group of low buildings	0.185	19.5
Area of luxurious buildings	0.185	20.5

roofing system was measured. For most of the period, the white elastomeric coating was close to the ambient temperature, while the black roof was to about 36 K warmer than the air at noon.

Taha *et al.* reported the surface temperatures of two-inch polyurethane roofs coated with various colours in Texas, USA, during August (Figure 11.1).[19] As shown, the surface temperature of the white coatings was close to 45°C, while the corresponding temperature of the black coatings was almost double this. Taha *et al.* also correlated the solar absorptivity of specific materials with their surface temperature when the materials were in horizontal position at noon on a clear, windless summer day in Austin, Texas (Figure 11.2). As shown, the temperature difference between white and black coatings is close to 45 K, and that between concrete and asphalt 20 K.

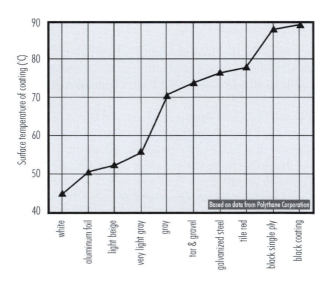

Figure 11.1. Surface temperature for various coatings (adapted from Taha et al.[19])

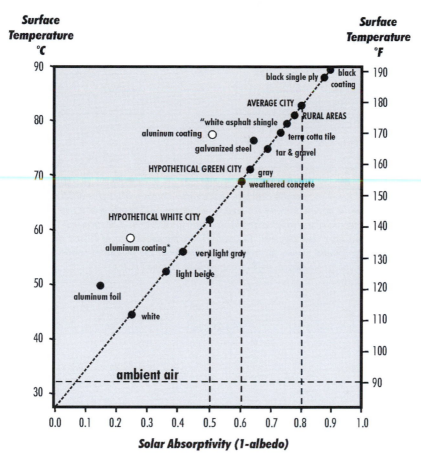

Figure 11.2. Solar absorptivity plotted against the surface temperature of horizontal surface for paints, roadways, roofing materials and cities; data refer to noon on a clear windless summer day in Austin, Texas (adapted from Taha et al.[19])

The figure shows the average surface temperature of a hypothetical green city with surface temperatures 17 K cooler than an average city as a result of the combination of white roofs, light streets and parking lots and urban vegetation. In the hypothetical 'white city', there is less urban vegetation and thus the surface temperature is reduced even further.

Santamouris *et al.* measured the distribution of the surface temperature of pavements and roads in seven different canyons in Athens, Greece, on a hourly basis and for a period of two to three days.[6] The results obtained are given in Figures 11.3 to 11.10. It was found that asphalt temperatures during the day period reach peak temperatures up to 57°C, while the corresponding maximum temperature for pavements made of white and dark slabs were close to 45°C and 52°C respectively. The mean temperature of all the materials during the night was close to 23–25°C.

The orientation of the streets, the *H/W* (height to width) ratio, together with the type of material used, defines the surface temperature of the materials. The effect of

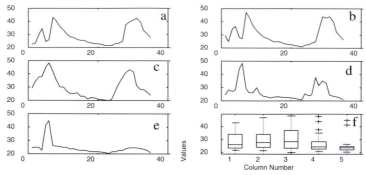

Figure 11.3. First canyon. Surface temperature measured at five different points on the pavements and the street: (a) on the SW facade and (e) on the NE facade of the street with (b), (c) and (d) between. The measurement points are close together. (a) and (c) are for pavements of White slab, while (b), (c) and (d) are for asphalt. (f) Boxplot of the measured temperatures for the whole period; 1–5 corresponds to plots (a) to (e)

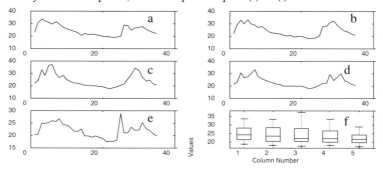

Figure 11.4. Second canyon. Surface temperature measured at five different points on the pavements and the street: (a) on the SE facade and (e) on the NW facade of the street with (b), (c) and (d) between. The measurement points are close together. (a) and (e) are for pavements of white slabs, while (b), (c) and (d) are for asphalt. (f) Boxplot of the measured temperatures for the whole period; 1–5 corresponds to plots (a) to (e)

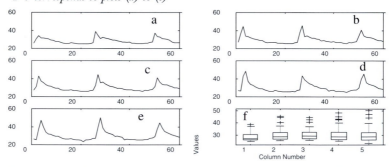

Figure 11.5. Third canyon. Surface temperature measured at five different points on the pavements and the street: (a) on the SW facade and (e) on the NE facade of the street with (b), (c) and (d) between. The measurement points are close together. For all points the material is dark slab. (f) Boxplot of the measured temperatures for the whole period; 1–5 corresponds to plots (a) to (e)

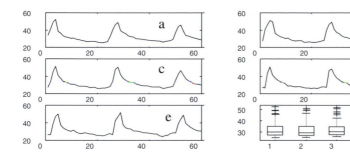

Figure 11.6. Third canyon – second cross section. Surface temperature measured at five different point in the street: (a) on the SW facade and (e) on the NE facade of the street with (b), (c) and (d) between. The measurement points are close together. For all points the material is dark slab. (f) Boxplot of the measured temperatures for the whole period; 1–5 corresponds to plots (a) to (e)

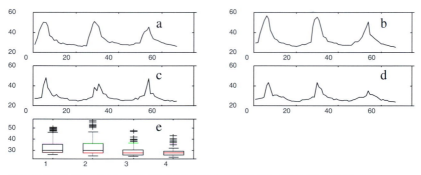

Figure 11.7. Fourth canyon. Surface temperature measured at four different points on the pavements and the street: (a) on the SE facade and (d) on the NW facade of the street with (b), (c) and (d) between. The measurement points are close together. (a) and (d) are for pavements of white slabs, while (b) and (c) are for asphalt. (e) Boxplot of the measured temperatures for the whole period; 1–5 corresponds to plots (a) to (e)

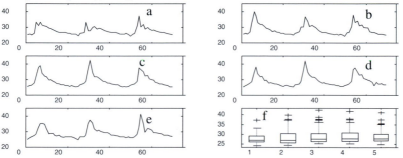

Figure 11.8. Fifth canyon. Surface temperature measured at five different points on the pavements and the street: (a) on the SE facade and (e) on the NW facade of the street with (b), (c) and (d) between. The measurement points are close together. For all points the material is light slab. (f) Boxplot of the measured temperatures for the whole period; 1–5 corresponds to plots (a) to (e)

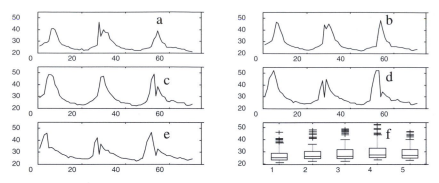

Figure 11.9. Sixth canyon. Surface temperature measured at five different points on the pavements and the street: (a) on the NW facade and (e) on the SE facade of the street with (b), (c) and (d) between. The measurement points are close together. (a) and (e) are for pavements of white slabs, while (b), (c) and (d) are for asphalt. (f) Boxplot of the measured temperatures for the whole period; 1–5 corresponds to plots (a) to (e)

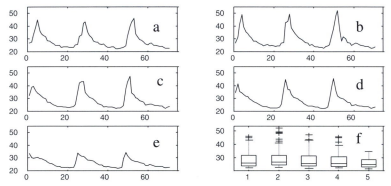

Figure 11.10. Seventh canyon. Surface temperature measured at five different points on the pavements and the street: (a) on the SW facade and (e) on the NE facade of the street with (b), (c) and (d) between. The measurement points are close together. (a) and (e) are for pavements of white slabs, while (b), (c) and (d) are for asphalt. (f) Boxplot of the measured temperatures for the whole period; 1–5 corresponds to plots (a) to (e)

the absorbed solar radiation on the increase in temperature of the materials used in pavements and roads was found to be very important. Table 11.4 summarizes the measured maximum temperature difference (DT) between pavement or road slabs located in the sunny part of the street and shaded slabs of the same material on the other side of the canyon. The table also shows the maximum difference in peak daily temperature for the same slabs in the sun and in the shade. As can be seen in the table, for white and dark slab pavements the instantaneous temperature difference across the street varies between 10 to 22 K as a function of street layout and orientation. For asphalt the instantaneous temperature difference goes up to 27 K.

During the same study, the temperature differences of these materials across a street have been compared. In particular, in all canyons the instant and daily peak temperatures of the white lab pavement and the asphalt were compared. The slabs were located on the same side of the street and had almost the same access to solar

Table 11.4. Maximum temperature difference and maximum difference in the peak daily temperatures between pavement and road slabs located in the sunny part of the street and shaded slabs of the same material, on the opposite side of the canyon

Canyon name	H/W	Orientation (° – relative to S)	Pavement slab		Asphalt	
			Max DT (K)	Max peak DT (K)	Max DT (K)	Max peak DT (K)
Giannitson	0.725–1	40	20	10	27	6
Kavalas	1	50	18	14	20	6
Omirou	1.75–2.62	35	19	22	20	7
Evrota	1–1.5	65	10	7	8	3
Mavromihali	2.47	60	22	8	21	14
Kordou	2.74	55	15	10	–	–
Valaoritou	2.47	30	20	13	–	–

Table 11.5. Maximum temperature difference and maximum difference in the peak daily temperatures between white pavement and asphalt road slabs located in the sunny part of the streets

Canyon name	Max DT (K)	Max peak DT (K)
Giannitson	14	9
Kavalas	10	7
Mavromihali	8	7
Evrota	6	5
Omirou	7	4

radiation. The observed differences are given in Table 11.5. As can be seen, asphalt shows a higher peak temperature 4–9 K during the day than a white slab. The instantaneous temperature differences range between 6 and 14 K. During the night no important differences between the various materials were observed.

Using infrared thermography, Santamouris *et al.* assessed the temperature of materials used in pavements and streets in Omonia Square in Athens during the summer period.[22] A typical picture is shown in Figure 11.11. As shown, the temperatures of unshaded and shaded asphalt were close to 52°C and 35°C respectively. At the same time, the temperature of shaded white slabs used in pavements was between 28°C and 31°C. During the experiment the ambient temperature was around 31°C.

Backenstow reported the daily variation of the surface temperature of panels covered with three membranes coloured black, beige and white.[23] The highest temperatures, measured under clear-sky conditions (Figure 11.12), were 81°C, 57°C and 51°C for the black, beige and white coatings respectively. Recorded temperatures under overcast conditions were much lower except when there was a break of the cloud cover (Figure 11.13). As expected, the effects of higher reflectivity are much less important for low solar radiation.

Appropriate materials used in the exterior part of a roof can contribute to decreasing its surface temperature and thus reduce the related air-conditioning load. Materials for 'cool roofs' should present a high albedo to solar radiation, close to 70%,

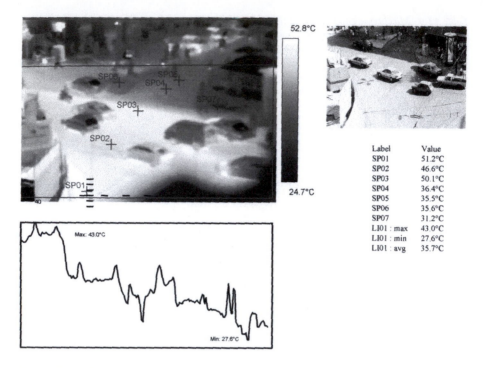

Figure 11.11. Infrared thermography of a section of Omonia Square in Athens, Greece. The picture was taken at 16:00 in September 1998

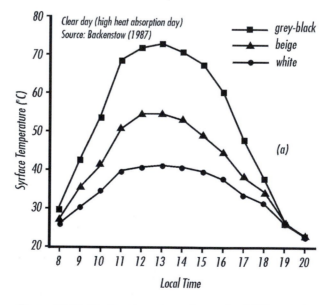

Figure 11.12. Surface temperature for panels of different colours in clear-sky conditions[23]

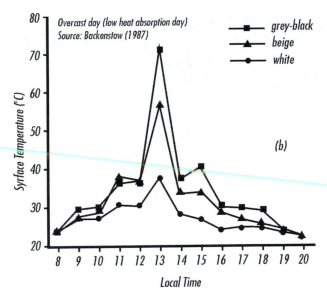

Figure 11.13. Surface temperature for panels of different colours in overcast-sky conditions[23]

while conventional roofing systems have a mean albedo around 20%. As reported in the next section, important energy gains can be expected when solar reflective materials are used in roofs. The group working on heat-island mitigation at the Lawrence Berkeley Laboratory has published data on the solar reflectance and thermal performance of various roof materials.[24] The group's database includes information on the characteristics of roof coatings, roof membranes, metal roofing and roofing tiles.

Roof coatings can be white, tinted or aluminium. White roof coatings include transparent polymeric materials, such as acrylic, and a white pigment, such as titanium dioxide or zinc oxide, in order to be opaque and reflective. Typical reflectivities of these materials are around 70–80 (Table 11.6).

Coloured coatings and, in particular, light colours are produced by adding tints to white coatings, thus reducing the solar reflectance (Table 11.7). For aluminium roof coatings an asphalt-type resin containing 'leafing' aluminium flakes is generally used. The term leafing 'refers to the tendency of the aluminum flakes to accumulate at the exposed upper portion of the coating, which is accomplished with specialized coatings on the flakes'.[24] The upper part is therefore an almost continuous aluminium layer, protecting the asphalt material from the ultraviolet radiation. This effect of this is to increase the solar reflectance to above 50% for the most reflective coatings (Table 11.8). However, this type of coating has a low infrared emmissivity.

Roofing membranes usually contain a fabric made from felt, fibreglass or polyester that is laminated or impregnated with a flexible polymeric material, which may range from bituminous hydrocarbon materials like asphalt to PVC or EPDM (synthetic rubber). In general, roofing membranes are fabricated from waterproof materials that are flexible and strong. They may consist of a single ply membrane or may be applied in multiple layers. The colour and the reflectance of the membrane is determined by

Table 11.6. Solar reflectance and thermal performance of white roof coatings[24]

Product	Solar reflectance	Infrared emittance	Temperature rise (°F)
White coating 1 (76.2 μm)	0.60	0.91	35
White coating 1 (191 μm)	0.72	0.91	23
White coating 1 (356 μm)	0.77	0.91	18
White coating 1 (660 μm)	0.79	0.91	16
White coating 2 (381 μm)	0.81	0.91	14
White coating 2 (889 μm)	0.81	0.91	14
White coating 2 (1143 μm)	0.82	0.91	13
White coating 3 (152 μm)	0.68	0.91	17
White coating 3 (254 μm)	0.75	0.91	20
White coating 3 (304 μm)	0.78	0.91	17
White coating 3 (381 μm)	0.77	0.91	18
White coating 3 (686 μm)	0.80	0.91	15
White coating 3 (1143 μm)	0.80	0.91	15
White coating 4 (127 μm)	0.68	0.91	27
White coating 4 (254 μm)	0.78	0.91	17
White coating 4 (304 μm)	0.80	0.91	15
White coating 4 (381 μm)	0.80	0.91	15
Koolseal Elastomeric on shingles	0.71	0.91	24
Henry White Coating on shingles	0.71	0.9	24
Aged Elastomeric on plywood	0.73	0.86	22
Flex-tec Elastomeric on shingles	0.65	0.89	30
Insutec on metal swatch	0.78	0.90	17
Enerchron on metal swatch	0.77	0.91	18
White Coating (1 coat, 8 mils)	0.80	0.91	15
White Coating (2 coats, 20 mils)	0.85	0.91	10
Triangle Coatings, Toughkote	0.85	0.91	10
Triangle Coatings, Trilastic	0.83	0.91	12
Triangle Coatings, high relectance	0.84	0.91	11
National Coatings, Acryshield	0.83	0.91	12
Utrecht acrylic, titanium white	0.83	0.91	12
Guardcoat, white	0.74	0.91	21
Koolseal elastomeric	0.81	0.91	14
MCI, elastomeric	0.80	0.91	15
Nexus/Visuron elastomeric	0.851	0.9	9

its upper surface, which is generally coated with a pigmented material or simply ballasted with roofing gravel. Table 11.9. gives the optical characteristics of some roofing membranes.[24]

Most metal roofs are of steel or aluminium construction. The mean solar reflectivity is close to 60%, while the emissivity is low. Cool white coatings can be used to increase the reflectivity of the roof. Table 11.10 summarizes the optical characteristics of some metal roofing materials.[24]

Finally, the optical characteristics of some roofing tiles are given in Table 11.11. Roofing tiles can be ceramic or concrete.

ALBEDO OF URBAN AREAS

Cities and urban areas in general are characterized by a relatively reduced effective albedo as a result of two mechanisms:

Table 11.7. Solar reflectance and optical performance of tinted roof coatings[24]

Product	Solar reflectance	Infrared emittance	Temperature rise (°F)
White Coating (1 coat, 8 mils)	0.80	0.91	15
White Coating (2 coats, 20 mils)	0.85	0.91	10
No pigment (1 coat, 18 mils)	0.36	0.91	59
No pigment (2 coats, 36 mils)	0.54	0.91	41
Raw Cotton (1 coat, 8 mils)	0.74	0.91	21
Raw Cotton (2 coats, 26 mils)	0.79	0.91	16
Grey Goods (1 coat, 8 mils)	0.40	0.91	55
Grey Goods (2 coats, 18 mils)	0.40	0.91	55
Desert Ridge (1 coat, 14 mils)	0.36	0.91	59
Desert Ridge (2 coats, 30 mils)	0.36	0.91	59
Red Pot (1 coat, 10 mils)	0.16	0.91	79
Red Pot (2 coats, 22 mils)	0.17	0.91	78
Green Gate (1 coat, 8 mils)	0.15	0.91	80
Green Gate (2 coats, 16 mils)	0.16	0.91	79
Charcoal Blue (1 coat, 16 mils)	0.12	0.91	83
Charcoal Blue (2 coats, 36 mils)	0.12	0.91	83
White Stucco (1 coat, 36 mils)	0.60	0.91	35
White Stucco (2 coats, 86 mils)	0.67	0.91	28
Raw Cotton (1 coat, 8 mils)	0.74	0.91	21
Brown on wood shingle	0.22	0.90	73

Table 11.8. Solar reflectance and optical performance of aluminium roof coatings[24]

Product	Solar reflectance	Infrared emittance	Temperature rise (°F)
Aluminium	0.61	0.25	48
Lomit on shingle	0.54	0.42	52
Premium nonfibred on black	0.56	0.41	50
Premium nonfibred on a rough surface	0.55	0.42	51
Nonfibered on black	0.52	0.42	52
Fibered on black	0.40	0.56	64
Premium fibred on a rough surface	0.37	0.58	67
Emulsion on a rough surface	0.30	0.67	72

- Darker buildings and urban surfaces absorb solar radiation.
- Multiple reflections inside urban canyons significantly reduce the effective albedo.

The typical albedo of European and American cities is close to 0.15–0.30, although much higher albedos (0.45–0.6) have been measured in some North African cities. Taha has compiled data on snow-free urban albedos for several cities and where possible has given the difference between the urban and rural albedo (Table 11.12).[25–37]

Variations in the albedo caused by the urban structure have been extensively studied. Cantat has shown that the albedo in Paris is about 16% lower than in the surrounding rural areas.[38] Aida has shown that absorption by the urban structure under clear weather conditions increases by about 20% compared to a flat surface of the same material, mainly because of the irregular building structure.[39] As shown in Figure 11.14, the

Table 11.9. Solar reflectance and optical performance of roofing membranes[24]

Product	Solar reflectance	Infrared emittance	Temperature rise (°F)
Grey EPDM	0.23	0.87	72
White EPDM	0.69	0.87	26
Black EPDM	0.06	0.86	89
Hypalon	0.76	0.91	19
T-EPDM	0.81	0.92	14
Firestone SBS bitumen on white	0.26	0.92	69
Smooth bitumen	0.06	0.86	89
White granular surface bitumen	0.26	0.92	69
Carlisle Syntec System Brite-Ply	0.77	0.90	18
Ecology Roof	0.80	0.90	15
Hypsam Roofing Systems, Hyload	0.75	0.90	20
Sarnafil Beige	0.435	0.9	51
Sarnafil Blue	0.617	0.9	33
Sarnafil White	0.831	0.9	11
Stevens Hi-Tuff EP White	0.78	0.9	16
Trocal Roofing Systems, White	0.77	0.9	18
Trocal Roofing Systems, White	0.83	0.9	12
Dark gravel on BUR roof	0.12	0.9	83
Light gravel on BUR roof	0.34	0.9	61
White-coated gravel on BUR roof	0.65	0.9	30

Table 11.10. Solar reflectance and optical performance of metal roofing[24]

Product	Solar reflectance	Infrared emittance	Temperature rise (°F)
New, bare galvanized steel	0.61	0.04	55
Aluminium	0.61	0.25	48
MBCI Siliconized Polyester White	0.59	0.85	37
Atlanta Metal products Kynar Snow White	0.67	0.85	28

Table 11.11. Solar reflectance and optical performance of roofing tiles[24]

Product	Solar reflectance	Infrared emittance	Temperature rise (°F)
Red clay tile	0.33	0.9	62
Red concrete tile	0.18	0.91	77
Unpainted cement tile	0.25	0.9	70
White concrete tile	0.73	0.9	22

increase in urban absorptance was greater than 20% at large zenith angles, i.e. when the sun is lower, but when the sun is lower the absorptivity increased by an average of 18% in urban areas relative to a flat area of the same material. In Figure 11.14 the curves represent the normalized increase in absorptivity for the urban case relative to a flat surface of similar material. The upper curve is for uniformly distributed building blocks, while the lower curve is for a long urban canyon.

Table 11.12. Selected urban albedo values[25–37]

Urban area	Albedo	Δ (urban – rural)
Los Angeles, CA, USA (city core)	0.20	0.09
Madison, WI, USA (urban)	0.15–0.18	0.02
St. Louis, MI, USA (urban)	0.12–0.14	–
St. Louis, MI, USA (centre)	0.19–0.16	0.03
Hartford, CT, USA (urban)	0.09–0.14	–
Adelaide, Australia (commercial)	0.27 (mean)	0.09
Hamilton, Ontario, Canada	0.12–0.13	–
Munich, Germany	0.16 (mean)	–0.08
Vancouver, BC, Canada	0.13–0.15	–
Tokyo, Japan	0.10 (mean)	–0.02
Ibadan, Nigeria	0.12 (mean)	0.03
Lagos, Nigeria	0.45	0.25

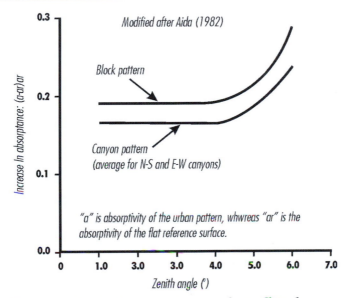

Figure 11.14 Relative increase in absorptance due to effects of geometry. The parameter 'a' is the absorptivity of the urban pattern, whereas 'as' is the absorptivity of the flat reference surface. (Adapted from Taha et al.,[19] based on the original work of Aida,[39].)

The decrease of the urban albedo as a result of the geometrical characteristics of urban canyons has been assessed by Aida and Gotoh,[40] who used a two-dimensional model developed by Craig and Lowry.[41] The authors evaluated the impact of different building width to canyon width ratios for a constant height-to-width ratio (Figure 11.15). As shown in the figure, as the irregularity increases, the effective albedo decreases. The role of building width to canyon width was studied by the same authors for a fixed building height (Figure 11.16).[40] As the streets get narrower, the effective albedo increases because of the reduced effect on the internal reflections within the canyons. An urban configuration with canyon width about twice the width of the buildings produces the lowest albedo at all zenith angles.

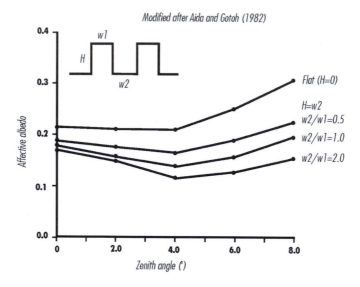

Figure 11.15. Effective albedo and urban geometry (adapted from Taha et al.,[19] based on the original work of Aida and Gotoh[40])

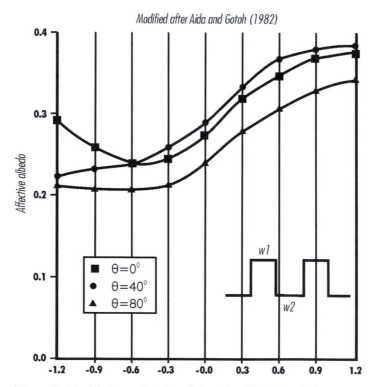

Figure 11.16. Albedo as a function of the ratio of building width to street width and zenith angle. (adapted from Taha et al.,[19] based on the original work of Aida and Gotoh[40])

THE ROLE OF MATERIALS IN THE ENERGY CONSUMPTION OF BUILDINGS AND CITIES

An increase in the surface albedo has a direct impact on the energy balance of a building. Large-scale changes in urban albedo may have important indirect effects at a city scale. Numerous studies have been performed to evaluate the direct effects of albedo change. All clearly demonstrate the benefits of using reflective surfaces. In all cases the temperature of roofs are significantly reduced, but the degree to which cooling load decreases depends on the structure of the roof and on the overall thermal balance of the building.

Using computer simulations and actual measurements, Bretz et al. reported that the increase of the roof albedo of a house in Sacramento, USA from 0.2 to 0.78 reduced the cooling energy consumption by 78%.[10] Simple calculations by Reagan and Acklam showed that when the reflectivity of a poorly insulated building is increased from 0.25 to 0.65, the heat gains through the roof are reduced by half.[42] When highly reflective materials are applied to the roof of an insulated building, heat gains through the roof were also reduced by half, although the absolute heat reduction is much lower.

Parker and Barkaszi measured the impact of reflective roof coatings on the use of air-conditioning energy in a series of tests on occupied buildings;[43] they used a before-and-after test protocol, in which the roofs were whitened at midsummer. The measured air-conditioner electrical savings in the buildings during similar pre-treatment and post-treatment weather periods averaged 19%, ranging from a low of 2% to a high of 43%. Coincident utility peak savings averaged 22%, with a similar range of values. A similar study was performed by Habel and Florence,[44] who reported important reductions in the roof temperature and heat flow through the roof, but insignificant reductions in the cooling load.

Simpson and McPherson, using scale models of residences in Arizona, found that white roofs (albedo ~ 0.75) were up to 20 K cooler than grey (albedo ~ 0.30) or silver albedo ~ 0.50) roofs and up to 30 K cooler than brown roofs (albedo ~ 0.10) roofs.[45] Measurements showed that simply increasing the albedo of a building surface may not be effective in reducing its temperature and heat gain if emissivity is reduced simultaneously. Regarding energy savings, reductions in the total and peak air-conditioning loads of approximately 5% were measured for insulated scale model buildings that had white roofs rather than grey or silver but were otherwise identical. When the ceiling insulation was removed, the air-conditioning reductions were much larger for white than for brown roofs, averaging about 28% and 18.5% for total and peak loads respectively.

Measurements of the indirect energy savings from large-scale changes in urban albedo are almost impossible. However, with computer simulations the possible change of the urban climatic conditions can be evaluated. Taha et al. used one-dimensional meteorological simulations to show that localized afternoon air temperatures on summer days can be lowered by us much as 4 K by changing the surface albedo from 0.25 to 0.40 in a typical mid-latitude warm climate.[46] Taha also used three-dimensional mesoscale simulations of the effects of large-scale albedo increases in Los Angeles to show than an average decrease of between 2 and 4 K may be possible by increasing the albedo by 0.13 in urbanized areas.[47] Akbari et al. showed that a temperature decrease of this magnitude could reduce the electricity load from air

Table 11.13. Direct and indirect savings from trees and light coloured roofing in Los Angeles, CA (after Rosenfeld et al.[9])

	Direct energy savings		Indirect energy savings		Smog benefits	Totals	
	Avoided peak power (MW)	A/C cost savings ($M/year)	Avoided peak power (MW)	A/C cost savings ($M/yr)	Avoided medical costs, 12% ozone reduction ($M/yr)	Total avoided peak power (MW)	Total cost savings ($M/year)
Cooler roofs	400	46	200	21	104	600	171
Trees	600	58	300	35	180	900	273
Cooler pavement	0	0	100	15	76	100	91
Total	1000	104	600	71	360	1600	535

conditioning by 10%.[48] Recent measurements in White Sands, New Mexico have indicated a similar relationship between naturally occuring albedo variations and measured ambient air temperatures. Taha *et al.* analysed the atmospheric impacts of regional scale changes in building properties, paved surface characteristics and their microclimates and considered the possible meteorological and ozone air-quality impacts of increases in surface albedo and urban trees in California's South Coast Air Basin.[49] By using photochemical simulations, they found that implementing high-albedo materials would have a net effect of reducing ozone concentrations and domain-wide population-weighted excess exposure to ozone above the local stand-ards would be decreased by up to 12% during peak afternoon hours.

Rosenfeld et al. simulated the impact of the use of increased urban albedo and tree planting for Los Angeles, CA.[9] The simulations were based on the assumption that the city albedo increased by 7.5% when 5% of the area was covered with ten million trees. The calculated direct and indirect energy savings, as well as some environmental benefits, are given in Table 11.13. Direct savings refer to the cooling effect on individual buildings. Indirect savings refer to cuts in air-conditioning load for all buildings as the temperature of the surrounding community drops.

The results show that the use of white roofs and shade trees in Los Angeles would lower the need for air conditioning by 18%, or 1.04 billion kWh, for the buildings directly affected by the roofs and shaded by the trees. Indirect benefits are also very important. If the temperature of the entire community drops a degree thanks to lighter roofs and pavements and to the evapotranspiration from the trees, then the air-conditioning load of all buildings is reduced, even if these buildings are not shaded or if they have dark-colour roofs. This indirect annual saving would be an additional 12% or 0.7 billion kWh.

REFERENCES

1.   Smith, M., J. Whitelegg and N. Williams (1988). *Greening the Built Environment*. Earthscan Publications, London.
2.   Anink, D., C. Boonstra and J. Mak (1996). *Handbook of Sustainable Building*. James and James Science Publishers, London.

3. Griggs, E.I., T.R. Sharp and J.M. MacDonald (1989). 'Guide for Estimating Differences in Building Heating and Cooling Energy Due to Changes in Solar Reflectance of a Low-Sloped Roof'. Oak Ridge National Laboratory Report ORNL-6527.
4. Yap, D. (1975). 'Seasonal Excess Urban Energy and the Nocturnal Heat Island – Toronto'. *Archives Meteorology, Geophysics, Biometeorology* Ser. B, Vol. 23, p.p. 68-80, 1975.
5. Robinette, 1977.
6. Santamouris M., N. Papanikolaou and I. Koronaki (1977). 'Urban Canyon Experiments in Athens. Part A. Temperature Distribution'. Internal Report to the POLIS Research Project. European Commission, Directorate General For Science, Research and Technology. Available through the authors: M. Santamouris, Physics Department, University of Athens, 15784 Athens, Greece.
7. Oke T.R, G.T. Johnson, D.G. Steyn and I.D. Watson (1991). 'Simulation of Surface Urban Heat Islands under 'Ideal' Conditions at Night – Part 2: Diagnosis and Causation'. *Boundary Layer Meteorology*, Vol. 56, pp. 339–358.
8. Akbari, H., S. Davis, S. Dorsano, J. Huang and S. Winnet (eds) (1992). *Cooling Our Communities – A Guidebook on Tree Planting and Light-Colored Surfacing*. US Environmental Protection Agency, Office of Policy Analysis, Climate Change Division.
9. Rosenfeld, A., J. Romm, H. Akbari H. and A. Lloyd (1998). 'Painting the Town White and Green'. Available through the Web site of Lawrence Berkeley Laboratory; URL:http://www.lbl.gov.
10. Bretz, S., H. Akbari, A. Rosenfeld and H. Taha (1992). 'Implementation of Solar Reflective Surfaces: Materials and Utility Programs'. LBL Report – 32467, University of California.
11. Baker, M.C. (1980). *Roofs: Designs, Application and Maintenance*. Multi-science Publications, Montreal.
12. Martien, P., H. Akbari and A. Rosenfeld (1989). 'Light Colored Surfaces To Reduce Summertime Urban Temperatures: Benefits, Costs, and Implementation Issues'. Presented at the 9th Miami International Congress on Energy and Environment, Miami Beach, FL, 11–13 December.
13. Stirling, R., J. Carmody and G. Elnicky (1981). *Earth Sheltered Community Design: Energy Efficient Residential Development*. Van Nostrand Reinhold, New York.
14. Edwards, D.K. (1981). *Radiation Heat Transfer Notes*. Hemisphere Publishing Corporation, Washington, DC.
15. Oke, T.R. (1987). *Boundary Layer Climates*. Cambridge University Press.
16. Asaeda, T., Vu Thanh Ca and Akio Wake (1996). 'Heat Storage of Pavement and its Effect on the Lower Atmosphere'. *Atmospheric Environment*, Vol. 30, No. 3, pp. 413–427.
17. Gustavsson, T. and J. Bogren (1991). 'Infrared Thermography in Applied Road Climatological Studies'. *International Journal of Remote Sensing*, Vol. 12, pp. 1811–1828.
18. Berg, R. and W. Quinn, (1978). 'Use of Light Colored Surface to Reduce Seasonal Thaw Penetration beneath Embankments on Permafrost'. *Proceedings of the Second International Symposium on Cold Regions Engineering*, University of Alaska, pp. 86–99.
19. Taha, H., D. Sailor and H. Akbari (1992). 'High Albedo Materials for Reducing Cooling Energy Use'. Lawrence Berkeley Laboratory Report 31721, UC-350, Berkeley, CA.
20. Cantat, O. (1984). 'Evolution du Climat et Urbanisation dans le Sud de l'Aglomeration Parisienne'. Report of Maitrise studies, University Paris – Sorbonne.
21. Boutell, C.J. and Y. Salinas (1986). *Building for the Future Phase I*, Volumes I and II. University of Southern Mississippi, Department of Construction and Architectural Engineering Technology.
22. Santamouris, M., Papanikolaou, N. and Georgakis, C. (1998). 'Study on the Ambient and Surface Temperature in Omonia Square, Athens, Greece'. Internal Report, Group Building Environmental Studies, Physics Department, University of Athens, Athens, Greece. Available through the authors: M. Santamouris, Physics Department, University of Athens, 15784 Athens, Greece.
23. Backenstow, D.E. (1987). 'Comparison of White versus Black Surfaces for Energy Conservation'. *Proceedings of the 8th Conference on Roofing Technology; Applied Technology for*

Improving Roof Performance, Gaithersburg, Maryland, 16–17 April. National Bureau of Standards and the National Roofing Contractors Association, pp. 27–31.

24. Lawrence Berkeley Laboratory (1998). 'Roof Coatings. Solar Reflectance and Thermal Properties of Roof Materials'. Available through the Web site of Lawrence Berkeley Laboratory; URL:http://lbl.gov

25. Taha, H. (1997). 'Urban Climates and Heat Islands: Albedo, Evapotranspiration, and Anthropogenic Heat'. *Energy and Buildings*, Vol. 25, pp. 99–103.

26. Taha, H. (1994). 'Aircraft-based Albedo Measurements over the South Coast Air Basin'. In Taha, H. (ed.), 'Analysis of Energy Efficiency of Air Quality in the South Coast Air Basin - Phase II', Report No LBL-35728, Lawrence Berkeley Laboratory, Berkeley, CA, Ch. 2, pp. 43–59.

27. Kung, E.C., Bryson, R.A. and Lenschow, D.H. (1964). 'Study of a Continental Surface Albedo on the Basis of Flight Measurements and Structure of the Earth's Surface Cover over North America'. *Monthly Weather Review*, Vol. 92, p. 543.

28. Dabberdt, W.F. and Davis, P.A. (1978). 'Determination of Energetic Characteristics of Urban–Rural Surfaces in the Greater St-Louis Area'. *Boundary Layer Meteorology*, Vol. 14, p. 105.

29. Vukovich, F.M. (1983). 'An Analysis of the Ground Temperature and Reflectivity Pattern about St Louis, Missouri, using HCMM Satellite Data'. *Journal of Applied Meteorology*, Vol. 22, p. 560.

30. Brest, C.L. (1987). 'Seasonal Albedo of an Urban/Rural Landscape from Satellite Observations'. *Journal of Applied Meteorology*, Vol. 26, p. 1169.

31. Coppin, P., Forgan, B., Penney, C. and Schwerdtfeger, P. (1978). 'Zonal Characteristics of Urban Albedos'. *Urban Ecology*, Vol. 3, p. 365.

32. Rouse, W.R. and Bello, R.L. (1979). 'Shortwave Radiation Balance in an Urban Aerosol Layer'. *Atmosphere-Ocean*, Vol. 17, p. 157.

33. Mayer, H and Noack, E.M. (1980). 'Einfluss der Schneedecke auf die Strahlungsbilanz im grossraum München'. *Meteorolische Rundsch*, Vol. 33, p. 65.

34. Steyn, D.G. and Oke, T.R. (1980). 'Effects of a Scrub Fire on the Surface Radiation Budget'. *Weather*, Vol. 35, p. 212.

35. Aida, M. (1982). 'On the Urban Surface Albedo'. *Japanese Progress in Climatolology*, pp. 4–5.

36. Oguntoyinbo, J.S. (197). 'Reflection Coefficient of Natural Vegetation Crops, and Urban Surfaces in Nigeria'. *Quarterly Journal of the Royal Meteorological Society*, Vol. 96, p. 430.

37. Oguntoyinbo, J.S. (1986). 'Some Aspects of the Urban Climates of Tropical Africa'. World Meteorological Organisation Report, pp. 110–135.

38. Cantat, O. (1989). 'Contribution à l'Etude des Variations du Bilan d'Energie en Region Parisienne'. PhD Thesis, University of Paris – Sorbonne.

39. Aida, M. (1982). 'Urban Albedo as a Function of the Urban Structure – A Model Experiment'. *Boundary Layer Meteorology*, Vol. 23, pp. 405–413

40. Aida, M. and Gotoh, K. (1982). 'Urban Albedo as a Function of the Urban Structure – A Two Dimensional Numerical Simulation', *Boundary Layer Meteorology*, Vol. 23, pp. 415–424.

41. Craig, C.D. and Lowry, W.P. (1972). 'Reflections of the Urban Albedo'. *Proceedings of the Conference on the Urban Environment and Second Conference on Biometeorology, Philadelphia, PA*. 31 Oct.–2 Nov. American Meteorological Society, pp. 159–164.

42. Reagan, J.A. and Acklam, D.M. (1979). 'Solar Reflectivity of Common Building Materials and its Influence on the Roof Heat Gain of Typical Southwestern USA Residences'. *Energy and Buildings* Vol. 2, pp. 237–248.

43. Parker, D.S. and Barkaszi, S.F. (1997). 'Roof Solar Reflectance and Cooling Energy Use: Field Research Results from Florida'. *Energy and Buildings*, Vol. 25, pp. 105–115.

44. Habel, M. and Florence, R. (1985). 'Design Methodology for the Analysis of a Solar Reflective Roof Coating and its Effect on the Cooling Load in Actual Environments'. Department of Mechanical Engineering, San Diego State University.

45. Simpson, J.R. and McPherson, E.G. (1997). 'The Effects of Roof Albedo Modification on Cooling Loads of Scale Model Residences in Tucson, Arizona', *Energy and Buildings*, Vol. 25, 127–137.

46. Taha, H. (1988). 'Site Specific Heat Island Simulations: Model Development and Application to Microclimate Conditions'. Lawrence Berkeley Laboratory Report No 26105; M. Geogr. Thesis, University of California, Berkeley, CA.

47. Taha, H. (1994). 'Meteorological and Photochemical Simulations of the South Coast Air Basin'. In Taha, H. (ed.), Analysis of Energy Efficiency of Air Quality in the South Coast Air Basin – Phase II, Rep. No LBL-35728, Lawrence Berkeley Laboratory, Berkeley CA, Ch. 6, pp. 161–218.

48. Akbari, H., Rosenfeld, A. and Taha, H. (1989). 'Recent Developments in Heat Island Studies: Technical and Policy'. *Proceedings of Workshop on Saving Energy and Reducing Atmospheric Pollution by Controlling Summer Heat Islands*, Berkeley, CA, 23–24 February, pp. 14–20.

49. Taha, H., Douglas, S. and Haney, J. (1997). 'Mesoscale Meteorological and Air Quality Impacts of Increased Urban Albedo and Vegetation'. *Energy and Buildings*, Vol. 25, pp. 169–177.

12

Applied lighting technology for urban buildings

P. Michel

Laboratoire des Sciences de l'Habitat, Ecole Nationale des Travaux Publics de l'Etat

ℛ

Daylight is a light source (see, for example, Figure 12.1). Occupant satisfaction and energy conservation may be improved if proper considerations are given to the following:

- view design;
- human factors;
- glare control.

Apertures to admit daylight to urban buildings must be designed with many elements taken into account. These include:

- thermal comfort and indoor air quality;
- acoustics;

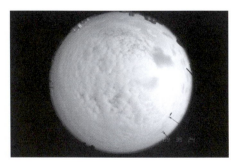

Figure 12.1. Fish-eye views of the sky[1]

- architecture;
- economics;
- energy consumption;
- security.

FUNDAMENTALS OF LIGHTING

Light

Light can be defined with reference to human beings as radiant energy that is capable of exciting the human retina and creating a visual sensation. Light can also be defined as a physical quantity in terms of its relative efficiency over the electromagnetic spectrum lying between 380 nm and 770 nm (Figure 12.2)

The characteristics of light can be well illustrated by different physical theories. In particular, some characteristics are best described by wave theory, while others need a corpuscular/quantum approach. However, it needs to be understood that light, exhibiting wave/particle duality, is radiation produced by electronic processes.

The different quantities that are used or evaluated when dealing with lighting engineering are given below. These quantities can be defined as both energetic and luminous quantities. For some quantities, such as exitance, the same physical definition leads to two different names. Similarly, the same physical quantities may be evaluated in different units, depending on whether these quantities are considered as energetic or luminous. The quantities are:

- *Luminous energy (Q)*. This is the visually evaluated radiant energy travelling in the form of electromagnetic waves. It unit is lumen seconds.
- *Luminous flux (Φ)*. This is the rate of flow of luminous energy over time. Its unit is lumen (lm).
- *Luminous intensity (I)*. This is the luminous flux density over solid angle in a given direction. Its unit is candela (cd).
- *Exitance (M)*. This is the total luminous flux density leaving a surface. Its unit is lm/m^2.
- *Luminance (L)*. This is the luminous flux per unit of projected area per unit solid angle leaving a given point in a given direction (brightness). Its unit is cd/m^2.

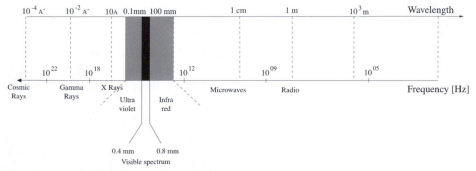

Figure 12.2. The electromagnetic spectrum[2]

- *Illuminance.* This is the incident luminous flux per unit area of the surface. Its unit is lux.

VISION AND COLOUR

Vision

The human eye can be considered as an optical system that creates an image of the environment on a photosensitive surface, the retina. The iris continuously adjusts the amount of light that enters the eye (adaptation). Focusing on objects is carried out by the ciliary muscles, which change the shape of the lens (accommodation).

Light passes through the vitreous humour to the retina, which contains two types of light-sensitive nerve fibre endings:

- *Rods.* Situated all over the retina, the rods function primarily during 'night' (scotopic) vision. Colour vision is not possible with rods.
- *Cones.* Located close to the optical axis, cones allow colour vision and are specialized for 'daylight' (photopic) vision.

Human beings are particularly sensitive to solar radiation, which has a peak of emission at about 505 nm. Our vision is at its best at about 555 nm during the day (yellow) and at about 510 nm at dawn or for night vision (blue). Light engineering both within buildings and outdoors has to take into account this relative spectral sensitivity curve, which is designated as $V(\lambda)$ for photopic vision and $V'(\lambda)$ for scotopic vision.

Colour

Colour temperature is a method for describing the colour of a light source by comparing it with the colour of a black-body radiator. As the colour of an incandescent lamp is similar to that of a radiator heated to 3000 K, the colour temperature of the incandescent lamp is said to be 3000 K.

A warm appearance (rich in red and orange) corresponds to a low colour temperature, while sources with a high colour temperature have a cool appearance (bluish light). A clear blue sky thus has a very high colour temperature (8000 to 30,000 K).

Another approach to colour measurement compares the light source to a reference one. Unlike colour temperature, which is concerned with the light produced, the Colour Rendering Index (CRI) deals with the appearance of coloured objects illuminated by the light source in comparison to the appearance when illuminated by a reference source. Incandescent lamps and daylight have a CRI of about 100. For some fluorescent or discharge light sources, the CRI can be very low (about 50).

The difference between these two approaches is that colour temperature is concerned with the additive principle, while CRI is about subtractive colour, i.e. the light that is reflected is the incident light less the light that is absorbed by the reflecting object.

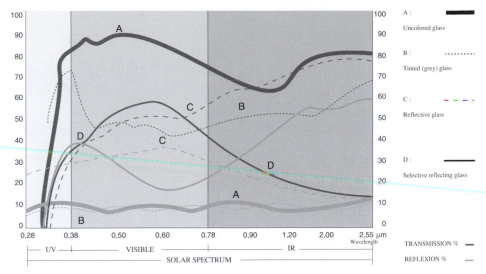

Figure 12.3. Transmissive and reflective properties of glazing materials[2]

DAYLIGHTING

Glazing materials

Glazing materials used in buildings for daylighting applications can be characterized by:

- optical properties: transmission, absorption, reflection, photometric properties (Figures 12.3 and 12.4);
- spectral information;
- thermal properties.

Glazing materials can basically be characterized by a very high transmittance in the visible (and near-infrared) spectrum and a very low transmittance in the (far) infrared.

Figure 12.4 (and below). Visualization of the photometric distribution of different glazing materials.[1]

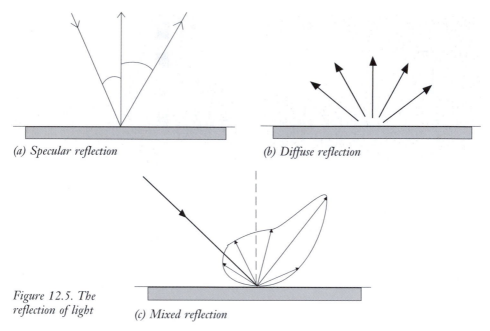

(a) Specular reflection   *(b) Diffuse reflection*

Figure 12.5. The
reflection of light   *(c) Mixed reflection*

However, optical properties may vary widely from one material to another and with the angle of incidence of the light.

Glazing materials typically have a high thermal conductivity, compared to the other building materials that are used in walls. In order to reduce heat transfer through the windows, manufacturers have developed various materials with high thermal performance, such as transparent insulation or low-emissivity glazing.

The reflection of light from the surface of a glazing material may be considered as either specular (mirror) or diffuse. Many surfaces may, however, be considered as mixed reflectors (Figure 12.5).

Glazing materials can be classified into six families according to their effect on light:

- transparent materials;
- light-scattering materials;
- materials that change the direction of the light;
- materials that absorb or reflect light depending on the angle of incidence;
- tinted materials;
- materials that have a variable transmittance depending on the application of a low voltage (electrochromic) or of solar input (thermochromic/thermotropic).

These families are considered in the following sections.

Transparent materials

These materials do not appreciably change either the direction or the colour of light, so that they are image preserving. These materials are thus used in applications where transparency either has no major drawback, for example with reference to security or confidentiality, or is of major importance because of, for example, the outside view, visual control or commercial considerations.

Figure 12.6. (top left) Light-scattering glazing materials; (top right) Sunny patch through clear glass; (bottom right) Light-scattering effect.

Sheet glass and moulded glass are examples of such materials, the approximate transmittance of which is about 85–90% (for normal incidence). These materials are now frequently used in multiple-glazing windows or as components to reduce heat transfer and/or noise transmission.

Transparent materials are (almost always) used in residential buildings; this can be a problem on lower floors and shutters or curtains may be required, especially for security reasons. Shading devices may be required on higher floors to prevent overheating.

These transparent materials are also very frequently used in commercial buildings, because they do not alter colour. This is especially important in shops where products are displayed and have to be seen from the street.

Light-scattering materials

The amount of diffusion in materials varies over a wide range, depending on the material and its treatment. The luminance of highly diffusing materials is almost constant for all viewing angles.

Light-scattering materials are of particular benefit when one has to reduce or prevent viewing from outside as well as from inside. At the same time, they prevent the generation of sunny patches on indoor surfaces, especially important in buildings such as sport halls, and, at the same time, preventing overheating.

Diffusing glass, prismatic materials and multilayer insulating and diffusing panels are examples of materials that provide light scattering (Figure 12.6).

A recent development is the use of aerogels which also provide high thermal insulation.

Materials that change the direction of the light

To avoid glare, a fraction of the incoming light is rerouted to the ceiling – preferably painted white – which reflects an additional amount of light into the room. The view outwards can be maintained while the brightness of the outdoor environment is attenuated.

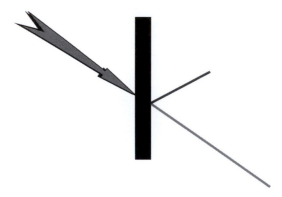

Figure 12.7. Incoming light after passing through built-in lamellae or holographic film

This kind of material – with a low transmittance coefficient – is thus of particular interest in cases where there is high direct radiation that can cause glare. This means that urban buildings are not so likely to require this technique because it leads to a reduction in the incoming light.

Built-in reflection lamellae, prismatic devices, polycarbonate panels and thin-film holographic materials (Figure 12.7) may offer such opportunity of a reduction of the contrast in the visual environment.

Materials that allow angular selection of the incoming light.

For specific angles of incidence, these materials either absorb light or reflect it to the outside. This can be achieved either with materials that preserve the direction of the transmitted light (for instance, micro lamellae) or those that change it; holographic materials are one example.

Such products can have aesthetic as well as thermal advantages in summertime, but they significantly reduce the light transmitted. Windows have, therefore, to be oversized to prevent a negative effect on overcast winter days. Moreover, this type of material has a variable efficiency that depends on the location (the latitude of the site) and on the time of the year (height and azimuth of the sun).

Materials that cause a colour shift

Tinted materials (glazing, polycarbonate) may be used to generate an appropriate luminous indoor environment, by increasing (bluish tint) or decreasing (yellowish tint) the colour temperature of the transmitted light (Figures 12.8 and 12.9). The choice of the tinted materials is carried out in conjunction with the choice of lamps and in accordance with the orientation of the facades. Tinted materials have to be carefully used in cases where colour rendering is fundamental.

Tinted materials are frequently used in the glazed facades of office buildings, offering the opportunity for aesthetic effects. However, architectural rules may limit their implementation in urban environments. Facades using these tinted materials are frequently overglazed, leading to thermal problems in both summer and winter.

Materials with variable transmittance

Developments on variable transmittance glazings have occurred in two major directions:

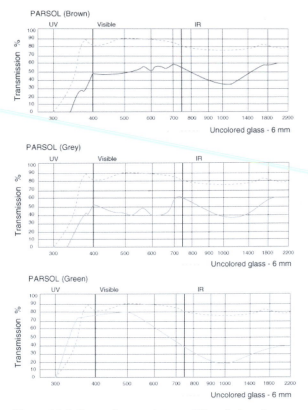

Figure 12.8. Spectral transmittance of Parsol glasses[2]

Figure 12.9. Bluish tinted material[1]

- *Electrochromic.* The transmittance properties are a function of an applied low DC voltage, which may be controlled manually or automatically. Glazing components can then be incorporated into a building management system.
- *Thermochromic/Thermotropic.* When a high solar input raises its temperature, a thermosensitive glazing material will switch to one of the following:

  – a scattering state (thermotropic material);
  – a coloured low-transmittivity state (thermochromic material).

These materials have several advantages, both preventing overheating and allowing good indoor daylight availability when needed. They allow a control strategy over the incoming light to be implemented in the facade. Recent technological researches and technical developments have allowed the industrial production of such materials at a 'reasonable' cost. At the same time, the optical and thermal properties of these glazing materials have been improved.

An additional benefit is the increase in the speed of the physical phenomena. Some electrochromic materials, available with large surface areas, can switch from very high transmittance to very low transmittance in a few seconds.

Components and technologies

Daylight components

Four major examples of building components that provide views of the outside and/ or penetration of daylight, while also controlling glare, are windows, clestories, roof apertures and atria:

Windows. Windows play a key role in buildings and must be considered as a source of heat as well as light. Moreover, psychosociological aspects have to be taken into account: windows are transparent to allow a view to the outside, but they must often prevent a view from the outside in.

A window may be considered as a potential area of energy conservation. It provides direct gain (especially in winter) and it can reduce the consumption of electricity for artificial lighting. A window must, however, be carefully designed to prevent heat excess losses in winter and to avoid visual discomfort.

The distance between the window wall and the inner wall of a room should be limited in order to avoid extreme daylight illuminances. Openings can be placed opposite one another to solve this problem, for example windows with their tops placed close to the ceiling. This, however, increases glare (the view of the sky).

Overhangs may reduce sunlight penetration during daytime in summer, limiting the range of illuminances. In the same way, venetian blinds can reduce sky glare and redirect light to the ceiling. For examples, see Figures 12.10 and 12.11.

Clerestory. Additional windows on the roof, with the same orientation as the main windows, can help in provide sufficient daylighting in case of unilateral sidelighting. While not appropriate for high buildings, clerestory windows may provide good indoor lighting environment in the case of deep rooms or (blind) corridors. They input light to the back of the room and may use the ceiling and the back wall as reflectors to produce a uniform environment. See Figures 12.12 and 12.13.

At the same time, depending on the size of the room or the corridor, clerestory windows provide a view onto the sky.

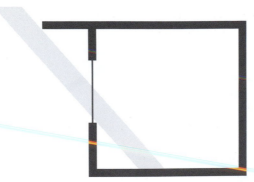

Figure 12.10. Window with overhang for unilateral sidelighting

Figure 12.11. Window with bilateral sidelighting[1]

Roof apertures. Roof apertures can be of many shapes, while special attention should be given to sunny patches, moisture and condensation. Openable roof apertures can be also used for ventilation and cooling.

Such components have been frequently implemented in commercial buildings, especially in educational buildings and in the retail sector, i.e. large and high one- or two-storey buildings. In such cases, roof apertures are a unique way of providing daylight inside the building (Figures 12.14 and 12.15).

Atria. Atria are more and more frequently used by architects in high buildings because of their aesthetic and architectural interest. Not only office buildings and hotels, but also educational establishments, are sectors where atria are commonly implemented.

The atrium may be used as a key element of the indoor climate, for lighting as well as for heating and ventilation. As a consequence, an atrium has to be carefully designed to avoid heat losses or overheating, glare, air movements, etc. This can be achieved, in particular, through smart control of openings and shading devices at the top of the atrium.

Shading devices

Many internal and external shading devices may be used to control the admission of direct radiation from the sun. Apart from vegetation, which may, in non-urban areas, be used as a protection against sun, shading devices may be fixed or movable.

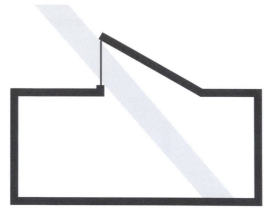

Figure 12.12. Clerestory (Drawing: Pierre Michel)

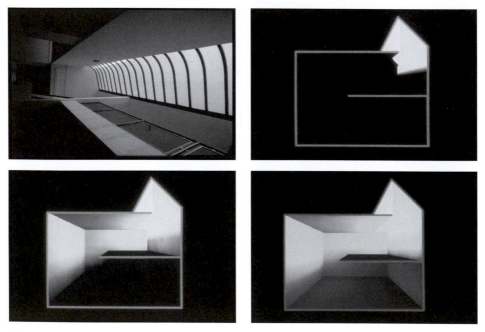

Figure 12.13. (Top left) Scale model of a clerestory. (top right) Simulated light coming in through a clerestory (one reflection). (Bottom left and right) Simulated light coming in through a clerestory (two and three reflections)[1]

Fixed devices need no control and almost no maintenance, but are unlikely to be efficient in any season and in any climatic condition. These devices always provide a reduction in the daylight penetrating the building. Movable devices may be adjusted depending on the season, climatic conditions and the time of the year (and of the day). Such devices thus have mechanical constraints and need to be connected to a controller for optimum efficiency and easy use.

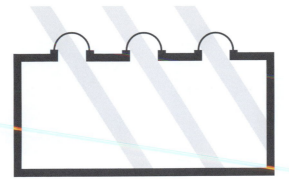

Figure 12.14. Roof apertures.

Figure 12.15. (top) Outside and inside views of a roof aperture. (bottom) Simulated light coming in through roof apertures[3]

DESIGNING WITH DAYLIGHT

Visual comfort

Glare can be evaluated in order to qualify the visual environment. Glare can have two major negative consequences:

- *Losses in visibility*. Reflection on a VDU screen and glare due to high-luminance sources are two examples where decreasing contrast (for example between the letters on a sheet of paper and the surrounding surface) leads to difficulty in reading or viewing objects (Figure 12.16). This effect increases with the age of the people involved.
- *Discomfort*. Discomfort glare is the sensation of pain caused by non-uniform distributions of luminance in the field of view (Figure 12.17). Discomfort can be caused by direct or indirect glare. Discomfort glare can be reduced by:
  - decreasing the luminance of the light source;
  - diminishing the area of the source;
  - increasing the background luminance around the source.

Different indexes have been proposed in order to define the comfort conditions inside a building:

Figure 12.16. Examples of losses of visibility due to glare. (Left) Unacceptable reflections on a VDU screen; (right) a sunny patch[3]

Figure 12.17. Tinted window panes[3]

- *The Bodmann-Söllner method.* For different activities, i.e. different lighting requirements, a minimum viewing angle is defined for the light sources (luminaires) in order to avoid glare.
- *The Visual Comfort Probability (VCP) method.* The VCP value represents the percentage of people who probably will not complain about the glare produced in the space. A luminaire-by-luminaire determination of discomfort leads to the calculation of an overall rating called the DGR (Discomfort Glare Ratio). VCP is then evaluated with the following equation:

$$VCP = \frac{100}{\sqrt{2\pi}} \int_0^{6.374-1.3227 \ln(DGR)} \exp(-t^2/2)\,dt. \tag{12.1}$$

- *The Unified Glare Ratio (UGR) method.* UGR is calculated using visual parameters in the field of view. Discomfort is then evaluated in terms of the position on a scale of discomfort. The formulation of UGR is:

$$UGR = 8\frac{0.25}{L_b}\sum \frac{L_s^2\omega}{p^2}, \tag{12.2}$$

where L_b is the background luminance, L_s is the luminance of each source, ω is the solid angle and p is the Guth factor.

Table 12.1. Illuminance values for various activities[3]

Type of activity	Illuminance category	Range of illuminances (lux)
Public spaces with dark surroundings	A	20–50
Simple orientation for a short temporary visit	B	50–100
Working spaces where visual tasks are only occasionally performed	C	100–200
Visual tasks of high contrast or large size	D	200–500
Visual tasks of medium contrast or small size	E	500–1,000
Visual tasks of low contrast of very small size	F	1,000–2,000
Visual tasks of low contrast and very small size over a prolonged period	G	2,000–5,000
Very prolonged and exacting visual tasks	H	5,000–10,000
Very special visual tasks of extremely low contrast and small size	I	10,000–20,000

Table 12.2. Lighting levels for various types of work spaces[4]

Task group and typical interior	Standard service illuminance (lux)
Minimum for a visual task	200
Rough work, machinery assembly, writing, reading	300
Routine work, offices, control room	500
Drawing office	750
Fine work, machining and inspection	1000
Very fine work	1500
Very demanding task in industry or laboratory	2000

Lighting requirements

Professional associations (for example, the Illuminating Engineering Society of North America – IESNA – and the Association Français d'Eclairage – AFE) have provided currently recommended illuminance categories and illuminance values for lighting design. Tables 12.1 to 12.3 illustrate the different categories and give values for some kinds of buildings.

Assessment criteria

Daylight Factor

The Commission Internationale de l'Eclairage (CIE) Daylight Factor (DF) at a particular point, used in daylighting design, is the ratio of the interior horizontal illuminance at that point to the simultaneous external horizontal illuminance. This method was developed in Europe for overcast sky conditions and direct sunlight is excluded from both interior and exterior values of illumination.

The Daylight Factor is thus an indicator of daylight performance of a room and its fenestration considered as a lighting system. Moreover, in strictly uniform overcast sky conditions, the daylight factor is a constant, i.e. the interior illuminance is a constant fraction of the exterior (variable) illuminance. However, if based on measured values, the daylight factor can be variable over time (even with overcast skies), because of the changing distribution of the sky luminance. Examples of daylight factors are given in Table 12.4.

Table 12.3. Illuminance categories[3]

Indoor activity	Category
Banks	
Lobby	
General	C
Writing area	D
Tellers' station	E
Health care facilities	
Anesthetizing	E
Corridors	
Nursing areas – day	C
Nursing areas – night	B
Operating areas	E
Eye surgery	F
Patients' rooms	
General	B
Observation	A
Critical examination	E
Reading	D
Toilets	D
Elevators	C
Exhibition halls	C
Cloth products industry	
Cloth inspection	I
Cutting or sewing	G
Pressing	F
Garages – service	
Repairs	E
Active traffic areas	C
Write-up	D
Museums	
Non-sensitive materials	D
General gallery areas	C
Restoration or laboratories	E
Residences	
General lighting	B
Specific visual tasks	
Dining	C
Ironing	D
Laundry	D
Reading	
in a chair – books	D
in a chair – poor copies	E

Illuminance

Illuminance is one of the basic quantities used in designing and evaluating lighting systems and their components. The calculation of illuminance may be carried out by considering a specific point (point-by-point method) or by determining the average uniform horizontal illuminance (lumen method) across the workplane.

Point-by-point method. The point-by-point method, which is used in non-uniform lighting layouts, is based on the Inverse Square Law of Illumination and the Cosine

Law of Illumination. It is used to calculate the direct component of illumination (as a proportion of the luminous intensity) under the assumption that the multiple internal reflections are negligible and the source of light (natural or artificial) can be considered a point source.

If the source cannot be considered as a point, it can be considered as several point-source-sized elements. Luminous flux transfer has to be considered if internal reflections cannot be considered negligible. The exitance of the source can then lead to the evaluation of illuminance through the calculation of geometric form factors or configuration factors.

Lumen method. The Illuminating Engineering Society (IES) Lumen method may be used either to design artificial lighting equipment or to predict illuminance from daylighting:

- *Artificial lighting.* The Lumen method is used to calculate the number of luminaires required for a uniform (general) lighting layout. The calculation of this number of luminaires and of the consequent average illuminance on the workplane depends on the characteristics of the room and of the luminaires, among which are:
  - Coefficient of Utilization (CU), which measures how a luminaire distributes light into a given room. The CU is an element of the photometric test report for a given luminaire.
  - Light loss factors, which take into account the Lamp Lumen Depreciation (LLD) – the decrease in light output of lamps over their life – and the Luminaire Dirt Depreciation (LDD) – the accumulation of dirt on the surface of a luminaire.
- *Daylighting.* The Lumen method, applied to daylighting, is used to predict the indoor illuminance at a specific point, with both direct and internally reflected components taken into account. The calculation of the illuminance depends on the following quantities:
  - the vertical illuminance incident on the window;
  - the area of the glazing;
  - the transmittance of the glazing;
  - the Light Loss Factor due to dirt accumulation;
  - Coefficients of Utilization K (for ceiling geometry) and C (for room geometry)

Table 12.4. Examples of daylight factor recommendations

Situation	Average daylight factor
Church	1% minimum
Dwelling	
bedroom	0.3%
kitchen	0.6%
living room	0.5%
Sports hall	3.5%
Hospital ward	1%
Office	2% to 2.5%
School classroom	2%

Contrast

Contrast is a measure of the ability of an observer to distinguish a minimum difference in luminance between two areas (an object and its background). Contrast is the basic distinguishing mechanism in vision:

$$C = \left| \frac{L_{obj} - L_{back}}{L_{back}} \right|. \tag{12.3}$$

Figures 12.7, 12.10, 12.12, 12.14, 12.16 and 12.17 were supplied by Piérre Michel.

REFERENCES

1. Fontoynant, M. *Et al.* (1993). *Properties of glazing materials for daylighting applications.* Final report (JOU2-CT940404). Ecole Nationale des Travaux Publics de l'Etat, Vaulx en Velin.
2. *Énergétique. Application au transfert de chaleur* (1999). Gérard Guarracino (Coordinates). Ecole Nationale des Travaux Publics de 'Etat, Vaulx en Velin.
3. *Lighting handbook* (1993). The Illuminating Engineering Society of North America, New York.
4. *Window Design* (1987). The Chartered Institution of Building Services Engineers, London.

FURTHER READING

European Passive Solar Handbook (1986). European Commission DG XII, Brussels.
Working in the City (1989). European Commission DG XII. Eblana Editions, Dublin.
Lighting Handbook (1993). Illuminating Engineering Society of North America, New York.

13

Active solar heating systems for urban areas

D. Mangold

Institute für Thermodynamik und Warmetechnik, University of Stuttgart

ဆာ

Urban areas both offer opportunities and present disadvantages for the usage of thermal solar energy. People who live in apartments generally do not have the possibility of installing their own solar thermal system, since their hot water supply is part of the central heating installation. Because urban areas have a high building density, and consequently the demand for energy as a function of building area is quite high, urban areas in particular offer possibilities for the economic application of central energy generation and delivery systems.

Central solar heating plants are the most economic way of providing thermal solar energy to housing estates for domestic hot water and room heating. Over 50% of the fossil-fuel demand of an ordinary district heating plant can be replaced by solar energy when seasonal heat storage is included in the plant. In this chapter the technology of central solar heating plants with diurnal and seasonal heat storage is presented and information is given on their design and economy. The first pilot plants with seasonal heat storage are described.

SYSTEMS FOR SOLAR ENERGY SUPPLY

Block or district heating systems are a prerequisite for the integration of economical large-scale solar heating systems. They consist of a central heating unit, a heat distribution network and, if necessary, heat transfer substations in the connected buildings. Centralized heat production offers high flexibility in terms of the choice of the heat generators and allows for seasonal storage of solar heat in summer to cover part of the heating demand in winter.

Solar-assisted district heating systems can be separated into systems with short-term or diurnal heat storage, designed to cover 10–20% of the total heat demand for room heating and domestic hot water with thermal solar energy, and solar systems with seasonal heat storage with solar fractions of more than 50%. The so-called solar fraction is that part of the yearly energy demand that is covered by solar energy. An example of a solar-assisted district heating system is shown in Figure 13.1.

Solar systems with diurnal heat storage are mainly used to supply heat to big multiple dwellings, hospitals, hostels and elderly people's homes or to the district heating systems of large housing estates. They are designed for a solar fraction of the domestic hot water supply of 80–100% in July and August, about 40–50% of the

Figure 13.1. South view of part of a housing estate with a solar-assisted district heating system in Hamburg, Germany. The collectors are mounted on the roofs (photo: *Aufwind, Hamburg*)

annual heat demand for domestic hot water (these figures apply to central and northern Europe). These solar systems are called *central solar heating plants with diurnal storage* (CSHPDS).

Central solar heating plants with seasonal storage (CSHPSS) aim at a solar fraction of 50% or more of the total heat demand for room heating and domestic hot water for a large housing estate (more than 100 apartments). The seasonal time shift between irradiance and heat demand is compensated by means of the seasonal heat storage. An example for a monthly energy balance is given in Figure 13.2 for the CSHPSS system in Friedrichshafen in Germany.

Most large-scale solar systems have been built in central and northern Europe, mainly in Sweden, Denmark, The Netherlands, Germany and Austria. Only a few systems exist in southern Europe; for example, in Greece there is only one system with seasonal heat storage (Lykovrissi in Athens).

Design guidelines for solar-assisted district heating systems in central and northern Europe are shown in Table 13.1. For comparison, guidelines for a small solar system for domestic hot water preparation are also given. The solar heat cost represents the investment required to save 1 KWh of end energy and is calculated with an amortization according to VDI 2067.[1]

It is vital for the optimum functioning of the solar system that it should be correctly integrated into the conventional heating system and that design quality should be as high as possible, both of the solar part and of all other components for heat supply: the district heating network, the heat-transfer substations and the HVAC systems.

Table 13.1. Design guidelines for solar-assisted district heating systems in central and northern Europe (FC = flat-plate collector)

System type	Small solar system for domestic hot water (for comparison)	Central solar heating plant with diurnal storage	Central solar heating plant with seasonal storage
Minimum system size	–	More than 40 apartments or more than 100 persons	More than 100 apartments (each 70 m²)
Collector area	1–1.5 m_{FC}^2 per person	0.8–1.2 m_{FC}^2 per person	1.5–2.5 m_{FC}^2 per MWh annual heat demand
Storage volume	50–80 litres/m_{FC}^2	50–60 litres/m_{FC}^2	1.5–2.5 m^3/m_{FC}^2
Solar net energy	350–380 kWh/m_{FC}^2 per annum	400–430 kWh/m_{FC}^2 per annum	280–320 kWh/m_{FC}^2 per annum
Solar fraction (new building code)			
domestic hot water:	50%	50%	–
total heat demand:	15%	20%	40–50%
Solar heat cost in Germany	€0.2–0.4/kWh	€0.08–0.15/kWh	€0.17–0.25/kWh

SOLAR TECHNOLOGY

Collectors

In Scandinavia the most economical placing of collectors for systems with very large collector fields is directly on the ground next to the central heating unit. Because of the high cost for building sites, this method usually cannot be realized in central Europe and collectors have to be installed on the roofs of buildings.

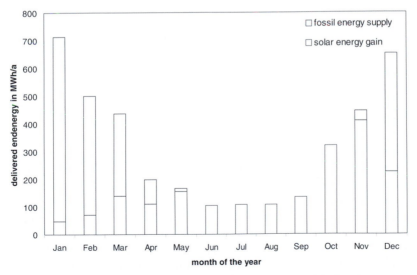

Figure 13.2. Energy balance of the CSHPSS system in Friedrichshafen, Germany (TRNSYS simulation)

There are three systems available:

- *on-site installation* with the system assembled directly on the roof;
- *large collector modules* industrially prefabricated with an area of 8–12 m²;
- *solar roofs* that make up complete roof modules.

With on-site installation the collector field can easily be adapted to the shape of the roof area. This approach is now seldom used because the assembly is dependent on the weather, while large collector modules can be mounted under almost any weather conditions. Today a number of collector manufacturers offer reliable sealing systems so that the collector system replaces the roof tiles and the collectors are mounted directly onto a subroof. In this case, a collector field of 100 m² or more costs between €200 and €300 per square metre of flat-plate collector area, including planning, assembly and piping on the roof (excluding VAT). If the collector field has to be mounted on a substructure, there will be additional construction costs of at least €50/m². The so-called solar roofs that have been developed in Germany and Sweden are complete roof modules, including rafters and heat insulation, which support a collector instead of the common roof tiles. Compared to the common roof, pilot systems have additional costs of about €150–250 per square metre of flat-plate collector area, completely mounted and with pipes connected (excluding VAT).

As a general rule, the whole collector area of a system should be installed in coherent fields that are as large as possible. The whole collector system is operated by a small, area-dependent flow of 10–15 litres/(m²h). This so-called 'low-flow' technology reduces the necessary pipe diameters, the amount of the heat-transfer fluid and the electrical energy required for the pumps. The collectors themselves are operated with high flow in order to obtain a high heat-transfer coefficient. For hydraulic reasons, as many collectors as possible are connected in series, with collector rows in parallel to each other. This requires collectors with absorbers connected in parallel and with a small pressure loss. In order to obtain a high system efficiency, stratified charging of the heat storage is necessary.

Seasonal heat storage

Various approaches can be used for seasonal heat storage (Figure 13.3):

- *Hot-water heat storage* is built as a steel tank or as a water-filled reinforced concrete tank, as a rule partially built into the ground. The roof area and the vertical walls of the storage are insulated against heat loss. The most recent pilot storage systems are in operation in Hamburg and Friedrichshafen (Germany) and have a watertight lining made of 1.2 mm stainless-steel sheets.
- In a *duct heat storage* system the heat is stored directly in the water-saturated soil. U-pipes, which are heat insulated near the surface, are inserted into vertical boreholes. By means of these U-pipes, heat is fed into and out of the ground. A pilot storage system for high temperatures up to 85°C is being built in Neckarsulm (Germany). The water-equivalent storage volume is 4–5 m³, meaning that 4–5 m³ of duct storage can store the same amount of energy as 1 m³ of water.

- In *gravel/water heat storage* a pit lined with water-tight plastic foil is filled with a gravel–water mixture as the storage material. The storage is insulated against heat loss at the sides and on top and – for small storage volumes – also at the bottom. Heat is fed into and out of the storage directly or indirectly. Storage of this kind has been in operation at the ITW of Stuttgart University since 1985 and in Steinfurt-Borghorst (Germany) since 1999. The water-equivalent storage volume is $1.3–2$ m^3.

- In *aquifer heat storage* systems, naturally occurring self-contained layers of ground water are used for heat storage. Heat is fed into the storage via wells and fed out by reversing the flow direction. The water-equivalent storage volume is $2–3$ m^3. A pilot storage system is being built in Rostock (Germany).

The decision to use a certain type of storage system mainly depends on the local conditions and, above all, on the geological and hydrogeological situation in the ground below the respective construction site.

For all types of storage, but mainly for aquifer and duct heat storage systems, a preliminary geological examination of the storage site is absolutely necessary. The approval of the water authorities must be obtained at an early stage. If different storage types are possible, an economic optimization should be carried out with the building costs for the different types of storage taken into account. This is because the storage types will be of different sizes according to the water-equivalent volume. So far, gravel/water heat storage systems have proved to be the least expensive.

ECONOMICS

In recent years, comprehensive energy concepts for the realization of economical energy-saving measures concerning the energy supply for residential areas have been developed. They result in a maximum reduction of fossil-fuel consumption at the lowest possible additional costs.

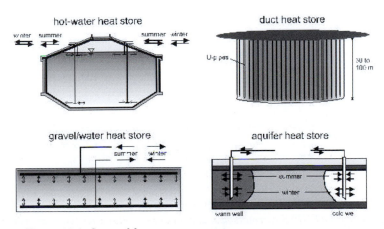

Figure 13.3. Seasonal heat storage systems

A comprehensive energy concept combines, as effectively as possible, insulation measures, rational energy conversion and supply, and active use of solar energy. Compared with the present levels of use, it can result in energy savings of more than 60%.[2]

An important part of these energy supply concepts is the use of solar thermal energy in large plants with and without seasonal heat storage. In Figure 13.4, both the cost–benefit ratio of large-scale solar systems – the solar heat cost as described above – and the additional investment cost as a function of living area are compared to those of small systems.

Because Scandinavian solar market prices are a little lower than German ones, Scandinavian CSHPDS systems can achieve solar heat costs of €0.06/kWh per annum.

Large-scale solar systems with diurnal heat storage (CSHPDS) come out best; the solar heat cost amounts to only €0.08 to €0.15/kWh per annum. Compared to small systems with solar heat costs of €0.2 to €0.4kWh per annum, CSHPDS systems are 2.5 times more cost-efficient. With additional investment costs of about 150–280 DM (€80–145) per square metre of living area compared to the normal building costs of residential buildings, a CSHPSS system with a solar fraction of more than 50% can be established.

As well as the question of finance, good project organization is the key to realizing this comprehensive energy concept. Before drawing up the development plan, local authorities, clients and planners should establish their common objectives. As a result, demands on buildings (orientation, construction, insulation standard) and heat-supply technologies can be integrated into the development plan and into subsequent stages, e.g. the purchase contract for the sites.

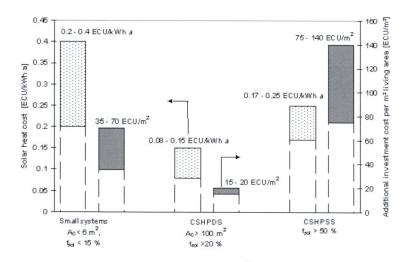

Figure 13.4. Solar heat cost (cost–benefit ratio) and the additional investment cost of CSHPDS and CSHPSS systems compared to those of small systems (German market prices, excluding VAT; A_c indicates the collector area, f_{sol} the solar fraction; ECU = €)

EUROPEAN OVERVIEW

In Europe, about 7 million m² of solar collectors are installed, corresponding to about 3500 MW of thermal power. Of this, 30 MW of thermal power is installed in large systems with collector areas of 500 m² or more. The ten largest plants are listed in Table 13.2. If not otherwise stated, the systems are built in housing estates.

In addition, there are 14 systems with individual collector areas of about 1000 to 1250 m² and 17 systems with collector areas between 500 and 1000 m². The interest in large-scale solar heating has increased during recent years, especially in Germany and Austria, where 15 new plants with collector areas over 500 m² have been put into operation since the beginning of 1997. For further information, see Dalenbäck.[3]

EXAMPLES OF GERMAN PROJECTS

As part of the research project 'Solarthermie 2000 – part 3: Solar assisted district heating' of the BMBF (Federal Ministry for Education, Science, Research and Technology), the first projects for solar-assisted district heating were realized in Germany. Scientific development and coordination is carried out by the Institut für Thermodynamik und Wärmetechnik (ITW) at the University of Stuttgart, Germany. In the first projects

Table 13.2. The ten largest European large-scale solar heating systems[3]

Plant, country	Year of initial operation	Collector area (m²)	Collector deliverer, country	System type with storage	Load size (GWh per annum)
Marstal, DK	1996	8040	Arcon, DK	CSHPDS, 2000 m³ water tank	28
Nykvarn, SE	1985	7500	TeknoTerm, SE; Arcon, DK	CSHPDS, 1500 m³ water tank	30
Falkenberg, SE	1989	5500	TeknoTerm, SE; Arcon, DK	1100 m³ water tank	30
Lyckebo, SE	1983	4320	TeknoTerm, SE	CSHPSS, 100,000 m³ water-filled rock cavern	8
Ry, DK	1988	3025	Arcon, DK	Directly connected to district heating	32
Friedrichshafen, D	1996	2700 (of 5200)	Arcon, DK; Paradigma, D	CSHPSS, 12,000 m³ water-filled concrete tank	2.2 (of 4)
Neckarsulm, D	1998	2700	Arcon, DK; Sonnenkraft, AU; SET, D; Paradigma, D	20,000 m³ duct storage	1.7
Hamburg, D	1996	2525	Wagner, D	CSHPSS, 4500 m³ water-filled concrete tank	1.6
Groningen, NL	1985	2400	Philips, NL	CSHPSS, 24,000 m³ duct storage	Unknown
Breda, NL	1997	2400	ZEN, NL	CSHPDS, 95 m³ water tank	1.2 (in industry)

(Ravensburg and Neckarsulm), the roof integration of large collector areas, as well as the system technology and safety devices for solar-assisted district heating systems, was tested extensively and improved as a result of the experience gained. In the following, some projects are described in detail to give ideas about possible system design and applications of active solar heating systems in urban areas.

CSHPDS systems

The central heating unit of the solar-assisted district heating system Ravensburg I (Figure 13.5) supplies 29 terraced houses with hot water and space heating by means of a four-pipe district heating network. The solar system supports the central domestic hot water heating with a solar fraction of 45%.

The solar system has been in operation since 1992. On the basis of measurements, the solar energy output is calculated to be 443 kWh/m² per annum. The investment cost for the complete solar system, including planning, amounted to €471 per square metre of collector area (excluding VAT). Figure 13.6 shows a view of the 115 m² collector area on the garages of the terraced houses.

If the buffer storage is also connected to the room heating network, then the stored solar energy can be used for room heating as well as for heating the hot water.

System design differs from case to case, especially in the way that the solar system is connected to the conventional heating system and in the design of the heat-transfer substations.

In the solar-assisted district heating system Neckarsulm I, solar energy is kept in the short-term buffer storage, which is directly connected to a two-pipe district heating network that supplies a whole residential area with heat for room heating and domestic hot water. The central heating unit, shown schematically in Figure 13.7, is situated in the cellar of the building that carries the collector field.

Up to now, only the first construction phase of the collector system with 360 m² of roof integrated collectors has been completed (Figure 13.8). The building cost of the solar system including planning, but without storage, came to €300 per square metre of collector area (excluding VAT).

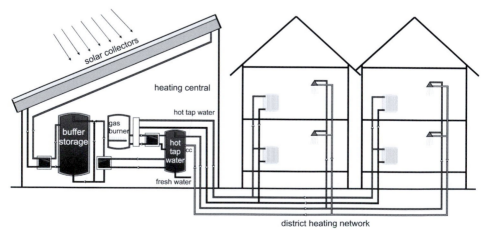

Figure 13.5. Schematic drawing of the CSHPDS system Ravensburg I, Germany

Figure 13.6. The collector field of the CSHPDS system Ravensburg I, Germany (on-site installation)

In its final stage, the solar system with its 700 m² collector area should satisfy 11% of the total heat demand of the housing estate. The simulated solar energy output amount to 508 kWh/m² per annum.

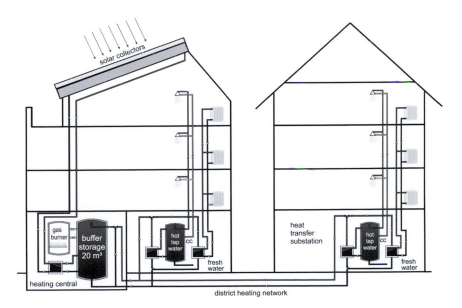

Figure 13.7. Schematic drawing of the central solar heating system with short-term heat storage Neckarsulm I, Germany

Figure 13.8. The collector field of the solar-assisted district heating system Neckarsulm I, Germany (large collector modules)

CSHPSS systems

The main technical data for current German CSHPSS projects is presented in Table 13.3.

Figure 13.9 shows a schematic drawing of the system in Friedrichshafen.[4] The heat obtained from the collectors on the roofs of the multi-family buildings is transported to the central heating unit via the solar network and is directly distributed to the buildings when required. The surplus heat in the summer period is fed into the seasonal

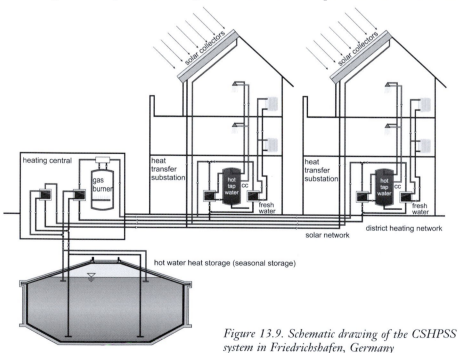

Figure 13.9. Schematic drawing of the CSHPSS system in Friedrichshafen, Germany

Table 13.3: Main technical data for current German CSHPSS systems

	Units	Hamburg	Friedrichshafen	Neckarsulm, Phase I	Chemnitz (partially built)
Housing area		124 terraced single-family houses	Eight multi-family houses with 570 apartments	Six multi-family-houses, school, commercial centre	One office building, one hotel and one warehouse
Heated living area	m²	14,800	39,500	20,000	not available
Total heat demand	MWh per annum	1610	4106	1663	1236
Solar collector area	m²	3000	5600	2700	2000
Heat storage volume	m³	4500 (hot water)	12,000 (hot water)	20,000 (duct)	8000 (gravel/water)
Heat delivery of the solar system	MWh per annum	789	1915	832	519
Solar fraction	%	49	47	50	42
Solar contribution per collector area	kWh/m² per annum	268	342	308	260
Total cost of the heating system	Million €	3.1	3.9	2.3	not available
Cost of the solar system (excluding subsidies)	Million €	2.2	3.2	1.5	1.2
Total heat cost (excluding VAT)	€/MWh	202	105	158	not available
Solar heat cost (excluding VAT and subsidies)	€/MWh	251	156	169	235

Figure 13.10. Views of the construction of the seasonal heat storage in Friedrichshafen: left: forming the concrete vertical walls; right: stainless-steel liner on the upper inside surface

heat storage to be used for space heating and domestic hot water supply in autumn and winter. Figure 13.10 gives pictures of the construction phase of the Friedrichshafen hot-water heat storage, made of reinforced concrete.

In Neckarsulm, a new heat collection- and distribution system has been built for the first time. Instead of two separate solar and heat distribution networks, a so-called three-pipe network has been realized. The collector arrays on each roof are connected by means of a heat exchanger to a single pipe, which is used both as heat network return and solar

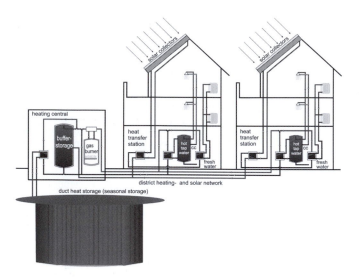

Figure 13.11. Schematic drawing of the CSHPSS system in Neckarsulm, Germany

*Figure 13.12. The solar roof on the primary school of the
CSHPSS system in Neckarsulm, Germany*

supply. In this way, a solar return-flow pipe can be saved and there is only water flowing in the network. In this case, antifreeze is obligatory only in the individual collector circuits, rather than in the whole network (Figure 13.11). Figure 13.12 shows a view of the primary school, having a so-called 'solar roof', within the CSHPSS system.

FUTURE PROSPECTS

In the upcoming years, further large-scale systems with seasonal heat storage will be built and not only in Germany. New concepts for seasonal heat storage systems will be applied and the storage technology, tested within the existing pilot plants, will be developed to reduce the building costs of storage. Efforts at cost reduction related to the collectors are also being made.[5]

Solar renovation of existing district heating systems will become more and more important as a way of reducing fossil energy demand and CO_2 emissions in existing urban areas.

For solar-assisted district heating systems with seasonal heat storage, the mid-term aim for solar heat cost is, at maximum, twice as high as the present conventional heat cost. Already, systems with short-term or diurnal heat storage can provide the best cost–benefit ratio of all solar thermal applications for heat supply. With the expected cost reduction in the collector industry, these systems should reach an economic level in the next few years.

REFERENCES

1. VDI 2067: *Berechnung der Kosten von Wärmeversorgungsanlagen* (1983). VDI-Verein Deutscher Ingenieure, Düsseldorf. (Calculation of the costs of heat generation systems, in German.)
2. Mangold, D., Schmidt, T., Wachholz, K. and Hahne, E. (1997). 'Solar Meets Business: Comprehensive Energy Concepts for Rational Energy Use in the Most Cost Effective Way', *Proceeding of the ISES Solar World Congress, Taejon, Korea*. Korea Institute of Energy Research.
3. Dalenbäck, J.-O. (1998). *European Large Scale Solar Heating Network*, hosted at Institutionen för Installationsteknik, Chalmers Tekniska Högskola, Göteborg, Sweden. URL: http://www.hvac.chalmers.se/cshp/.
4. Mahler, B., Schulz, M. and Hahne, E. (1997). 'Central Solar Heating Plants with Seasonal Storage in Hamburg-Bramfeld and Friedrichshafen-Wiggenhausen (Germany) – First Results of Monitoring Program'. *Proceedings of North Sun 1997*, Espoo-Otaniemi, Finland, Vol. 2, pp. 675–682.
5. Hahne, E. *et al.* (1998). *Solare Nahwärme – ein Leitfaden für die Praxis, BINE-Informationspaket*. TÜV Verlag Rheinland. (Solar assisted district heating – a manual for practice, in German.)

14

Indoor air quality and ventilation in the urban environment

P. Michel
Laboratoire des Sciences de l'Habitat, Ecole Nationale des Travaux Publics de l'Etat

ဢ

Ventilation has a key role in providing an optimum indoor climate in buildings. Indoor air quality (IAQ) is directly determined by the performance of the ventilation system, while air movements inside the building may be of particular importance for thermal comfort, especially in summer conditions. Moreover, ventilation may strongly affect the energy efficiency of the building.

The specific characteristics of the urban environment, in terms of temperature, air quality or wind speed, have thus to be taken into account to ensure a good performance of the ventilation system.

INDOOR AIR QUALITY

Sick building syndrome and building-related illness

The term 'Sick Building Syndrome' (SBS) generally refers to situations in which building occupants experience acute health and/or discomfort effects that are apparently linked to time spent in a building, while at the same time no specific illness or cause of these effects can be identified. The complaints may be localized in a particular room or zone, or may be widespread throughout the building. The occurrence of SBS specifically corresponds to the following signs:

- Building occupants complain of symptoms such as lethargy, headaches, lack of concentration, runny nose, dry throat, eye and skin irritation.
- The causes of these symptoms are not clearly identified.
- Symptoms frequently disappear soon after leaving the building.

The term 'Building-Related Illness' (BRI) is used when symptoms of a diagnosable illness are identified and can be attributed directly to airborne building contaminants.

A 1984 World Health Organization Committee report suggested that up to 30% of new and remodelled buildings worldwide may be the subject of excessive complaints related to indoor air quality (IAQ).[1]

Frequently, problems result when a building is operated, managed or maintained in a way that is not in accordance with its original design or prescribed operating procedures. Sometimes, IAQ problems are a result of poor building design or occupant activities. BRI is said to occur in the presence of the following aspects:

- Building occupants complain of symptoms such as cough, chest tightness, fever, chills and muscle aches.
- The symptoms can be clinically defined and have clearly identifiable causes.
- Complainants may require prolonged recovery times after leaving the building.

It is important to note that complaints may of course result from other causes. These may include an illness contracted outside the building, acute sensitivity (e.g. allergies), job-related stress or dissatisfaction and other psychosocial factors. Nevertheless, studies show that symptoms may be caused or exacerbated by IAQ problems.

Many parameters have been investigated in order to understand the causes of sick buildings, focused in four major areas:

- *Ventilation systems*: ventilation rates, air distribution, filtration, maintenance of the duct system, etc.
- *Building contaminants*: asbestos, carbon dioxide, dust, humidity, odours, smoke, etc.
- *Occupants*: age, gender, state of health, occupation.
- *Miscellaneous*: lighting, noise, stress, electromagnetic radiation, psychological factors, etc.

Research and observations have shown that ventilation thus represents only one aspect of a very complex problem. However, it seems that ventilation has a more significant role in these problems when ventilation rates are very low. Airtight buildings result from energy conservation measures that lead to a reduction of ventilation rates. As a consequence, the World Health Organization reported in 1986 that 'energy-efficient but sick buildings often cost society far more than it gains by energy savings', by increasing absenteeism rates and decreasing employee productivity.[2]

INDOOR POLLUTANTS AND POLLUTANT SOURCES

Indoor pollutants result from both outdoor and indoor sources. Indoor pollutants may either have a natural origin or result from human activities. Moreover, air pollutant sources may differ from one building to another, according to the building design and purpose.

Outdoor sources

Outdoor sources are of particular importance in urban areas and result in a poor quality of the air entering the building, air cleaning being inefficient or ineffective in buildings that are naturally ventilated or are ventilated by mechanical extraction systems. Major outdoor sources are the following:

- industrial emissions;
- traffic pollution;

- nearby sources;
- soil-borne pollutants.

Industrial emissions

Local as well as distant industrial installations may be responsible for high concentrations of:

- oxides of nitrogen and sulphur;
- ozone;
- lead;
- volatile organic compounds (VOCs);
- smoke, particles and fibres (asbestos, dust, etc.).

The effects of these pollutants can be enhanced by specific climate conditions. This is especially the case in urban areas, with the conjunction of, for example, the heat-island effect and the air-flow distribution around buildings.

Traffic pollution

Especially in urban areas, traffic emissions can explain a major part of outdoor air pollution close to streets, highways, tunnel entrances and parking areas, the pollutants being:

- carbon monoxide and dioxide;
- carbon dust;
- lead;
- oxides of nitrogen.

Nearby sources

Combustion emissions from nearby stacks and contaminated emissions from cooling towers can cause problems when located close to air sources.

Soil-borne pollutants

Soil-borne sources of pollutants can penetrate buildings through the foundations and may be enhanced by extract ventilation. Soil-born pollutants include:

- radon (a naturally occurring radioactive gas);
- methane (the product of organic decay);
- moisture.

Indoor pollutants

Indoor pollutants can arise from three main sources:

- human and animal metabolism;

Table 14.1. CO_2 production rate for various activities[2]

Activity	Metabolic rate [W]	CO_2 production rate [litre/s]
Sedentary work	100	0.004
Light work	150–300	0.006–0.012
Moderate work	300–500	0.012–0.020
Heavy work	500–650	0.020–0.026
Very heavy work	650–800	0.026–0.032

- occupant activities;
- building materials and equipment.

Major indoor air pollutants are the following:

- *Carbon dioxide.* CO_2 is a product of metabolism and is also given off by carbon-containing materials. It can affect breathing and cause nausea at high concentrations. In Table 14.1 are indicated the metabolic rate and the CO_2 production rate for different activities.
- *Carbon monoxide.* CO is a highly toxic, undetectable – odourless, colourless and tasteless – gas, resulting from incomplete combustion. CO sources include poorly vented heating appliances, motor vehicle emissions and tobacco smoke.
- *Environmental tobacco smoke (ETS).* ETS is the mixture of different kinds of smoke that come from the burning end of a cigarette, pipe or cigar, and smoke exhaled by the smoker. It is a complex mixture of several thousands of compounds, more than 40 of which are known to cause cancer and many of which are strong irritants.
- *Formaldehyde.* This colourless pungent-smelling gas can cause burning sensations in eyes and throat, nausea and difficulty in breathing. Major sources are building materials, smoking, household products and the use of unvented fuel-burning appliances.

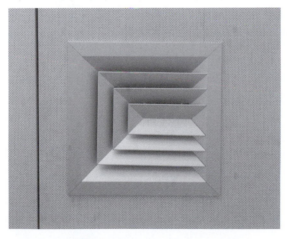

Figure 14.1. Ventilation exhausts in the showers at a swimming pool[4]

- *Moisture*. This is principally generated by occupant activities (cooking, washing, drying of washing in the residential sector). Condensation of vapour can cause considerable damage to a building through mould growth and fabric decay. Figure 14.1 shows one possible source.
- *Odour*. Generated from metabolism or emitted from furnishings or fabrics, odours cause discomfort and often justify ventilation.
- *Ozone*. Ozone is produced by office appliances (e.g. copiers and laser printers) and it may cause respiratory difficulties. Ozone can also be produced in the outdoor air by industrial activities such as thermal power plants
- *Particulates*. Dust, organic fragments, fibres and smoke particles can have varying levels of toxicity according to their type and size. Long-term exposure to these can cause bronchitis, emphysema and lung cancer.
- *Volatile organic compounds (VOCs)*. WHO defines VOCs as organic compounds with boiling points in the range 50–260°C. VOCs are emitted from furnishing fabrics and household chemical products. VOCs have a typical odour and some of them are considered as toxic.

INTERNATIONAL STANDARDS FOR INDOOR AIR QUALITY

Some years ago, the International Energy Agency produced a review of international standards for Indoor Air Quality, including the World Health Organization standards. Three different concentration levels are listed in this review:

- *MAC*: Maximum Allowable Concentration at the work space for an eight-hour period (occupational health criteria);
- *ME value*: Maximum Environmental value;
- *AIC*: Acceptable Indoor Concentration – 'For concentrations below the AIC, the negative health effects are either negligible or, if no threshold is known, are at least tolerable.'

Tables 14.2 to 14.6 present these different standard concentration limits for major indoor air pollutants.

Table 14.2. International standards for CO_2 concentration levels

Level	MAC (ppm)	Peak limit (ppm)	ME value (ppm)	AIC (ppm)	Remarks
Country					
Canada	5000			1000–3500	
Germany	5000	2 × MAC		1000–1500	
Finland	5000	5000		2500	
Italy				1500	
The Netherlands	5000	15,000		1000–1500	
Norway	5000	MAC + 25%			
Sweden	5000	10,000			
Switzerland	5000			1000–1500	
UK	5000	15,000			
USA	5000			1000	
Columbus Space Station				4000	

Table 14.3. CO concentration levels for various areas

Zone	Concentration range (ppm)
Natural base level	0.044–0.087
Rural areas	0.175–0.435
Industrial areas	0.87–1.75
Downtown levels	up to 40

Table 14.4. International standards for CO concentration levels

Level	MAC (ppm)	Peak limit (ppm)	ME value (ppm)	AIC (ppm)	Remarks
Country					
Canada	50	400		9	
Germany	30	2 × MAC	8–43	1–18	Depending on the duration and the type of room
Finland	30	75		8.7–26	Depending on the duration
Italy	30				
The Netherlands	25	120		8.7–35	Depending on the duration
Norway	35	+50%			
Sweden	35	100		12	
Switzerland	30		7		
UK	50	400			
USA	50	400		9	
WHO		9–87			Depending on the duration
Aeroplanes	50				

Table 14.5. International standards for NO_2 concentration levels

Level	MAC (ppm)	Peak limit (ppm)	ME value (ppm)	AIC (ppm)	Remarks
Country					
Canada	3	5		0.3	Offices
				0.052	Homes
Germany	5	2 × MAC	0.05–0.1		
Finland	3	6		0.08	Daily average
				0.16	Hourly average
The Netherlands	2			0.08–0.16	
Sweden	2	5		0.15–0.2	
Switzerland	3		0.015–0.04		
UK	3	5			
USA	3	5		0.3	
WHO			0.16	0.08–0.21	

Table 14.6. International standards for HCOH concentration levels

Level	MAC (ppm)	Peak limit (ppm)	ME value (ppm)	AIC (ppm)	Remarks
Country					
Canada	1	2		0.1	
Germany	1	2 × MAC		0.1	
Finland		1		0.12	New buildings
				0.24	Existing buildings
The Netherlands	1	2		0.1	
Norway	1	+100%			
Sweden	0.5	1		0.01–0.1	
Switzerland	1			0.2	
UK	2	2			
USA	1	2		0.1	
WHO				0.08–0.1	Depending on effects considered

REDUCING INDOOR POLLUTANT CONCENTRATION

Controlling outdoor air pollutants

Filtration: Mechanical supply ventilation systems can incorporate filters to prevent pollutants from the outside entering the ventilation system. Dust filters are frequently used in ventilation systems, while special applications (hospitals, clean rooms, etc.) may require high-specification filters able to remove gaseous pollutants or particles.

Siting air intakes: Buildings in an urban environment may be located near important pollutant sources (streets with high traffic density, car parks, exhaust stacks, cooling towers, etc.). Building design has to take these local sources into account in order not to introduce high concentrations of pollutants from the outside. Air intakes must thus be located away from these specific pollutant sources.

Fresh air dampers controlled by air quality: Temporarily closing fresh-air dampers during traffic pollution peaks can improve air quality at these times. This is especially the case during the commuting periods.

Building airtightness: A building has to be effectively sealed from the outdoor environment in order to prevent contamination through air infiltration.

Controlling indoor air pollutants

Source control: Source control aims at eliminating avoidable pollutants by restricting pollutant emissions (VOCs and formaldehyde from furnishings, tobacco smoke, etc.).

Ventilation at source: Highly localized sources, such as water vapour generated by washing, drying and cooking, or specific pollutants emitted at the workplace should be directly vented to the outside in order to prevent pollutant action. Local mechanical extractors can be used for this, to prevent these pollutants from diffusing through the whole building.

Figure 14.2. Duct network of a ventilation system in a swimming pool

Dilution:
General ventilation is needed to dilute and remove residual pollution from unavoidable contaminant sources. The minimum need for ventilation is the necessity to contain metabolic pollution at acceptable levels. The ventilation system has to be designed to control the dominant pollutant, e.g. water vapour or CO_2, depending on the use and occupation of the building. Figure 14.2 shows an example.

VENTILATION STRATEGIES

Ventilation appears to be the process to improve the indoor air quality of a space through the control and/or the evaluation of pollutants. Thus ventilation should do the following:

- prevent disorders from occurring in the building envelope and building equipment;
- ensure good indoor conditions for the health and comfort of the occupants.

Two types of ventilation can be defined:

- Natural ventilation, which includes the movement of outdoor air through intentional openings (doors, windows) and through unintentional openings in a building's shell. These result in infiltration and 'exfiltration'.
- Mechanical ventilation, which is intentional ventilation supplied by electric fans, usually included in a building's HVAC system.

NATURAL VENTILATION

Many buildings in the world, especially old buildings, are naturally ventilated, using wind pressure as well as stack pressure (Figure 14.3).

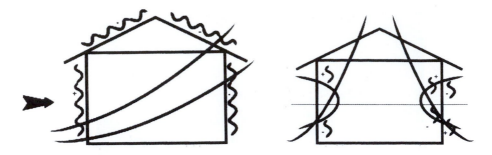

Figure 14.3. Natural driving mechanisms: (left) wind-driven flow; (right) stack driven flow

Driving forces

Wind pressure: Wind that blows against a rectangular-shaped building produces a positive pressure on the windward face and a negative pressure on the leeward face, so that the air enters openings and passes through the building from the windward side to the leeward side (Figure 14.4).

Stack pressure: The temperature differences between the interior and exterior of a building induce air-density differences that cause pressure differences. When the indoor temperature is higher than the outdoor temperature, indoor air goes out of the building at its top while outdoor (cooler) air enters through openings near the basement.

The directions are reversed when the temperature balance changes. These two forces have to be well combined to ensure that they interact and are not in opposition. This can be achieved through the building architecture and the design of openings to improve airflow in natural ventilation configurations.

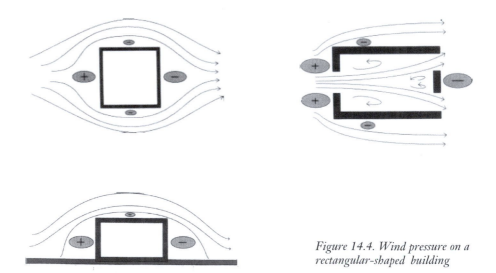

Figure 14.4. Wind pressure on a rectangular-shaped building

Building architecture and openings

Different elements have to be focused to ensure good natural ventilation performance.

Building airtightness: The building envelope must be as airtight as possible so that air flows specifically through designed openings.

Air-quality reservoir: The volume of the enclosed space is a key element of natural ventilation design. Under certain circumstances a building can play the role of an air-quality reservoir, compensating for the variable nature of the natural ventilation and ensuring outdoor air quality without a constant ventilation rate.

Openings: Naturally ventilated building design has to combine permanent openings, for background ventilation, with controllable openings to adjust ventilation to users' needs. Natural ventilation components include:

- *Openable windows*. These are the principal component of natural ventilation. However, they often cause discomfort and energy losses during the heating season, because they are often uncontrollable and used to decrease indoor temperature in the case of overheated spaces.
- *Air vents*. These replace window openings in winter conditions. They offer a permanent moderate area of opening and provide limited uncontrolled ventilation.
- *Automatic inlets*. Air inlets can be designed to be sensitive to air quality or climate parameters, such as temperature, humidity and pressure. The area of opening of such automatic inlets is automatically (mechanically) adjusted through the variation of a design parameter.

Natural-ventilation strategies

Cross-flow ventilation: A clearly defined airflow path is established from the incoming to the outgoing air stream, through the zone of occupancy. The interior space has to be as open as possible and the distance between the openings has to be maintained below a practical limit, which depends on the ceiling height.

Single-sided ventilation: True single-sided ventilation (small openings along one side of a room) is almost uncontrollable because it is driven by turbulent fluctuations. A better situation is provided by multiple openings above one another or by a large opening, so that air can enter and be driven out by wind and stack forces. This is illustrated in Figure 14.5.

Passive stack and atrium ventilation: The airflow is driven through a stack by a combination of stack pressure and wind pressure. A separate stack is needed for each room, in order to avoid cross-contamination. This ventilation principle can be applied to atria. The atrium is used as a passive stack itself. Top openings have to be automatically controlled

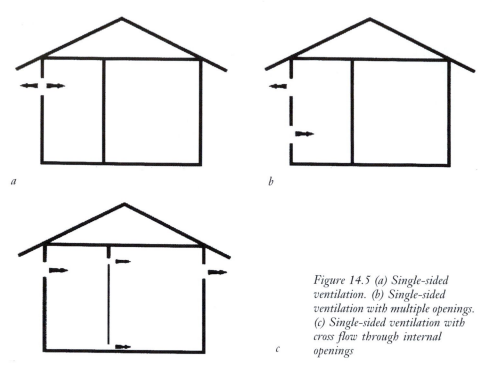

a

b

c

Figure 14.5 (a) Single-sided ventilation. (b) Single-sided ventilation with multiple openings. (c) Single-sided ventilation with cross flow through internal openings

and building joints must be well sealed to prevent uncontrolled airflows. Examples of stack effects are shown in Figures 14.6 and 14.7.

Control

Airflows in natural ventilation depend on climatic conditions as well as on the positions of openings. Only this latter parameter can be controlled and the following multiple constraints have to be taken into account:

- For security and safety reasons, especially in urban areas, openings sometimes have to be kept closed.

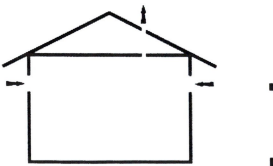

Figure 14.6. The stack effect in a dwelling

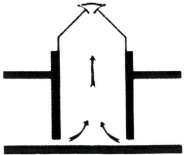

Figure 14.7. The stack effect in an atrium

- In a similar way, keeping the openings closed helps to keep the inside noise at an acceptable level.
- Internal door opening cannot be easily controlled.
- Shading devices may reduce or alter airflow rates.

Natural ventilation control, based on CO_2 concentration and/or indoor temperature, induces controllable automatic openings. The risk of discomfort, due to high-speed airflow, can be avoided by implementing specific openings of reduced area above doors and windows.

Advantages and disadvantages in an urban environment

The major advantages of a natural ventilation strategy are the following:

- Natural ventilation is especially suitable in a mild or moderate climate.
- It can be used during warm periods as a passive-cooling technique.
- It is usually inexpensive, compared to mechanical solutions.
- A minimum level of maintenance is required for controlled openings.
- High flow rates are possible when needed.

The major drawbacks are the following:
- Natural ventilation is unsuitable for noisy and/or polluted locations.
- There are possible security risks in urban areas.
- Indoor air-quality problems and excessive heat loss may occur for some (uncontrolled) ventilation rates.

It is noticeable that, because cleaning of the incoming air is not really practical in this case, natural ventilation can hardly be implemented in urban environments with bad outdoor air quality.

MECHANICAL VENTILATION

The purpose of mechanical ventilation is to provide controlled ventilation of a building, if possible not – or hardly – affected by outdoor climatic conditions (Figure 14.8). The required air movement is supplied by electric fans, linked to different system components. Polluted air has to be extracted close to pollutant sources (e.g. kitchen or bathroom). Air can be mechanically extracted or/and supplied.

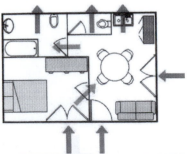

Figure 14.8. The ventilation principle for a dwelling

System components

Fans: Fans are used to provide the driving force for mechanical ventilation, operating by consuming electrical energy (Figure 14.9). They are classified as centrifugal or axial, depending on the direction of the airflow through the impellers.

Ducts: Ducting is used to transfer air (Figure 14.10). The duct resistance to airflow depends on the air-flow rate and the length, the cross-sectional area, the geometry and the surface roughness of the duct. Ducts should be insulated to prevent heat losses and

Figure 14.9. Ventilation fans

have to be well sealed to prevent loss of conditioned air. Moreover, a ducting system has to be periodically maintained (cleaned) in order to prevent any effects on health.

Air-diffusing equipment:

- *Diffusers*. These discharge mechanically supplied air into the ventilated space (Figure 14.11). The position and the design of diffusers are major elements in

Figure 14.10. Duct system

determining the thermal comfort inside the building. Specific applications may require low or very low air speeds.

- *Air intakes*. These are openings through which outdoor air is collected for ducting to a ventilation system (Figure 14.12). They have to be located away from pollutant sources.
- *Air grilles*. Air grilles are used to capture exhaust air from a space (Figure 14.13).

Figure 14.11. Diffusing elements

Figure 14.12. Air intakes

Figure 14.13. Air grilles

Mechanical ventilation configurations and strategies

Mechanical extract ventilation: A fan is used to mechanically remove air from a space. Fresh air enters through vents provided for this purpose (Figure 14.14). Air also enters extraction ducts from wet or polluted zones. The underpressure provided by the fan has to be slightly greater than the weather-induced pressure, so that the flow process is dominated by the mechanical system.

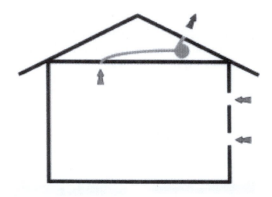

Figure 14.14. Mechanical extract ventilation system[3]

Extract systems include:

- *Local extraction*. In this case (usually for small buildings), pollutants (often moisture) are extracted near the source of production. Wall, window or cooker-hood fans are such local extract ion systems that vent the polluted air directly to outside.
- *Controlled ducted extraction*. Often used in single-family and apartment dwellings, such systems are operated by a central control connected through ducting to extraction grilles (Figure 14.15). Extraction is carried out in polluted or wet zones, while air inlets are placed in living rooms.

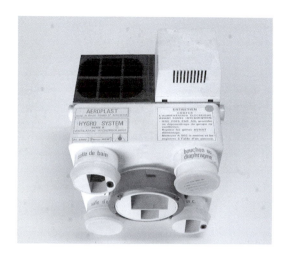

Figure 14.15. Mechanical extraction ventilation system

Mechanical supply ventilation: Outdoor air is mechanically introduced into the building. This process induces a positive pressure in the building. Indoor air is driven through purpose-provided and/or infiltration openings. Extraction ventilation reduces the risk of infiltration and thus all the incoming air can be precleaned and thermally conditioned.

This process has several important applications where a building has to be pressurized and/or the outdoor air is polluted. Examples are downtown buildings, where the outdoor air can be conditioned prior to distribution, industrial clean rooms and rooms for allergy control.

Mechanically balanced mixed ventilation: The process combines extraction and supply systems, using separated ducted networks (Figure 14.16). A heat exchanger is frequently implemented to transfer heat from the extraction air stream to the supply air stream (Figure 14.17). The building must be perfectly sealed for optimum performance and all duct networks have to be in conditioned spaces.

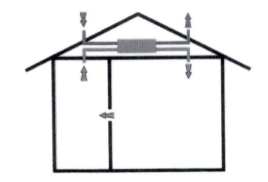

Figure 14.16. Mechanically balanced mixed ventilation system

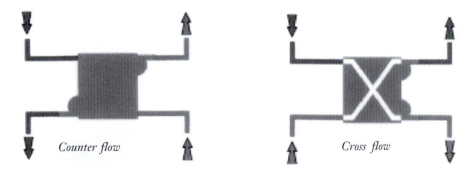

Counter flow   *Cross flow*

Figure 14.17. Heat exchangers

Mechanical balanced 'displacement' ventilation: In this case of balanced ventilation, the supply air displaces, rather than mixes with, the room air. Supply air is introduced at 2 to 3 K below the ambient room temperature and will thus displace the air already present within the space, which will be forced out via ceiling extraction (Figure 14.18).

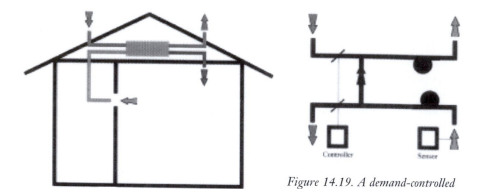

Figure 14.19. A demand-controlled ventilation system in an office building

Figure 14.18. A mechanical balanced 'displacement' ventilation system in a classroom

Demand-controlled ventilation (DCV): With DCV, the rate of ventilation is automatically controlled in response to variations of indoor air quality. DCV can be extremely effective where a dominant pollutant is identified, for example CO_2, moisture or particles. An example is shown in Figure 14.19.

A DCV system includes:

- a group of sensors to monitor the dominant pollutants;
- a control system or a building management system for adjusting the ventilation rate in response to need;
- a ventilation system.

Heat recovery

Various ventilation heat-recovery systems can be used to recover thermal energy from exhaust air for use within the building and/or to precondition incoming fresh air. Different systems can be used for ventilation heat recovery, depending on the target:

- air-to-air heat recovery;
- heat pumps;
- dynamic insulation;
- ground preheat recovery.

These systems may be applied to recover sensible and/or latent heat. The net benefits of these techniques have to take into account the thermal efficiency (lost heat usefully recovered) as well as several other factors:

- increased investment and running costs;
- air infiltration;
- auxiliary energy consumption.

Air-to-air heat recovery: Air-to-air heat-recovery techniques, implemented in mechanical balanced mixed ventilation systems, transfer heat from the exhaust 'hot' air to the supply 'cold' air. Various systems can thus be used to ensure the energy exchange, among them:

- *Plate heat exchangers*. Often used in dwellings, where ducts are close together, these are static materials made of layers of separated elements in which both supply and exhaust air flow, while exchanging heat. These products are very simple and avoid contamination, but they lead to extra energy consumption by fans and have to be by-passed in summer conditions.
- *Heat recovery systems from flue gases*. Principally used in dwellings with gas-fired central heating systems, this technique needs a plate air-to-air heat exchanger to transfer energy from the flue gases to the supply airflow of a balanced ventilation system.

Heat pumps: Air-to-air/liquid electrical heat pumps may be used for the heating or preheating of supply air, domestic hot water or heating fluid. The evaporator of the heat pump is placed in the exhaust airflow, while the condenser is located either in a reservoir tank or in the supply air stream.

Although the energy balance of the heat-recovery system may be positive, a careful energy and financial analysis has to be conducted, taking into account extra capital and operational costs.

Figures 14.3, 14.10 and 14.12 to 14.19 were provided by Pierre Michel.

REFERENCES

1. World Health Organization (1984). Air Quality Guidelines for Europe. WHO, Geneva.
2. Liddament, M. (1996). *A Guide to Energy Efficient Ventilation*. AIVC, Coventry.

FURTHER READING

ASHRAE (1996). *HVAC Systems and Equipment*. ASHRAE, Atlanta.
ASHRAE (1997). *Fundamentals*. ASHRAE, Atlanta.
INRS (1996). *Principes Généraux de Ventilation*. INRS, Paris.
Judet de la Combe, A. and Sesolis, B. (1991) *Aéraulique*. AICVF, Paris.
Judet de la Combe, A. and Sesolis, B. (1992). *Ventilation*. AICVF, Paris.
Lester, J.N., Penny, R. and Reynolds, G.L. (eds) (1992). *Quality of the Indoor Environment*. IAI, Publications Division, London.
Mansson, L. (1992). *Demand Controlled Ventilation Systems*. IEA Annex 18 Report. Swedish Council for Building Research, Stockholm.
Martin, A. (1995). *Control of Natural Ventilation*. BSRIA,
Pont, P. du and Morrill, J. (1989). *Residential Indoor Air Quality and Energy Efficiency*. ACEEE, Washington, DC.
Recknagel, H., Sprenger, E, Hönmann and Schramek, E.R. (1995). *Manuel Pratique de Génie Climatique*. Pyc Edition, Paris.
Stroebel, R., Berthelot, V. and Charré, B. (eds) (1995). *La Qualité de l'Air en France en 93–94*. Ademe, Paris.

15

Urban settlements

P. Michel

Laboratoire des Sciences de l'Habitat, Ecole Nationale des Travaux Publics de l'Etat

ℰↃ

The site and the location of a city are the two basic elements that are considered in urban geography and they differ in scale.

THE SITE AND THE LOCATION

The site

The site of a city can be analysed in terms of the major features of regional and national geography, as discussed in the following sections.

Figure 15.1. Lyons is located at the confluence of the Rhône and the Saône

Figure 15.2. The actual centre of Lyons is still located at the confluence of the Rhône and the Saône

Topographic features

Below some examples of major topographic features are described and it is noticeable that water plays a key role in the definition of these topographic features. Most middle-sized and large French cities are close to water, a river, a lake, the sea or a canal. The first phase in the development of cities around the world corresponded to 'river valley civilizations', for example the Nile in Egypt, the Yellow River in China, the Tigris and the Euphrates in Mesopotamia and the Ganges in India. Figure 15.1 shows the site of Lyons.

The following are important among the many topographic features:

- *Major river confluence*. Examples are Belgrade, Lyons, Coblenz, Namur and Turin. In each case the old city was often established in the area of the confluence but has been extended in directions that depend on the topographic neighbourhood, e.g. hills or mountains, marshes. Examples of cities in the neighbourhood of a major river confluence are Avignon, Grenoble, Bratislava and Warsaw.
- *Estuary or tidal river*. Examples of cities situated on estuaries or tidal rivers are Hamburg, Antwerp, Bordeaux, London, Shanghai and Seville.
- *Strait*. Cities have been established on one or both sides of a strait depending on either topographic or non-topographic features, such as a country border. Examples are Gibraltar, Istanbul and Copenhagen.
- *Lake*. Cities sited on lakes include Geneva, Zurich and many other alpine cities, as well as Stockholm.
- *Island*. Some cities, like Mexico City or Paris, were first established on an island in a river or in the sea, close to the shore, and have subsequently been extended.

*Figure 15.3. Fourvière — a hill above the Rhône-Saône confluence —
was the first location of Lyons*

Other features

Elements other than physical criteria also have to be taken into account in explaining the site of a city. A settlement may have been created for one or more of the following:

- political reasons (e.g. the border of a country);
- religious reasons;
- linguistic reasons;
- ethnic reasons;
- economic reasons (a connection between several economic areas).

An example is Amsterdam, which is at the centre of the area of which it is the capital.

The site can be considered as the basic environment within which the location of the city is defined.

The location

The location of a city is the place where the city was founded. This will be a key element in the structure of the city and it can also be a symbol for the city. A river, a mountain, an island or an estuary may provide a city with a strong image that can be preserved through the ages.

The original location was always defined on the basis of essential needs:

- *Protection and defence*. Many original locations are well protected by natural barriers, such as water (river, lake or sea) or mountains.

- *Microclimate*. A quiet location by the sea (with, for example, a natural harbour) or a sunny location protected from winds in the mountains may have been the original reason for the location.
- *Water* (in desert regions).
- *The economy*. The city had to be placed at a connection location where business could be developed. This was often inconsistent with protection needs (military as well as natural protection).

The original location may have been a compromise between or a combination of these different elements. As an example, a south-oriented slope close to a river confluence may constitute a good location in terms of military constraints, climatic advantages and economic interest.

The site and the location represent important elements in explaining the foundation of a city. However, other major factors have to be included in the analysis in order to understand the development of towns.

The impact of climate

Climate can be considered as one of the major elements in the settlement, morphology and dynamics of a city.

Architecture everywhere has been strongly influenced by local climatic conditions. Precipitation, availability of sunlight, temperature and wind have clearly influenced the design of houses and buildings. Aspects that have been affected include choice of the raw materials, opening areas, the slope of the roof and the mass of the building.

Urban morphology has also been defined by taking into account natural considerations. Matters affected include the width and orientation of the streets as a function of wind direction and urban design so that natural water flow is allowed. Urban design in hot climate is characterized by high-density housing in narrow streets, so that radiation received is minimized, while in colder climates solar radiation is more efficiently collected in open spaces.

URBAN MORPHOLOGIES

Urban system

Large cities often do not have a clear structure, but are the result of the work of several different designers. Although suburbs often have a quite different structure, three different structures can however be identified for modelling the centre of a city:

- *Concentric structure*. The city is the centre of a star-shaped network of streets. Successive belts can often be visible, some of them being 'green belts' or pheripheral highways. Paris, Milan and Brussels are representative of such a structure, which can also frequently be identified in cities dating from the Middle Ages.
- *Semi-concentric structure*. The concentric structure is altered when the city was located on a riverside, a seaside or at the edge of a forest. Vienna and Frankfurt

Figure 15.4. The orignal morphology of the old centre of Nice has been well preserved

are based on this plan. Amsterdam also presents such a structure, which is determined by semi-concentric canals.

- *Orthogonal structure.* This structure is the result of a clear decision by the urban designers. It is possible where no major natural obstacle impedes such a development. Two grids may sometimes coexist, linked by a major avenue. Turin and the new town of Barcelona show this structure, which is very common, especially in North America.

Other structures may sometimes be identified, with combinations or elements of those above. They always are based around several centres. They can also be the result of successive urban retrofitting operations.

Urban functions

As well as the original requirements and considerations, urban structure and urban landscape are closely linked to major urban functions, which can be summarized in four main categories:

- the industrial function;
- the trade function;
- the business function;
- the administrative function.

In addition, there are functions that are specific to certain cities, such as harbours, mining, railways and tourism.

Figure 15.5. The design of the Lyons suburbs has been strongly influenced by the development of chemical and mechanical industries

The industrial function

The removal of industry to the periphery of cities is a very significant process, so that transport becomes a key factor in the choice of the location of an industry.

The design of many industrial cities has been strongly influenced by the siting of the manufacturing industry and of the housing estates for workers in the factories. Mulhouse, Roubaix and Saint-Étienne in France and Sarrbrücken in Germany are examples of the many cities shaped and designed by the development of industries, especially during the eighteenth and the nineteenth centuries.

The new distribution of industrial activities has had several impacts on the morphology of these 'old' industrial cities:

- Large companies wish to be close to large cities, i.e. to administrative, financial and political power centres. An important consideration is therefore that they stay in or close to these towns.
- In contrast, small and medium-size companies can easily find opportunities to establish themselves in suburban areas or in the country. These two combined effects tend to be important distinguishing factors between large and small cities.
- The decentralization of factories means that large free spaces are becoming available in the centre or close to the centre of industrial cities. These spaces offer city managers a good opportunity to create green parks in the central area, when none have previously existed.
- At the same time, urban retrofitting operations can be take place so that buildings located near the released zones can be restored and upgraded.
- Finally, 'peripheral' cities are creating industrial centres for manufacturers in specialized areas.

The trade function

Four major steps can be identified in the dynamics of the trade function in cities:

- *Concentration.* All trades and services are concentrated in centres with high densities of population.
- *Suburbanization.* The retail sector develops activities in suburban areas. Implementation is an element in segregation, as it is statistically linked to average incomes.
- *Degradation.* Slum areas develop in city centres, as a result of which high-quality shops close and migrate to peripheral zones. These shops often thus benefit from being sited at commercial centres, or 'Great Leasing Areas' (GLA), close to peripheral highways.
- *Renovation.* Renovation and retrofitting operations, both on the urban structure and on the buildings, offer shopkeepers the opportunity to move back to the centre and create an appropriate environment for their operations.

The business function

All major cities, both in the developed nations and in some developing countries, have business, financial and trade activities organized in dedicated areas called Central Business Districts (CBD). Manhattan in New York, the 'City' in London and 'La Défense' in Paris are some famous examples of these large CBDs.

CBDs concentrate administrative, trade and financial leaders of large companies into relatively small areas. This concentration leads to the development of a typical architecture and urban structure, made up of high buildings. The residential potential of these areas decreases as business activities increase. This can lead to the creation of a 'ghetto' of new and 'high-tech' buildings around the CBD (Figure 15.6).

Figure 15.6. La part-Dieu is the CBD of Lyons

The administrative function

In all major cities the development and rapid growth of the administrative sector has taken place. This is especially true in national capitals, but it is also the case in large regional capitals. This increase leads to social segregation between a small high-level class and the growing huge population of employees.

The administrative sector of the city becomes a high-class sector. High-level civil servants and administrative leaders use their cars to travel from and to their houses, which are located in peripheral areas. Employees either live in the city or outside and around the administrative sector.

The administrative function is the basis of the urban structure in 'new' capitals like Washington, Brasilia and Bonn. However, even in 'old' administrative cities like Paris or London, is it possible to identify a specific area dedicated to this activity.

Brasilia, created in 1957, is an example of an *ex nihilo* city. The plan of the city has nothing to do with topographic considerations, historical references or economic aspects. The architects focused on the symbolic significance of the structure: Brasilia is designed in the shape of an aeroplane and only light, 'clean' industries were allowed.

Specific functions

- **Network nodes**
  - *Harbour cities*. Many big cities have been created as a harbour; London, Hamburg, Amsterdam, Copenhagen and Gothenburg are examples. The basic structure of the city is linked to harbour installations and activities. Nowadays, harbours and cities are developed separately and the activities of the cities are more and more detached from the original ones.
  - *Railway stations*. Many cities, big and small, have been created around or because of the railway, especially in old industrial countries. In these cities, the main station often constitutes the centre of the town and the station area has been fully developed as a trade or administrative sector. Several activities were developed in conjunction with the railway activity: road transportation, agricultural markets, etc. At the same time, the railway constituted a break in the urban space, especially when there was heavy traffic. This situation often led to the development of two separated towns, sometimes each one with its own plan. Major infrastructure has often been necessary to link the two parts. In the case of large cities and simultaneous growth of the railway, medium-sized cities have appeared along the railway network, each corresponding to a secondary station.
  - *Mining cities*. In many cases, such cities are not single entities, but group-ings of several small or medium-sized cities. These often have a typical monotonous alignment of identical buildings, with streets organized in a clear geometrical structure. These cities retain the impression of the company that created and planned the city. In most cases, the economic key role of the company carried over in a significant way into social and political aspects of the city, especially concerning housing, which was provided to workers as well as to foremen and management.

Figure 15.7. Nice, main city of the French Riviera [ENTPE]

- *Cultural or touristic cities.* Many medium-sized and some large cities can be characterized in terms of a specific cultural or touristic function that often has a strong impact on its urban development, depending on the socio-economic situation of its – permanent or transitory – population. Examples are:
  - seaside resorts, e.g. Brighton, Cannes, La Grande Motte and Nice (Figure 15.7);
  - mountain resorts, e.g. Chamonix, Courchevel and Zermatt;
  - cities with specific cultural activities, e.g. Bayreuth, Salzburg, Vittel and Marienbad;
  - cities with universities, e.g. Heidelberg, Cambridge and Leuwen;
  - cities with international functions, e.g. Geneva and Brussels.

URBAN DYNAMICS

Urban space

Urban space can be easily and propitiously substituted for the concept of city. This space represents a contrast between two opposite elements:

- the centre of the city, often inherited from previous centuries;
- the suburbs, recent (less than 100 years) and rapidly growing.

More and more people are inhabitants of these cities or urban spaces, but live outside the city for a significant part of their time (holidays, telecommuting, time spent in travel, etc.). Table 15.1 gives details of the largest conurbations in the world, both in 1995 and predicted for 2015.

The increase in the average income level induces a general improvement in indoor comfort conditions (heating, domestic hot water, cooling, etc.), as well as a development

Table 15.1. *Population (in millions) of the largest urban agglomerations in 1995 and 2015 (data from the United Nations)*

City	Population (1995)	City	Population (2015)
Tokyo, Japan	26.8	Tokyo, Japan	28.7
São Paulo, Brazil	16.4	Bombay, India	27.4
New York, USA	16.3	Lagos, Nigeria	24.4
Mexico City, Mexico	15.6	Shanghai, China	23.4
Bombay, India	15.1	Jakarta, Indonesia	21.2
Shanghai, China	15.1	São Paulo, Brazil	20.8
Los Angeles, USA	12.4	Karachi, Pakistan	20.6
Beijing, China	12.4	Beijing, China	19.4
Calcutta, India	11.7	Dhaka, Bangladesh	19.0
Seoul, South Korea	11.6	Mexico City, Mexico	18.8

Table 15.2. *Government expenditures (water supply, sanitation, drainage, garbage collection, roads and electricity) (Data from the United Nations)*

Income grouping	US $ per capita	Regional grouping	US $ per capita
Low-income countries	15.0	Sub-Saharan Africa	16.6
Low–middle-income countries	31.4	South Asia	15.0
Middle-income countries	40.1	East Asia	72.5
Middle–high-income countries	304.6	Latin America and Caribbean	48.4
High-income countries	813.5	Eastern Europe, Greece, North Africa and Middle East	86.2
		Western Europe, North America and Australia	656.0

of urban facilities (appliances, telephone, etc.). In Table 15.2 are listed government expenditures for the different income and regional groupings.

At the same time, the required urban space per capita rises, as a result of the development of urban facilities as well as the growth of road networks.

Two major elements have played and continue to play a key role in the definition and evolution of urban space:

- Technological advances lead to new architectural and urban designs and the creation of very high buildings and superposed urban areas.
- Road traffic increases so that city centres become saturated.

In general, rural and urban spaces are more and more connected and they interpenetrate. The car has reduced the distance between the city and the country and private housing is being built in the suburban areas around cities.

Urban theories

Many theories have been developed to model the urban space and its dynamics. Some of them are concerned with specific aspects, although a multidisciplinary approach has to be used in order to understand all interrelations and flows between the various elements.

Global economic theories

These theories aim at modelling the city from a holistic point of view, considering internal economic relationships. One theory represents the city as an economic unit, like a company or a state and it is then possible to define inputs and outputs of this unit as well as internal financial flows, investments, etc. In this matrix can be included elements related to households, private companies and public institutions.

Space theories

Urban dynamics can also be modelled by theories that analyse the spatial structure of urban spaces. There are various approaches used:

- *Mathematical approach*. Residential densities are linked and correlated with the distance to the centre. This supposes that this centre can be easily identified in time and space.
- *Physical approach*. Based on gravitational theory, this approach assumes that the attraction of a point depends on its weight and the squared distance from another point. Therefore, it is possible to consider residential zones as well as work reservoirs.
- *Financial approach*. This micro-economic theory is only based on the financial value of space and its evolution.
- *Sociological approach*. Each urban zone reflects the power of a social class. This theory therefore models the city and its dynamics through a social segregation of space.

Urban dynamics

Urban space characteristics

Urban space may be considered from different points of view:

- *Geometric space*. Urban space is defined through geometrical considerations: lengths and heights, surfaces, densities, etc.
- *Physical space*. The location can be characterized by natural elements: topography, climate, lakes and rivers, vegetation, etc.
- *Time space*. Each point in this space can be represented through isochrons. This space is determined by the distance (in time) between two points.
- *Economic space*. Each element has an economic weight depending on its situation and its intrinsic value.
- *Social space*. Segregations are especially highlighted on the basis of housing, facilities and employment.
- *Actual space*. Each inhabitant has their own view of the urban space, depending on their age, income level, housing, type of work, etc.

Table 15.3. Population density of seven cities compared to the population density of the countries they are in (data from the United Nations)

City	Density (capita/km²)	Country	Density capita/km²)
Beijing	34,177	China	122
Dhaka	11,019	Bangladesh	825
Mexico City	13,925	Mexico	45
Toronto	6,391	Canada	3
Tokyo	13,925	Japan	328
London	7,299	UK	235
Los Angeles	2,878	USA	26

Table 15.3 gives the population densities of seven major cities in comparison with the densities of the countries as a whole.

Urban expansion

Different factors have to be integrated when the development of the cities is analysed:

- *Physical factors*: topography, waterflows, etc. The weight of these factors is reduced by technological improvement.
- *Land factors*: the sizes and shapes of estates.
- *Economic factors*: the financial values of land.
- *Political factors*.

At the same time, this development can be analysed in terms of the dynamics of the urban morphology. Transportation networks are in this case the major element. These networks define and organize the urban space. The growth of a city is directly linked to the development of transport facilities such as railroads and highways. The out-migration of inhabitants from the centre to the suburban areas has been possible because of the improvement of transport networks and because of the reduction in the length of the working day.

The financial value of land increases from the suburbs to the centre. This value is also directly linked to parameters such as:

- the distance to roads;
- the distance to major crossings;
- the distance to railway stations.

The urban dynamics and the expansion of the urban space then have several major consequences. These include:

- Plot sizes decrease and shapes become more geometric.
- House moves become more frequent.
- Roads and highways become more and more developed.
- Buildings become higher and higher.

Figure 15.8. A newly retrofitted ward in Lyons

Urban retrofitting

In addition to urban expansion, retrofitting is another key element in the life of a city. Facilities and buildings have to be upgraded on a regular basis in order to keep them operational.

Urban centres often are very old zones, which have to be rehabilitated for improvement of public health and comfort as well as for economic reasons. Earthquakes, fires and war damage are other – extraordinary – causes of reconstruction and redesign.

From social, cultural and economic points of view, (electoral) wards are, in large and medium-sized cities, the reference scale for many inhabitants. They provide people with readable and visible distances and spaces. These wards are also social identifiers and social segregation can be revealed through the size of the wards; upper class wards are larger and more open.

BUILDING DESIGN

The basic structure of urban spaces was, until recently, based on a very irregular framework in old wards and on a geometric quadrangular design in other areas. There are three phenomena that can be used to explain and/or justify changes in this structure:

- Private houses are now essentially located in peripheral areas, while the city is more and more occupied by high residential or office buildings.
- Space functions tend to be mixed in order to avoid the negative effects of zoning, i.e. empty residential buildings and zones during the day and empty office buildings and zones during the night.

- Development of green parks and water features (e.g. artificial lakes) that may change microclimates.

Some important transformations can be seen in buildings in urban areas (in particular residential buildings):

- Buildings and dwellings are better and better equipped. With, for example, running water, electricity, heating, sanitation, telephone and electric appliances.
- The overall performance and quality of buildings and dwellings have improved and are continuing to do so, for examples thermal and acoustic characteristics, daylighting and artificial lighting, indoor air quality. At the same time, both dwelling area per capita and ceiling heights are increasing.
- Economic considerations lead to a more and more standardized design of new dwellings.

The development of retrofitting operations and the transformation of the buildings and dwellings in urban areas have, however, various consequences:

- New or retrofitted buildings are more comfortable and, as a consequence, often more expensive. This improvement in the quality induces some substitution of population. Former occupants are sometimes unable to stay in retrofitted buildings, or cannot purchase new dwellings.
- Renovation operations in urban areas sometimes lead to substitution of office buildings for residential buildings, so that the *intra muros* population of the cities decreases.
- If they are not retroffited, (old) wards may often deteriorate progressively. The departing residents are replaced by people on lower incomes with no means to improve, or at least maintain, the quality of the dwellings. Only public operations can change this movement by renovating these areas.

Part 2

16

———

Guidelines for integrating energy conservation techniques in urban buildings

N. Chrisomallidou
Laboratory of Building Construction and Physics, University of Thessaloniki

ℰℐ

INTRODUCTION

Energy strategies in urban buildings – problems encountered

The implementation of energy-saving strategies in all categories of buildings – residential, offices, shopping centres, hospitals or schools, either new or existing – in an urban environment appears to be a serious problem, the solution of which demands careful management. This difficulty is caused by the current condition of the urban fabric of cities, with respect to their planning and their frequently chaotic expansion, which has taken place without taking into account the principles of energy-conscious design on an urban scale.

Some of the factors that usually have a negative effect on the design and construction of urban buildings of low-energy consumption are:

- The layout of the basic road network with a specific orientation. This layout affects the siting of the buildings on either side of the road, giving them an orientation that, in most cases, is not suitable for the implementation of solar and energy-saving techniques.
- The relationship between the height of a building and the width of the road, which causes overshadowing and thus prevents access to direct sunlight in living spaces.
- The relationship between plot frontage and depth, which can determine how many internal spaces will have a southern aspect.
- Densely built urban centres, which result in the obstruction of the airflow and of sunlight by the walls of tall buildings.
- The lack of greenery, which has been replaced by concrete and tarmac.
- The overshadowing caused by adjacent buildings and other landscape features, which is difficult to avoid.
- Building regulations and codes that in most cases determine the dimensions of a building and thus its geometrical form and its position on the plot.

Examples in Thessaloniki are shown in Figure 16.1 to 16.3.

Figure 16.2. View of a section of Thessaloniki

Figure 16.1. The centre of Thessaloniki, Greece

Figure 16.3. General view of the centre of Thessaloniki

In spite of all the negative aspects of the urban environment and its buildings, there are still many possibilities for energy-saving intervention, which also mitigate the greenhouse effect, filter pollutants and mask noise. These interventions, which are the subject of the guidelines in this chapter, start with the site layout and the design of the building and then move on to the implementation of energy-saving measures.

Furthermore, it must be noted that urban centres present milder climatic conditions than suburban or rural sites (higher temperatures, less intensive winds) as a result of the heat-island effect. So, in spite of the negative effects of the above-mentioned factors, in particular of the difficulties that areencountered in the widespread implementation of bioclimatic architecture in urban buildings, the mild urban microclimate compensates in the winter period for the negative elements (inappropriate orientation, high density, shading, etc.). If we intervene correctly in the design layout, better climatic conditions will be achieved not only in the winter but, especially, during the summer, when serious overheating problems can occur in hot climates. A large number of air-conditioning appliances are used, especially in the southern urban areas, to solve the problems created by the improper energy design of buildings. This unfortunately leads to increased cooling loads during the summer and to overconsumption of electric energy, which also increases peak energy demand and creates failures and black-outs in the energy transport network. These strategies lead to an artificial, isolated and mechanically controlled indoor environment, which is obviously inhospitable for the residents.

Solar architecture or energy-conscious design in urban buildings?

In recent years there has been much talk of the need to save energy in the building sector by using simple and financially expedient methods that generally utilize conventional building materials.

During this period, in which numerous researchers have concerned themselves with energy issues, various terms have made their appearance, such as 'passive solar design', 'bioclimatic' or 'solar architecture', 'rational use of energy', 'energy-conscious design' etc., which describe a tendency, a philosophy or a method relating to architecture and building practice. In reality, however, over and above their conceptual differences, all these terms have the following in common: they relate to a type of architecture that respects the environment and applies various methods and techniques to reduce the amount of energy consumed in heating, cooling, lighting or hot-water use, provided of course that these do not affect the thermal comfort conditions in the interior of the building concerned. A key role in the effort to achieve this, if one isolates the 'building' from the built environment of which it forms part, is played by the building's shell, because it forms a 'filter' between the external and internal environments. Thus the majority of energy-saving interventions are concerned with the correct design of this 'filter' and aim to render it effective in exploiting the positive influences on it and in nullifying the negative ones, in accordance with the season, regardless of whether these influences stem from the external environment or the interior. Of course, it would be a serious omission not to mention the importance of energy design at the town-planning level, where one has to piece together a puzzle consisting of groups of buildings of different categories (industrial buildings, homes, offices, recreational buildings, hotel complexes, etc.), the design of open spaces and the lay-out of the streets. It is clear that the earlier and more effectively energy problems are dealt with on this large scale, the easier it is to implement various energy-conservation strategies on the small scale, i.e. in the buildings themselves.

As for the energy-saving techniques that can be applied in a building, it appears – although it is by no means clear – that two main categories exist (Figure 16.4). The first category includes the simple methods of conserving energy, involving the application of thermal insulation in the external building elements and interventions in the building's mechanical systems, while the second category includes strategies for passive solar heating, natural cooling and lighting. It should be noted, of course, that the second category of techniques can only be applied if certain conditions are fulfilled, such as a south orientation for the main facade, the avoidance of shading by adjacent buildings etc., while, as a general rule, the simple methods of energy conservation must also be applied at the same time. The reverse is not true. A combination of both categories, of course, will provide the best possible energy-saving solution.

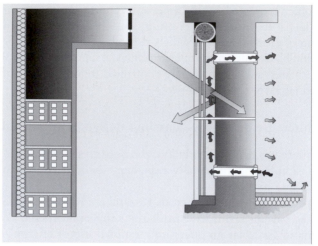

Figure 16.4. Energy saving or passive solar technology?

A great deal has been written about traditional architecture, which has a lot to teach us about the way in which it has dealt with thermal and visual comfort. It is certain that the anonymous constructors of the past had excellent knowledge of how to solve their problems by working together with nature, since the technological means did not exist to enable them to do otherwise. As is well known, they would position their buildings in relation to the wind, sun, shade, vegetation, lie of the land, etc. Thus, with very limited materials and technology, they erected buildings that were well adapted to the climatic conditions of the local area. Nowadays, similar techniques are being employed in modern architecture and these are also the subject of an applied scientific field that is attempting to become part of a wider architectural practice.

Unfortunately, even today, there are a lot of doubts about the efficiency of conscious energy design and of passive methods, in both architectural and financial terms, as well as fear that a bioclimatic or solar building will have to sacrifice aesthetics for the sake of its 'solar' character. However, there is so much proof of the opposite, along with all the experience and the know-how, that the slightest hesitation is unjustifiable.

A building designed to satisfy certain energy requirements does not have to be 'solar'. There is no need for it to differ from a 'conventional' building, as long as it is designed and built according to certain energy principles.[1]

DESIGN GUIDELINES

During the design process, the architect should start with the site layout, the positioning of the building in the plot, the landscaping, the form of the building, which in turn is very closely related to the functional arrangement of internal spaces, the construction of the external elements, the choice of materials, etc. The first and most important stages, especially of the design of buildings in an urban environment, are the study of the site, the spacing between buildings to avoid overshadowing and the arrangement of the exterior by planting, windbreaks, the creation of wind channelling etc.

Site layout

The first basic problem the designer has to face in the urban context is that he/she has to place the building in an already _existing built environment_ (Figure 16.5(b)). The case of an empty block is very rare, so it is difficult to propose solutions like the one in Figure 16.5(a), in order to ensure that the buildings will be facing south, and to avoid overshadowing by adjacent buildings, as indicated by energy-conscious design, so as to permit solar access in living spaces.

In new settlements there are fewer problems and greater possibilities for implementing energy-saving strategies (Figure 16.6).

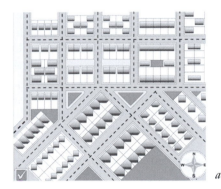
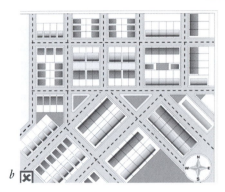

_a_   _b_

Figure 16.5 (a) Proposed arrangement to avoid overshadowing and to allow winter solar access; (b) the usual arrangement of building rows on plots, parallel to the roads

Figure 16.6 (a) Example of site layout of a workers' apartment block with no energy-saving criteria.[2] (b) Simple site replanning and reorientation of the dwellings in a workers' apartment block, in order to facilitate the implementation of passive solar strategies, can almost halve the energy demand per house[2]

_a_

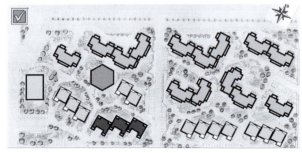

_b_

The roads

The second basic problem that very often appears in the already formed fabric of cities is related to the road layout. This, as a rule, influences the main orientation of the buildings and the spacing between them. It therefore affects natural daylighting and solar access in the internal spaces, as well as the shading of the buildings and the natural ventilation, depending on the airflow characteristics in the canopy layer (Figure 16.5 and 16.6).

Generally all solutions concerning the positioning of the buildings on a site presuppose a 'sun lighting–shading' study, and, as has already been mentioned, an equivalent treatment of all the buildings in the block, taking of course into account the circulation axes and the orientation of the plot (Figure 16.6(b)).

The parallel arrangement of the roads along the east–west axis is advisable. This solution provides the buildings with the possibility of south orientation (Figure 16.6(b)).

In urban sites the risk of *overshadowing* from adjacent buildings is fairly high, cancelling the positive impact of south orientation (solar gains, daylighting). If possible take into account the minimum distance between buildings so as to avoid overshadowing and permit the access of winter sun into internal spaces (Figure 16.7).

According to Yannas,[3] the rule is: For any point on a surface, the view of the sun becomes obstructed when the altitude angle of the obstructed object – the obstruction angle q exceeds the solar altitude angle. The parameters that define the obstruction are its height and the distance from the affected surface (Figure 16.8).

The distance required between buildings in order to avoid overshadowing increases in proportion to geographical latitude of the location, where the solar altitude decreases correspondingly (Figure 16.8(a)). The required distance also increases when the height of the nearby buildings is increased. For example, in southern latitudes we could locate the buildings closer to each other or we could keep the same distances as in the northern latitudes and increase the height of the building (Figure 16.8).

If some storeys in south elevations are overshadowed by other buildings, then these floors should serve for secondary uses (workshops, storehouse etc.) that will have a short daily operation schedule and perhaps low temperature demands (non-residential). In this case, it is recommended that the two different uses should be thermally separated by insulating the floor between them, in order to eliminate the heat flow from the main use to the secondary one. A different heating plant should also be provided for each use (Figure 16.9) so that the user can stop the heating of the spaces for periods when the space is not in use.

For floors with no winter solar access problems, consider passive solar systems for heating, cooling or daylighting (see the appropriate sections later in this chapter).

The following site planning considerations, along with the internal organization of individual buildings, can be significant in reducing energy demand.

Landscaping to optimize microclimatic conditions

During the early design stage of buildings in the urban environment, it is necessary to include the landscaping of the surrounding area, having as a basic criterion the improvement of the external climatic conditions, both in winter and in summer. The shading and evaporative cooling that planting can offer, as well as water surfaces and

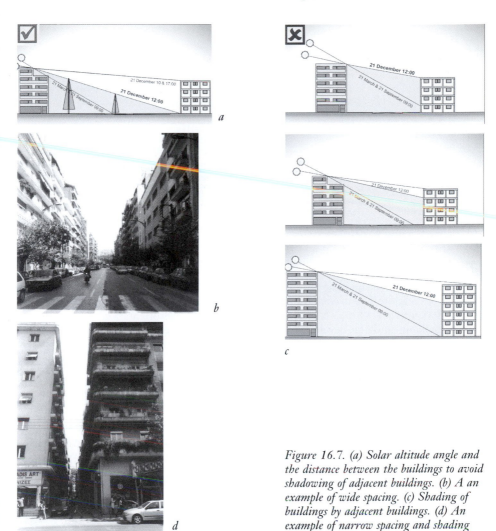

Figure 16.7. (a) Solar altitude angle and the distance between the buildings to avoid shadowing of adjacent buildings. (b) A an example of wide spacing. (c) Shading of buildings by adjacent buildings. (d) An example of narrow spacing and shading

Figure 16.8. (a) The obstruction angle (north latitude 52°). (b) The obstruction angle (south latitude 38°)

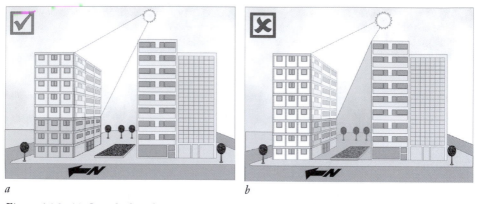

a   b

Figure 16.9. (a) Overshadowed storeys are put to secondary use. (b) Avoid this solution

☑
- One elevation of the building must have a south orientation. The maximum acceptable deviation is ±25°.
- The building must be developed along the east–west axis in order to maximize the length of the south elevation (Figure 16.6(b)).
- If both tall and low buildings are proposed in a housing project, the tall buildings should be positioned on the north side so as to avoid shading the lower ones.

Figure 16.10 demonstrates these principles.

- Locate the buildings as close as possible to the north side of the building plot. In this

☒
- Avoid the types of building plot that, in the urban environment, can cause problems in the orientation of the building. Usually these types of plot are small in area, irregular in shape and located in densely built areas.
- Avoid plots on narrow streets with tall buildings (Figure 16.11).
- Avoid plots on north–south road axes.

Figure 16.10. View of the solar village in Lykovrissi designed by Tombazis[1]

Figure 16.11. Narrow streets with tall buildings.

way the space opposite the urban building is increased, solar access to southern spaces is made easier and, of course, the surface of the garden is bigger on the favourable, in terms of the microclimatic conditions, side of the site.

- Try to find the solar altitude angle for 21 December for your region and calculate the shortest acceptable spacing between the buildings in order to limit overshadowing as much as possible and to allow the access of the winter sun into internal spaces (Figures 16.7 and 16.8).
- The width of the streets, the urban planning and the built environment should be taken into account in the relevant national regulations and codes and restrictions on the maximum height of buildings should be imposed.

As mentioned above, in most cases, in an urban context we have to place the building in an already existing built environment and the possibilities for interventions are extremely limited. So, if you have to deal with an east–west orientation, see the sections below on energy conservation strategies and passive solar cooling strategies, while for a north orientation only the energy conservation strategies apply.

---

wind channelling through natural or artificial barriers, reduce the effect of solar radiation in summer, while in winter they shelter the building.

Siting and site layout

In the metropolitan centres of big cities, where buildings of all kinds, office buildings, shopping centres, entertainment and hotel facilities etc., are needed, the building density is at its greatest level and space is overexploited. It is therefore impossible to meet the needs for free spaces, squares and parks, while the shading and related problems due to tall buildings and narrow streets are greatly increased.

Although the possible interventions in very dense metropolitan areas are rather limited, it is useful to use the experience gained in projects carried out in expanding urban areas and in newly designed building complexes so as to avoid incorrect approaches and ineffective procedures.

Buildings close to each other in a linear development reduce, in most cases, the positive impact of unobstructed wind movement, while layouts in the form of a chessboard decrease the wind shadow and so can increase the potential for natural cooling. The final choice depends on the dominant external climatic conditions (Figure 16.12).

The arrangement of buildings in an irregular pattern is preferable and short distances or narrow streets between them should be avoided. Uninterrupted building arrays form a solid wall, which dramatically affects air circulation and air renewal. This phenomenon also increases the heat-island effect, creating discomfort in the indoor environment during the summer, as a result of the decrease in the natural cooling of the buildings.

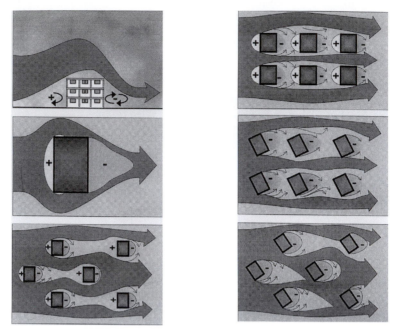

Figure 16.12. Airflow patterns (after Bowen[3])

Buildings of similar height, oriented towards the south, together with a proper urban planning and site layout, contribute remarkably to successful energy and environmental interventions, giving great potential to a properly designed energy-efficient building.

Vegetation and windbreaks

Vegetation. For the reasons discussed above, the lack of planted spaces in urban areas affects more than just the view. Temperatures monitored in areas without vegetation are 5 to 12 K higher in the warm period compared to nearby suburban or rural areas. A study at the University of Athens concluded that the cooling-energy demand of a building in the centre of Athens is almost double that of a building in the nearby countryside (see Chapter 11). Successful solutions to such problems include the creation of many small-sized planted areas, which are more effective than an extended planted area.

In the area close to the building the increase in the area planted helps to form better microclimatic conditions and positively affects:

- the annual energy balance of the building itself, increasing solar and wind protection, reducing the temperature of the urban environment and thus improving comfort indoors during summer and mitigating the greenhouse effect;
- the pollutant concentration levels;
- the indoor air quality (IAQ);
- the sound pollution levels.

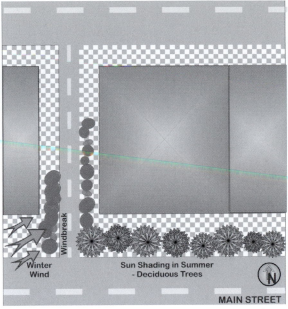

Figure 16.13. Vegetation operates to provide both wind-breaks and sunshading

In the urban context, the right choice of vegetation, including the type, shape and location of the flora as well as the correct location and size of the planted areas, is therefore important, producing a positive effect during both summer and winter. Some care in the choice and placement of vegetation on or near buildings should, however, be taken to avoid structural damage (Figure 16.13).[4]

Evergreen trees placed near the northern facade of a building increase the thermal protection, in accordance with the wind environment and the thermal characteristics of the building envelope.

Deciduous trees on the south side of a building offer natural protection from solar radiation and evaporative cooling in the summer, while at the same time they allow solar access to the internal spaces in winter. Furthermore, trees and grass, or other ground cover plants, positively influence the microclimatic conditions because they absorb large amounts of solar radiation, helping to keep the air and ground beneath them cool, while evapotranspiration leads to further reduction of the external temperatures (Figure 16.14).

Windbreaks. The design of the area around a building (the existence or planting of trees, bushes and other plants and barriers in relation to the openings in the building) creates areas of high and low pressure, thus affecting the flow pattern and the speed of the wind (Figure 16.15). It is possible to increase the speed of the airflow towards the inside of the building or over it, or to divide the air current and drive part of it through the building while another part goes above it, by constructing a fence or a hedge around the building. The combination of different windbreaks (low or high walls or trees) and their distance

Figure 16.14. Trees offer a windbreak in winter and shading from the sun in summer

from the building can produce different results (improve cross-ventilation, create a calm sheltered zone behind it etc.).[5]

Gaps in windbreaks, openings between buildings and openings between the ground and a canopy of trees can create wind channels, increasing wind speeds by about 20%.[6]

Aqueous surfaces

In addition to vegetation and the different techniques that affect the outdoor temperature and define the route and intensity of the wind, other landscaping techniques include the use of ponds, streams and cascades for evaporative cooling (Figure 16.16). These techniques should be implemented in warm and dry climates.

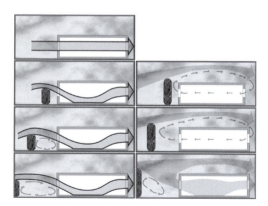

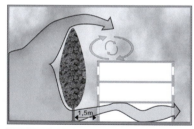

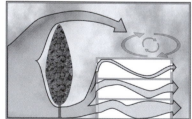

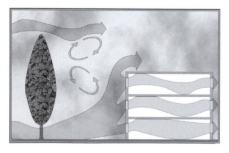

Figure 16.15. A windbreak produced by vegetation, affecting the flow pattern and the speed of the wind

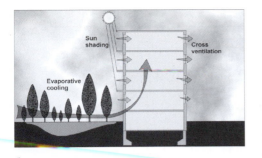

a

b

Figure 16.16. (a) Natural cooling by different techniques (evaporative cooling, sun-shading devices, cross ventilation). (b, c) Evaporative cooling

c

The implementation of outdoor evaporative cooling strategies improves the external comfort conditions by decreasing the temperature of the air surrounding the building. As a direct result, internal air temperatures, and consequently indoor cooling loads, are lower.

Surface albedo

Surface materials with a high albedo index (reflectivity) to solar radiation reduce the amount of energy absorbed through building envelopes and urban structures and keep their surfaces cooler (Table 16.1). Materials with high emissivities are good emitters of long-wave energy and readily release the energy that has been absorbed as short-wave radiation. Lower surface temperatures contribute to decreasing the temperature of the ambient air, as heat convection intensity from a cooler surface is lower. Such temperature reductions can have significant impacts on cooling-energy consumption in urban areas, a fact of particular importance in cities with hot climates (see Chapter 11).

The form of the building

The choice of a defensive (closed) or aggressive (open) building

The decision between these two choices (Figure 16.17) depends on the main orientation of the urban building, the climatic conditions, the use of the building (office, residential

Table 16.1. The albedos of various materials

Material	Albedo	Emissivity
Concrete	0.3	0.94
Red brick	0.3	0.90
Building brick	–	–
Concrete tiles	–	0.63
Wood (freshly planed)	0.4	0.90
White paper	0.75	0.95
Tar paper	0.05	0.93
White plaster	0.93	0.91
Bright galvanized iron	0.35	0.13
Bright aluminium foil	0.85	0.04
White pigment	0.85	0.96
Grey pigment	0.03	0.87
Green pigment	0.73	0.95
White paint on aluminium	0.80	0.91
Black paint on aluminium	0.04	0.88
Aluminium paint	0.80	0.27–0.67
Gravel	0.72	0.28
Sand	0.24	0.76

buildings, shopping centre, etc.) and other design criteria, such as the view, safety, noise, construction cost, etc. Both strategies could lead to the same amount of energy consumption by the implementation of energy saving and/or passive solar techniques for heating, natural cooling and daylighting. The envelope of an aggressive form opens to allow visual and physical contact with the exterior and controlled interaction with natural seasonal changes.

The following considerations are necessary in making the choice.

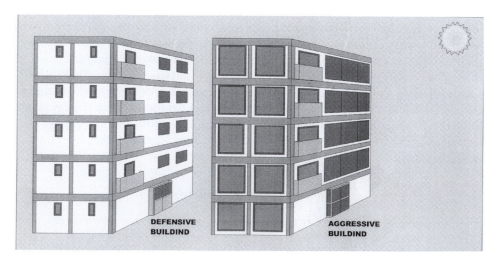

Figure 16.17. Defensive and aggressive buildings

- When the main orientation of the facade of the building is within ±25° of south, then an *aggressive–open form* with large glazed surfaces (windows or solar walls) should be chosen in order to take advantage of the thermal solar gains.
- If for some reason (an east, west or north orientation, the view, safety etc.) a *defensive–closed form* is chosen, then it is necessary to compensate for the limited thermal solar gains by reducing the transmission losses of the building shell (increase the insulation of the building elements, construct a compact building with a small area-to-volume ratio).

- An *aggressive form* should not be chosen when a building has a north orientation that does not allow solar access or when it creates overheating problems during summer (east or west orientation).
- It is well known that an aggressive building costs more than a defensive one (the cost of glazing is three times as high as that of the wall structures). If solar energy is not utilized, or if the building's operating schedule is not continuous, which means fewer energy savings, then the payback period of the additional cost is increased dramatically.

In all the aforementioned cases choose a defensive form, and determine the size of the openings, taking into account the demand for daylighting and natural ventilation.

The positive and negative aspects of an aggressive form of building (Figure 16.18) are as follows.

Figure 16.18. Example of an aggressive urban building[1]

☑

- Decide on this form only if a south orientation is guaranteed (Figure 16.18).
- Maximize the south elevation for solar collection and minimize the remaining external surface areas.
- This form should be applied in urban buildings with a regular operating schedule.
- Study the overshadowing from buildings on the opposite side of the road (see the earlier section in this chapter).
- Consider the possibility of integrating into the building shell active and/or passive solar systems in combination with building elements with a high thermal capacity.
- Plant deciduous trees in front of the facade. The first three–four floors could be shaded in summer. In addition, use fixed horizontal louvres or overhangs to control sunlight in the south-oriented windows.
- Provide night-time insulation of glass panes for the winter period and for the passive solar system, if one exists.
- Use advanced glazing (low-emittance glazing and windows with low U-value and high total solar energy transmittance g) and strong thermal insulation in the building envelope, in order to eliminate heat losses and to conserve the solar heat gains in internal spaces for a longer period of time.
- Arrange internal spaces accordingly, as discussed later in this chapter.
- Clerestories could be provided in the roof for daylighting and for natural cooling ventilation during summer (see below).

☒

- Avoid openings with single glass panes.
- Avoid glass panes of dark colour and of reflecting or absorbing type. This kind of glazing is recommended for west elevations because it offers protection from solar radiation during summer and thus helps to avoid overheating (Figure 16.19).
- Avoid overshadowing.
- Avoid thermal bridges in the external building envelope.
- Avoid the building having a great depth in order to ensure daylighting of all interior spaces.

Figure 16.19. Reflecting glass in urban buildings (usually offices)[7]

The positive and negative aspects of a defensive form of building (Figure 16.20) are as follows.

Figure 16.20. A defensive building

☑

- Reinforce the thermal insulation of the external building elements in order to eliminate thermal losses and find a form with a small area-to-volume ratio.
- Use double glazing or, as widely as possible, special thermal insulation glazing with a low *U*-value for the openings.
- Size the openings so that the needs for daylight and for natural-cooling cross-ventilation are satisfied in the interior.

☒

- Avoid single glass panes in the openings.
- Avoid thermal bridges in the external building envelope.

Architectural composition

For a given heated volume and surface floor area there can be numerous building forms (Figure 16.21), depending on the designers and their architectural ideas, as well as on various external factors (plot area, privacy, view, etc.)

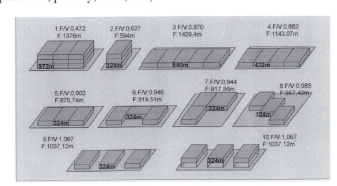

Figure 16.21. Various arrangements of apartments of 108 m² in area[8]

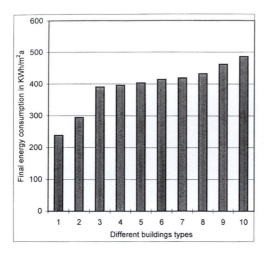

Figure 16.22. Final energy consumption of the uninsulated buildings in Figure 16.21 (simulations carried out for the climatic conditions of Thessaloniki, Greece[8])

From an energy-saving point of view, the different forms that the apartments take determine the different energy behaviour of each building if it is assumed that all the other parameters remain constant (heated floor area, heated volume, temperature of the interior, construction, operation schedule, etc.), Thus, the building's form constitutes one of the most important factors, in determining its final energy consumption (Figure 16.22). Good and bad arrangements are shown in Figure 16.23. The form of the building in relation to the position and the size of its openings can greatly affect the thermal balance of the building, since it determines the thermal solar gains, as well as the thermal losses through the external envelope. The general design objectives are as follows.

☑

- Reduce through the form the exposure of the building to the dominant winds in order to minimize their negative effects.
- Keep the form of the building as compact as possible, so as to minimize exposure of the external building elements to weather effects during both winter and summer.
- Use sheltering and buffering as additional means of heat-loss control through the building's form.
- The building should be in direct contact with the ground to benefit from the thermal capacity of the ground.
- In detached units, and generally in units that have a large external surface, the building's insulation should be increased. The aim is to compensate the high heat losses of the shell and reach the same energy consumption levels as in compact buildings (Figures 16.21 and 16.22, building type1).

☒

- Avoid forms or shapes of buildings that cause overshadowing.
- Do not allow the form to constrain the solar access to the living spaces achieved by the site layout.
- Do not neglect to protect against direct sunlight urban buildings with high internal gains, presenting a high cooling load (offices, commercial centres etc.).
- Omit the internal corridor; with an internal corridor the back half of the north side of the apartments will not have winter solar access.
- The apartments in the middle face the problem of overheating during summer, because of the single-sided ventilation (Figure 16.23(b)).

- Extend the building on the east–west axis to maximize its south façade (Figure 16.23(a)).
- Try to expose all apartments to the sun. With this arrangement, all apartments will also be cross-ventilated and therefore naturally cooled during the summer months.

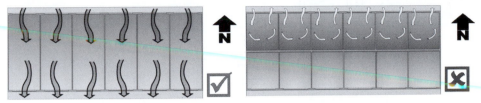

Figure 16.23. (left) Proposed arrangement of the apartments and (right) a wrong arrangement

The composition of four buildings. Figure 16.24 shows various arrangements of buildings. Figure 16.25 shows an example of a courtyard, The following suggest some of the choices.

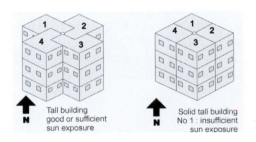

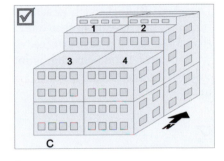

a   *b*

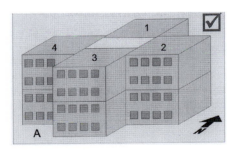

c   *d*

Figure 16.24. (a) Two arrangements of four buildings. (b) Most of the units are exposed to the sun. (c) By the creation of an internal atrium or courtyard the first unit is exposed to the sun. (d) Only two of the four buildings are exposed to the sun in the right way

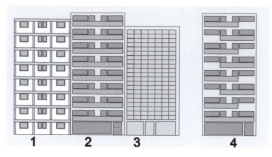

Figure 16.25. An internal courtyard

Figure 16.26. The relation of a building to other buildings

The relation of a building to other buildings

Detached buildings and those at the end of a row of buildings or flats in multistorey urban buildings that have more than one external wall will obviously have greater heat losses than those with only one external wall (Figure 16.26).

This can be compensated by increased insulation or solar gains through south-oriented windows or passive solar systems. What to do and what not to do are described in the following.

☑
- In the case of the detached Building 4 in Figure 16.26, the insulation of the shell should be increased, whereas if it faces south it should be reinforced by passive solar systems (see the later section in this chapter). The aim is to reduce the increased heat losses and therefore the building's energy consumption so that it reaches the same levels as those of Building 2, which should be used as a reference case.
- Buildings 1, 2 and 3 need milder measures than building 4, while buildings 1 and 3 need stronger energy interventions than Building 2. Building 2 can be the reference case in relation to the final energy consumption.
- Reinforce the insulation of the flat roof in all cases and provide independent heating control in every apartment. In this way energy consumption can be minimized.

☒
- In terms of energy saving, buildings with two or more external walls should not be dealt with in the same way as buildings with one external wall.
- Surfaces that are in direct contact with the external environment should not be left without insulation. Thermal bridges are created and heat losses are increased. There is also a real danger of condensation inside the mass of the materials, affecting the building elements.

The relation of a building to the ground

In areas where the temperature of the ground is higher than the temperature outside ($t_{ground} \sim 11°C$), buildings that are in direct contact with the ground have better thermal behaviour and therefore consume less energy. The internal climatic conditions of these buildings are also better during the summer.

Studies have shown that in buildings on pilotis (i.e. with an open ground floor) the consumption of energy is increased by only 5%.[9] Figure 16.27 shows various arrangements.

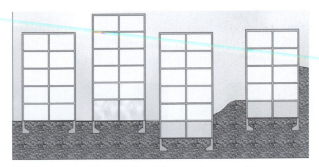

Figure 16.27. Relation of buildings to the ground

Single-storey and multistorey buildings

As a rule, urban buildings are multistorey buildings. These buildings consume less energy per square metre of floor area than a building that is lower in height (Figures 16.28 and 16.29).

Figure 16.28. Buildings with one to eight floors having the same floor area for a typical floor plan

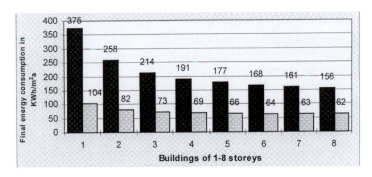

Figure 16.29. Heating energy consumption of buildings with one to eight storeys. The dark bar refers to uninsulated buildings, while the lighter bar refers to the same building but this time insulated (simulations carried out for climatic conditions in Thessaloniki)[8]

There are two alternative situations:

- The lower the building, the stronger must be the insulation of the external envelope – constructed of high-quality materials – so as to overcome the high thermal losses in comparison with tall buildings; or
- The stronger the insulation of the building envelope, the lower the impact of the building form. The differences in energy consumption between buildings are limited (Figure 16.29).

In low buildings the insulation of floor and roof should be a top priority. The opposite is true in the case of taller buildings, where the thermal protection of vertical surfaces (walls and windows) should be reinforced (Figure 16.30).

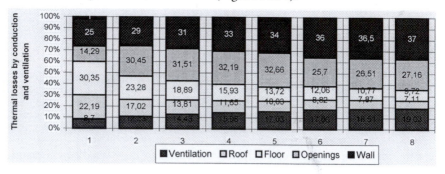

Figure 16.30. Buildings with one to eight storeys. The contribution of the external building elements and ventilation to the total thermal losses by conductivity and ventilation[8]

Undoubtedly, in many conventional tall urban buildings that have a central heating plant, a higher level of energy consumption than expected has been recorded.[10] Apart from irresponsible energy behaviour on the part of users, this is mainly due to insufficient, or a complete lack of, thermal insulation on the roof and therefore greater conductivity losses. As a result, the lower temperatures on the top floor lead the occupants to increase their energy consumption to ensure thermal comfort, but leading to simultaneous overheating of the lower floors.

If Figure 16.30 is used as a baseline, priorities in terms of interventions in external building elements can be determined in relation to their contribution to the total thermal losses (conductivity and ventilation).

For reasons of comfort, condensation and hygiene, predetermined air changes per hour are necessary in every category of building (residential, school, office, etc.). The thermal losses caused by the air changes cannot, therefore, be reduced below a certain level. What should be avoided, however, is the increase of ventilation losses due to uncontrollable air changes and/or to insufficient airtightness of the shell. Space ventilation and thermal losses are particularly important in tightly insulated shells and bioclimatically designed urban buildings. In the heating period, heat recovery systems should be considered. In summer, ventilation is the most efficient method of natural cooling, if of course the external temperature permits.

Building orientation – the design of openings

The most important thing to keep in mind when solar energy is used to heat buildings during winter and to avoid overheating during summer is the orientation of a building's openings.

South orientation

South openings can be used as collectors of solar radiation during winter and direct radiation can be avoided during summer if appropriate shading devices are used.

The benefits from a south orientation are as follows:

- better distribution of solar gains compared to other orientations (Figures 16.31 and 16.32);
- energy savings in heating (Figures 16.33 and 16.34);
- reduced risk of overheating compared to east and west orientations in summer;
- simple horizontal sun-shading devices (overhangs, balconies);
- sufficient thermal solar gains to cover the limited thermal load during fall and spring, in southerly regions with a mild climate, thus shortening the heating period.

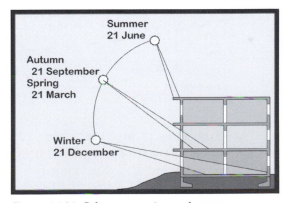

Figure 16.31. Solar access to internal spaces

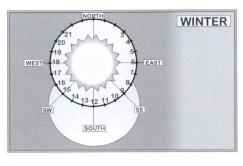
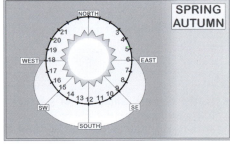

a   b

Figure 16.32. Sun paths (a) in the winter ; (b) in spring and autumn

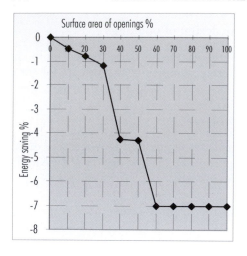

Figure 16.33. Energy saved by increasing the number and size of the south openings. Simulations carried out in a four-storey urban building in the climatic conditions of Thessaloniki[11]

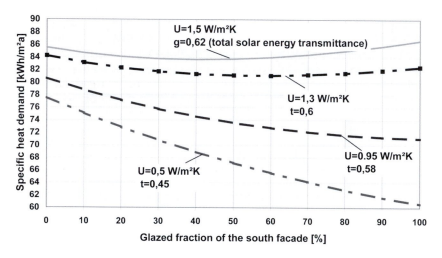

Figure 16.34. The impact of various qualities of glass and the glazed fraction of the south facade on the specific heat demand (KWh/m² per annum). Calculations carried out for the climatic conditions of north Germany[12]

What to do and what not to do regarding openings in south orientations are as follows.

☑

• In the climatic conditions of southern European, increase the size of the south openings in order to increase the thermal solar gains (direct gains) and therefore to decrease the final energy consumption (Figures 16.33 and 16.35). In other regions it is recommended that the energy

☒

• Avoid planting evergreen trees, or anything that would cause overshadowing, in front of the facade.
• Do not reduce the thermal capacity of the floor that directly absorbs the solar radiation by insulating its internal side or by laying carpets and putting furniture on it.

balance of south openings or, even better, of the whole building (solar gains, thermal losses) should be simulated in order that the optimum surface area of the openings can be determined (Figure 16.34). Generally, a surface area of 50% of the south facade constitutes a reasonable result.

- In order to secure privacy, safety and protection from noise, direct gain can be combined with other passive solar systems (mass wall, trombe wall, sunspace, etc.).
- Arrange interior spaces according as discussed later in this chapter.
- If the size of the openings is increased, reinforce the insulation of the envelope in order to conserve thermal solar gains in interior spaces.
- Provide night-time insulation of the openings.
- Ensure that there is controlled ventilation of the spaces.

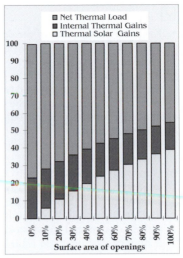

Figure 16.35. Increasing the number and size of south openings results in a reduction of the net thermal load and an increase of thermal solar gains. These results are valid for the climatic conditions of Thessaloniki[9,11]

North orientation

North openings provide spaces with a better quality of lighting because they only allow diffused rather than direct light. They are more useful during summer, but they should be of a limited size or they will cause high thermal losses during winter (Figure 16.36). What to do and what not to do regarding openings in north orientations are as follows.

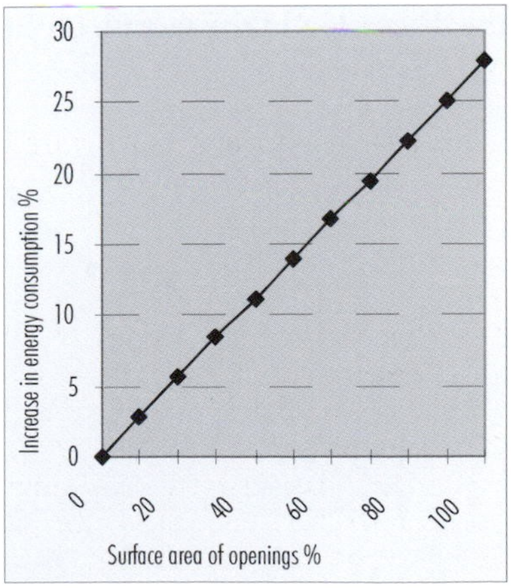

Figure 16.36. Energy consumption in relation to north openings. Simulations carried out in a four-storey building in the climatic conditions of Thessaloniki[9]

☑

- Keep the surface area of the openings as large as necessary for lighting and natural ventilation purposes.
- Increase the insulation of the building's envelope by ~20% compared to the usual standard, in order to attain the same level of energy consumption as a building with a south orientation.
- Use double glazing or specially insulated glass panes.
- Generally, and especially in north orientations that are affected by the wind, control the airtightness of the envelope. Reduce uncontrolled air infiltration by proper design and good workmanship, sealing cracks, pipe and cable holes and selecting continuous finishes rather than jointed ones where possible.
- Use draught lobbies for all entrance doors.
- Teach occupants to develop energy-saving ventilation habits.
- Consider the possibility of installing a controlled ventilation system incorporating an air-to-air heat exchanger.

☒

- Do not enlarge the openings.
- Do not forget the measures that should be taken outside the building against cold winter winds (see the previous discussion in this chapter).

East and west orientations

East and west openings have very few advantages at any time of year, which is why it is advisable to have them only if absolutely necessary for lighting reasons or in order to provide a view (Figure 16.37). West openings, in particular, increase the temperature, and therefore the cooling load, of interior spaces during summer, as they allow direct radiation in the afternoon (Figure 16.38). Generally, if there are east or west openings, then the shading devices that should be used must be external and vertical in order to be effective. In order to deal efficiently with the overheating problem during summer, see the discussion in the section on passive cooling strategies later in this chapter.

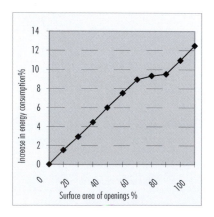

Figure 16.37. Energy consumption in relation to the size of east or west openings. Simulations carried out in a four-storey building in the climatic conditions of Thessaloniki[9]

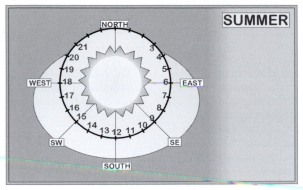

Figure 16.38. The sun path in summer

What to do and what not to do regarding openings in east and west orientations are as follows.

☑

- Provide the proper shading devices to avoid overheating during summer (see discussion later in this chapter).
- Provide cross-ventilation in the rooms for natural cooling (see discussion later in this chapter).
- In offices, shopping centres, etc. use special types of glass panes, such as reflecting or absorbing glass, in order to reduce solar radiation.

☒

- Avoid large areas of windows because the risk of overheating is increased in the summer period.

Functional arrangement of internal spaces (thermal zones)

Suggested arrangements of rooms are shown in Figure 16.39.

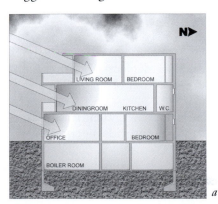
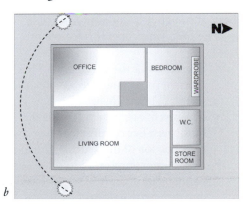

Figure 16.39. Functional organization of interior spaces. Creation of thermal zones: (a) cross section; (b) floor plan

Suggestions as to what to do and what not to do are as follows.

☑

- Spaces that are used for long periods and also require high indoor temperatures (living room, kitchen, dining room, office) should be placed on the south side.
- Spaces that are used for short periods and/or spaces that require lower indoor temperatures (WC, bedrooms) should be placed in an intermediate thermal zone.
- Auxiliary spaces, if they exist, should be placed on the north side in order to operate as a buffer thermal zone between the heated spaces and the outside environment (storage spaces, garages, etc.).

☒

- Do not place bedrooms on a side that overlooks a road. Operational problems are created in these rooms because of the noise.

ENERGY CONSERVATION STRATEGIES

Thermal insulation of opaque external building elements

In buildings in urban environments where the incoming solar radiation and solar gains are reduced as result of improper orientation or overshadowing by other buildings, the implementation of energy-saving measures probably provides the only solution for decreasing the consumption of conventional fuels.

Even if the reduction of thermal loads is possible by increasing the solar gains, the implementation of energy-saving measures – by the use of heavy insulation – is essential in order to retain in the internal spaces the heat gained.

The energy goal set for an urban building (measured in KWh/m^2 per annum) can be achieved by combining the use of solar energy with energy-saving measures, or by implementing heavy insulation in the envelope only. The first option is inevitably the best, although it is not always feasible.

The following general rules are proposed:

- The stronger the insulation of the external building elements, the fewer the thermal losses and the smaller the energy consumption of the building in winter (Figure 16.40).
- The first five centimetres of insulation saves much more energy than is achieved by the next five centimetres. A cost–benefit analysis can provide the criteria for the selection of the width of the insulation.
- The more complex a building's architectural form is (having high area-to-volume ratio), the heavier the envelope insulation must be, in order to balance out the increased thermal losses. The designer should always consider the implications of his decisions. If incorrect decisions are made during the design and construction process, this leads to increased problems for future

Figure 16.40. Winter

users, burdening them with the consequences and increasing the cost of energy consumption.

- Use the same means to deal with the insulation of the envelope, the ventilation, the thermal solar gains and the heating, because they are the basic factors of the building's thermal balance and therefore affect the internal climatic conditions and energy consumption.
- Insulate the structure uniformly, avoiding thermal bridging.
- Position the insulating layer so as to provide a thermal capacity appropriate to the heating system, the use of the building and the expected solar gains.

The thermal insulation of the building shell will achieve the following:

- It increases the thermal comfort in interior spaces.
- It reduces the possibility of vapour concentration on building surfaces, provided that there are no thermal bridges.
- It increases the initial construction cost, but reduces the running costs through the savings in energy consumption.
- It ensures that thermal solar gains will be conserved for a long period in the interior of the building.

Energy saving by insulating the external construction elements

External walls. The external wall structures can be thermally insulated on the outer side, in the cavity of a double wall or on the inner side. The ventilated wall structure method can also be implemented. Figure 16.41 shows the energy saving as a function of insulation thickness.

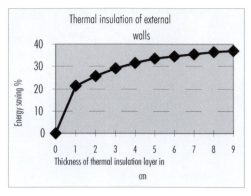

Figure 16.41. Energy saving by insulating the external wall of a non-insulated urban building; climatic conditions in Thessaloniki[9]

Particularly in urban multistorey buildings, the thermal insulation of the external walls is the key factor in the final energy consumption (see also Figure 16.30).

Flat roof. The roof, the most important structural element of the building shell, always used to be inclined, but has developed gradually and has finally been replaced to a large extent – in particular in urban buildings – by a flat roof. However, problems that are practically non-existent with inclined roofs began to develop as the form of the roof changed.

The requirements that a flat roof must satisfy in order to avoid damage and to adjust the internal climatic conditions can be summarized as follows:

- It must be waterproof and damp-proof both from external rain and internal relative humidity.
- It must have the slopes necessary to facilitate and achieve, in a short period of time, the removal of rainwater.
- It must provide satisfactory thermal protection – in winter and in summer – for the spaces it covers.

The above must be taken into account in the design stage and in the construction of the building in order for the result to be functionally and structurally sound.

From the energy point of view the roof plays an important role in low buildings (one to three storeys). Figure 16.42 shows energy savings as a function of insulation thickness.

Floor over the ground. By insulating the floor over the ground, energy savings (Figure 16.43) are significantly smaller than those of other building elements. This is because the temperature of the ground is higher than the external temperature.

Studies in the urban area of Thessaloniki in Greece have shown that the lack of thermal insulation of floors over the ground facilitates thermal dissipation through the floor and thus the internal conditions during summer can be better in adjacent rooms.

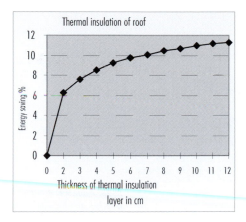

Figure 16.42. Energy saving by insulating the roof of an uninsulated urban building; climatic conditions of Thessaloniki[9]

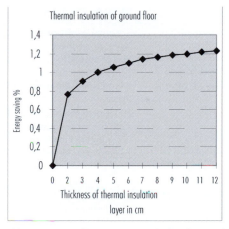

Figure 16.43. Energy saving by insulating the floor over the ground of an uninsulated urban building; climatic conditions of Thessaloniki[9]

Floor over a pilotis. With the insulation of a floor over a pilotis (an open space on the ground floor), not only are important energy savings achieved, but sound insulation of the spaces above is also ensured. Figure 16.44 shows the savings as a function of insulation thickness.

Openings

The glazing elements of a building are normally its weak points and ones that concern the energy-conscious designer more than the opaque building elements. Their particularity lies in a number of functions that they are supposed to perform. Each one of these functions demands a different surface (see Table 16.2). So, in order to achieve good living conditions and according to the function that is examined each time (lighting, ventilation, sound protection, etc.), the required surfaces vary and can be small, medium or large.

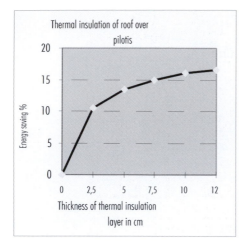

Figure 16.44. Energy saving as a result of insulating the roof over the pilotis of an uninsulated urban building; climatic conditions of Thessaloniki[9]

Table 16.2. Requirements of openings

Basic requirements – large surfaces	Secondary requirements – small surfaces
Selective maximization of solar heat gains	Thermal insulation (in winter and summer)
Daylighting	Wind protection
Ventilation	Sun protection
Sunlighting	Sound protection
Functional and visual connection of the	Protection from unwanted viewing
interior and exterior spaces	Privacy
	Functional requirements of the space; closed
	surfaces for the positioning of furniture
	Cleaning and maintenance

While the building is being designed, in order that all the requirements can be satisfied, aesthetics, morphology and function should be taken into account, as well as energy. Based on these parameters, the best size of the surface of the openings can be determined.

In recent years, and according to the new 'energy-conscious logic', much criticism and many doubts have been expressed regarding every kind of urban building in which lightweight glass facades dominate, irrespective of the building's orientation, and the use and normal function of the internal spaces in terms of climatic conditions (thermal comfort).

The overheating of spaces in summer, the increased thermal losses and consequently the low temperatures in winter force tenants to install air-conditioning to make the place habitable in terms of climate. This results in a dramatic increase in the operational cost of these particular urban buildings.

In terms of energy, when the dimensions of the openings are decided, thermal losses should be considered as well as the anticipated thermal solar gains, which mainly depend on the orientation of the opening. Thermal losses can be divided into transmission and ventilation losses.

One should take into account the fact that, in comparison with solid walls, the transmission losses through conventional glass can be five to seven times higher. The

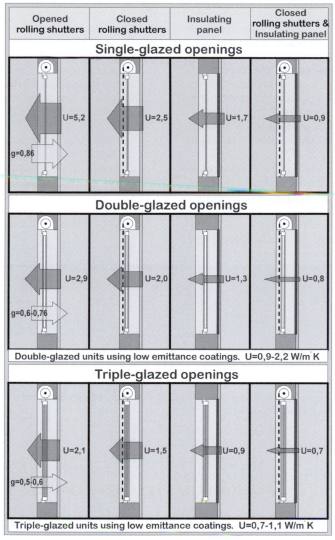

Figure 16.45. U-values of different types of glass panes with or without night-time insulation. Lower U-values of 0.7 or 1.1 W/m²K have been achieved by using advanced glazing; g is the total solar energy transmittance

rate decreases in advanced glazing materials (low-emittance coatings, aerogels, transparent insulation, evacuated glazing etc. – Figure 16.45), where U-values of 0.7–2.2 W/m²K have been achieved.

Ventilation losses depend on the way in which the opening in the wall is implemented and generally on its airtightness. Old openings, and openings that do not close tightly, present high ventilation losses. When the windows are used for supply of fresh air, then the amount of heat that is lost is very large.

Suggestions as to what should be done and what should not be done as far as windows are concerned are as follows.

- Provide for weather-proofing on windows in order to reduce the unwanted ventilation losses significantly.
- Construct the connection between the frame of the window and the wall very carefully, so as to avoid or reduce thermal bridges.
- Choose double-glazing with low emissivity and low *U*-value to reduce transmission losses (Figure 16.45).
- Choose the size and the position of the opening in an external wall according to the orientation and therefore the quantity of thermal solar energy that you want to collect, as well as in relation to sunlight being directed inside the space (Figure 16.46).

- Do not place large windows on the east or west sides. They will cause overheating in the summer.
- Do not forget to use the proper sunshading device on the openings in each orientation and keep in mind the total cost of the openings, including also the cost of solar protection.
- Do not place large openings on the north side. The gains are very limited and result only from diffuse radiation, while there are large thermal losses.

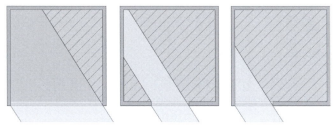

Figure 16.46. The size and location of windows affect solar gains and natural lighting

PASSIVE SOLAR HEATING STRATEGIES

The implementation of passive solar heating strategies in an urban building presupposes the implementation of energy-saving measures, so as to increase the efficiency of the strategies and to ensure the conservation of heat in the internal spaces for a long period. Moreover, everything in the *Design guidelines* section (siting and site layout, overshadowing by adjacent buildings, the form of the buildings, etc.) should be taken into account.

Heating strategy

The heating strategy that uses passive solar systems is based on solar collection, storage and distribution of heat and undoubtedly on heat conservation.

Passive solar systems are simple constructions, usually integrated into the building's shell. The materials used are very often common building materials. Their basic goal is to exploit solar energy to the full and to provide a passive form of heating, in order to achieve thermal comfort conditions with the lowest possible energy consumption.

Solar collection

In urban buildings, provided that the right orientation is secured and overshadowing from adjacent buildings is avoided, applying the well-known principles of passive solar systems is a simple matter (Figure 16.47).

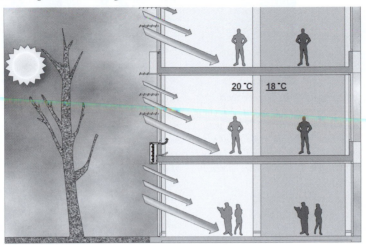

Figure 16.47. Solar collection through south openings

The systems that can be applied are as follows:

- direct gain through windows, clerestories or skylights (Figure 16.48);
- the mass or Trombe wall (Figures 16.49 and 16.50);
- the thermosyphonic panel (Figure 16.51);
- the attached sunspace (Figure 16.52);
- the atrium (Figure 16.53).

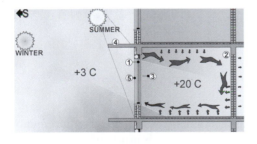

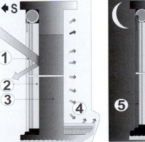

Figure 16.48. Operation of direct solar gain during morning in winter and summer.
1.Double glass pane – use of the greenhouse effect. 2. Thermal storage by means of the solid shell of the building. 3. Moveable elements used for night-time insulation. 4. Sun protection. 5. Insulated awnings

Figure 16.49. Operation of a mass wall during the day in winter and at night.
1.Double glass pane. 2. 10–15 cm air gap. 3. Mass wall. 4. Thermal radiation. 5. Night-time insulation or daytime sun protection

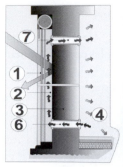

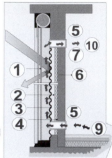

Figure 16.50. Operation of a Trombe wall during the day in winter and at night. 1.Double glass pane. 2. 10–15 cm air gap. 3. Mass wall. 4. Thermal radiation. 5. Night-time insulation or daytime sun protection. 6. Lower vent, closed at night during winter. 7. Upper vent, closed in the summer

Figure 16.51. Operation of a thermosyphonic panel during the morning in the winter and at night. 1.Double glass pane. 2. 2 + 2 cm air gap between layers 1 and 3 and between layers 3 and 4. 3.Absorbent surface. 4.Insulation. 5.Pipe of diameter 15 cm. 6.Wall. 7.Thermal radiation. 8.Night-time insulation or daytime sun protection. 9. Lower vent, closed at night during winter. 10. Upper vent, closed in the summer

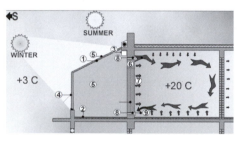

Figure 16.52. Operation of the attached sunspace in summer and during morning in winter. 1. Double glass pane. 2. Insulated floor. 3. Ventilation vent. 4. Fresh-air inlet. 5. External insulated awnings for night-time insulation or daytime sun protection. 6. Solid thermal storage element heated by radiation. 7. Heat transfer through the balcony door. 8. Air circulation vents. 9. Dampers for the vents

Figure 16.53. Atrium

If for reasons of noise, privacy or security, or on account of the functional requirements of the internal spaces, south openings cannot be used extensively in urban buildings, then a combination of the other opaque passive solar systems could be the solution.

The mass and Trombe walls and the thermosyphonic panel can be applied in existing urban buildings through the use of very simple additional structures in place of or close to conventional walls.

To avoid overheating of the interior during the summer, particularly in urban buildings that have high-density use (office blocks, schools, commercial buildings, etc.), it is necessary to close down any of the systems by providing a suitable form of shading, opening the glass collector in front of the system and closing the upper and lower vents in the case of the Trombe wall and the thermosyphonic panel.

Suggestions as to what should be done and what should not be done concerning solar collection are as follows.

☑

- Aim at a south orientation of the glass collector surfaces. The orientation is the most important factor for achieving maximum solar gain. A ±25° deviation to the east or the west reduces the effectiveness of horizontal shading devices and also slightly reduces the solar heat gains. This should be taken into serious consideration in the design of urban buildings, especially in the Mediterranean area, where summer over-heating is a common phenomenon.
- It is recommended that vertical solar collector panes be used in order to deal more effectively with problems such as cleaning, shading and night-time insulation. If the vertical slope of the solar collector pane is changed in order to be closer to 90° to the angle of the solar rays in winter, the amount of solar energy collected during winter will be increased, depending on the latitude of the site. In this case, overheating is also likely to occur in summer if there are problems in fitting solar shading devices.[13]
- In order to increase the thermal solar gains, use reflectors in front of the solar collector pane, either outside the building or in the interior, in order to direct solar radiation towards specific locations where thermal storage mass is placed. Adjust the angle of the reflectors and investigate any glare problems. The materials used for solar reflectors are aluminium sheets, thin metal foils, glass reflectors or mirrors.
- Use external moveable insulation on over-cast days and during the night, combined with internal curtains where possible, in order to reduce heat loss from the solar collectors.
- The quality of the glass used in solar collector panes significantly affects the amount of energy collected, as well as the amount of thermal loss. It is therefore important to choose glass panes of high quality and high standards, which will ensure high total solar energy transmittance g as well as low heat loss (U-value).

☒

- Avoid shading the south collecting surface with external obstacles, such as nearby constructions, buildings, trees, vertical shading devices, etc. if there is a heat demand in the building (and not just cooling demands).
- Avoid the use of reflective glazing in a south orientation as this reduces the admittance of solar irradiation into the interior and therefore is not compatible with maximizing passive solar gain.
- Do not forget to clean external collecting surfaces frequently. Dirt and dust decrease the effectiveness of the surfaces, because they reduce the transmittance of solar energy.
- In temperate and hot climates avoid a large amount of solar heat gain, especially in non-residential buildings.

Heat storage

Positive and negative considerations concerning heat storage are as follows.

☑

- Use heavyweight materials with increased heat storage capacity for the construction of walls and floors (Figure 16.54). Metal tanks filled with water, reinforced concrete and solid brick or stones are some of the most commonly used building materials that have high thermal storage mass (see the section on *Thermal capacity* later in this chapter).
- Place the thermal-storage mass elements mainly in the area that receives the greatest amount of solar radiation and consequently has the greatest solar gains.
- If the existing thermal mass appears to be insufficient, consider solutions such as gravel bedding under the floor between the ground and the ground floor or an extra thermal mass wall placed inside for extra thermal storage capacity.
- Select dark colours for the external surfaces of heat storage elements. It is well known that dark-coloured surfaces absorb solar radiation better than light-coloured ones. For example, a grey-coloured surface absorbs a large quantity of solar radiation by primary storage (75%) while a brick-red wall absorbs only 55%.
- Investigate the need for heat storage in a building for both heating and cooling periods.
- The construction elements around the heat storage facilities should be externally insulated so as to avoid heat loss and to direct heat to the living spaces.

☒

- Do not decrease the thermal storage capacity by using coverings and fabrics like carpets, furnishings or other obstacles where direct sunlight falls onto the building elements, because it is in these surfaces that most of the heat-storing process is taking place.
- Avoid the use of light-coloured floors when they are used as thermal storage elements, because the storage efficiency is significantly decreased.
- It is not recommended that high heat storage capacity should be used in buildings that are not operated continuously, because it means a great time lag when the heating of the interior is begun. In these cases the thermal insulation of the envelope of the building must be a priority.
- Any floor heating system is incompatible with the use of the floor as a heat storage element because the floor is already thermally charged with heat from the heating system and it is therefore impossible for it to receive most of the heat from the incoming solar radiation.

Figure 16.54. Heat storage in elements with high thermal inertia

Heat distribution

Heat distribution is illustrated in Figure 16.55.

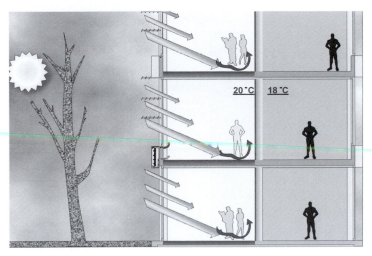

Figure 16.55. Solar heat distribution

Positive and negative considerations concerning heat distribution are as follows.

☑

- Natural distribution of heat is preferable to hybrid or forced distribution in which mechanical means (e.g. fans or pumps) are used.
- Forced (mechanical) heat distribution must be used when immediate distribution of heated (or cooled) air within a space is needed, or when the heat storage area is far from the target space.
- Locate the storage area near the heated area in order to use heat distribution by natural means only. In this way thermal losses are decreased as well as the time lag in the heat transfer.
- Use automatic control mechanisms to turn fans on if and when they are required to distribute the stored heat in a living space.

☒

- If the thermal insulation of the floor is required, this must not be carried out on the upper side of the floor, but underneath, in order to avoid a decrease in the thermal capacity of the floor.

Heat conservation

The principles of heat conservation are shown in Figure 16.56.

Positive and negative considerations concerning heat conservation are as follows.

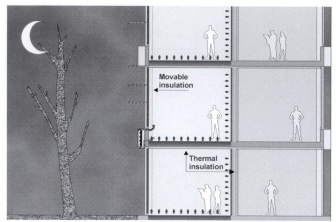

Figure 16.56. Operation of the building by night, when heat conservation is applied in the building's envelope

☑

- A sheltered location, the microclimatic conditions, the compact form of the building, the arrangement of internal spaces (thermal zoning, buffers between heated spaces and outside) are the main design measures of heat conservation (see discussion of thermal design earlier in this chapter).
- Moreover, high insulation levels of opaque and transparent elements (double or triple glass panes), reflective coatings on inside surfaces, moveable shutters and controlled ventilation, to provide fresh air only as and when required, conserve the collected, stored and distributed heat for a long period inside the heated spaces of a building.

☒

- Internal insulation of the external construction elements is not recommended. It causes a decrease in the thermal capacity of the elements and an increase in thermal bridges.
- Do not enlarge the openings in the north, east or west elevations. This causes an increased thermal load in winter (see the earlier discussion in this chapter on the design of openings) and a corresponding increased cooling load in summer.

PASSIVE SOLAR COOLING STRATEGIES

In urban buildings most of the overheating problems during the summer are caused by heat-producing facilities and installations, by a high level of artificial lighting and by the presence of a large number of occupants or clients. Such buildings are office buildings, stores, shopping centres, entertainment halls, etc. In residential buildings, the cooling needs are significantly lower – with the exception of those in the Mediterranean area. The main reasons are the lack of large heat-producing facilities and easily applied natural cooling strategies such as ventilation and shading.

Control of overheating

External gains control

As mentioned earlier in this chapter, several techniques provide cooling through the use of trees, natural vegetation, neighbouring buildings and water near building surfaces, either for shading and evaporative cooling or by wind channelling in summer.

Sun shading

A combination of external and internal shading devices can offer efficient solar control. An ideal solution is to have shades that can be moved, either seasonally or on a daily basis.

Positive and negative considerations concerning shading are as follows.

☑

- Divert the solar radiation from the opaque solid elements of the building envelope by applying efficient strategies. Special care should be taken to shade the external transparent elements of the building envelope in order to reduce the incoming heat flow and to reduce the risk of overheating.
- Use fixed or moveable sunshading arrangements or find the appropriate combination and arrangement for the most efficient solar protection. Moveable shadings arranged by the inhabitants are generally a good solution.
- External shading devices are preferable and more effective than internal ones (Figures 16.57 and 16.58). This category includes devices fixed to the outside of the window jamb or attached to the building envelope. Operable units include jalousies made of wood or metal, exterior venetian blinds, shutters, awnings and fixed or moveable overhangs. Fixed horizontal louvres or overhangs in general are effective in the control of sunlight in south-oriented windows (Figure 16.58), while in west- and east-oriented windows the shades should be vertical to counter the path of the summer sun (Figure 16.59). The most important advantage of exterior-mounted

☒

- Avoid the exclusive use of internal sunshading devices, such as roller shades, venetian blinds, curtains and drapes. Combine internal and external shading devices. The major disadvantage of shades, drapes and the like is that, regardless of how reflective they are, they trap heat on the interior of the glass, so it remains indoors. Internal and external shading must be designed as a whole.
- Do not block the daylighting and the natural cooling ventilation by using the internal shading devices as obstacles near the openings in the building.
- External shading devices, and especially horizontal ones, should not be in direct contact with the external envelope. Holes or openings should be kept or constructed in order to assist and enhance air circulation and to prevent heat from getting trapped in the upper space of the shading devices.
- Glazing materials with a low transmission coefficient, for aesthetic reasons, are not recommended for houses designed as full-time residences, especially for use in south-oriented openings. With this type of glass, solar gains are reduced not just in the summer but throughout the year.

sunshades is that they block the heat on the outside of the building.

- In urban buildings with specific uses, such as office buildings, shopping centres, stores etc., special glazing with a low solar transmittance coefficient or, if required, a high shading coefficient, should be used.
- When designing the external shading devices, take into account the need to control solar gains in summer, as well as the heating and the lighting performance of the building.

Figure 16.57. Interior shading devices (roller shades, venetian blinds, curtains and drapes)

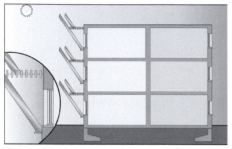

Figure 16.58. Exterior horizontal shading devices (fixed or moveable overhangs)

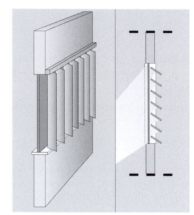

Figure 16.59. An exterior vertical shading device

Examples of sunshading are shown in Figures 16.60 and 16.61.

Figure 16.60. Sunshading of a pedestrian zone in a school complex

Figure 16.61. Sunshading by a reed roof in a market place in Morocco

Thermal capacity

Thermal inertia is an important parameter, especially during the summer when temperature fluctuations during a 24-hour period are very significant. It helps avoid the daytime heat and keep the night-time coolness inside the building for a longer period. The thermal capacity of a building's elements delays the heat transfer to the interior of the building by soaking up excessive heat for several hours. During the night, when the external temperature is lower, the stored heat is slowly expelled to the environment by radiation and by convection (Figure 16.62).

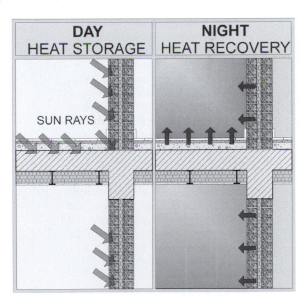

Figure 16.62. Operation of the thermal mass of a floor during day and night

Positive and negative considerations concerning thermal capacity are as follows.

☑

- In warm climates such as those of the Mediterranean, materials of high thermal capacity should be used in order to help the building operate as a thermal storage bank.[13] This will improve the building's performance during both summer and winter.
- Special care should be taken in providing night-time natural ventilation during the summer period, which is essential for the rejection of the stored heat by convection. This is the best (if not the only) way of passive cooling during the night. In regions with great temperature differences between day and night the night-time ventilation reduces both the cooling load of the building and the maximum daytime internal temperature by 1–2 K.
- Always keep in mind that the increased thermal capacity helps to reduce cooling loads during the summer and to avoid excessive temperatures – overheating – caused by high solar and internal thermal gains. Thermal mass delays the appearance of the maximum temperature values (time lag) and also reduces their absolute values.
- Thermal mass should be combined with adequate insulation of the external envelope of the building. This combination leads to reduced mean internal temperatures and to satisfactory thermal comfort conditions during the summer.
- In order to place the thermal mass within the building and distribute it correctly, a series of parameters should be taken into consideration. These are: the orientations of the building and its facades; the thermal insulation of the construction elements; the potential and the characteristics of the type of ventilation provided; the external climatic conditions; the operation schedule and the control mechanisms of the installed heating and cooling systems; and the thermophysical characteristics of the building materials.

☒

- Thermal mass placed on northern facades or elements generally does not play a very important role, except in urban buildings and especially in office buildings with large internal gains due to artificial lighting, electric installations and office equipment.
- Do not underestimate the role of thermal mass in the east or mainly the west elements and facades, especially during the summer period. Both the external envelope of the building and the internal spaces become thermally charged by solar radiation.
- Do not neglect the contribution of thermal mass combined with adequate insulation of the roof of the building. The roof receives solar radiation almost vertically during the summer period, causing overheating problems in the spaces directly below.

- In existing urban buildings of a given design and mass distribution it is very important to combine the elements that already exist and their thermal capacity with the correctly placed insulation of the envelope.
- Thermal mass with no insulation effectively cancels out any possible solar gains during the winter.

External surfaces of the building

Coatings. Positive and negative considerations are as follows.

☑

- In order to select the colours of the external surfaces, the heating and the cooling loads should be estimated. In cold climates where the heating needs are greater, dark colours should be preferred for the building's elevations (Figure 16.63), while light colours (Figure 16.64) should be used in warm sunny areas with long hot summers in order to increase the surface reflectance and to avoid high surface temperatures.
- The use of highly heat-reflective materials is necessary on surfaces exposed to the summer sun. Special attention should be given to east- and west-oriented surfaces and to the roof.

☒

- Do not use dark colours on external surfaces in buildings in warm climates. Because of the higher absorbency of the dark colours, the temperature of the building envelope, and consequently the cooling load, is increased.

Figure 16.63. Buildings painted in light colours

Figure 16.64. Buildings painted in dark colours

Glazing. Positive and negative considerations are as follows.

☑
- Control the solar heat gains by sensible sizing and positioning of the glazed surfaces, taking the orientation into consideration.
- East- and west-facing openings should be minimized. They increase the summer overheating risk and therefore the cooling load.
- In urban buildings with a given orientation, east or west, when there is a demand for large openings for architectural or functional reasons, choose a special glazing type that reduces solar heat gains during the summer period, while allowing for natural lighting and unobstructed views (Figure 16.65). The most effective glazing types in these cases are the reflecting or the absorbing ones.

☒
- Do not neglect to implement the proper sunshading devices at every external opening of the building according to its orientation (see the previous section). The larger the unprotected openings are, the greater is the solar gain and therefore the cooling load and the danger of overheating.
- Reflective glazing is not recommended for use in residential buildings, because it is rather expensive and architecturally inappropriate. It is therefore more suitable for office and commercial buildings.

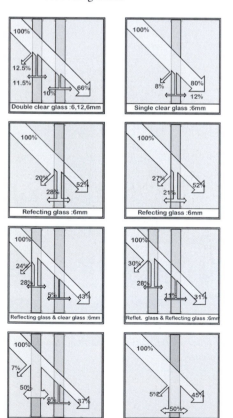

Figure 16.65. Different types of glass panes[14]

Planting next to building skin. Positive and negative considerations are as follows.

☑

- Place plants, such as climbing ivy, next to the building skin (Figure 16.66). The most obvious benefit of vegetation is its summer shading ability.
- Bear in mind that a dense cover of plants will intercept the sun's radiation before it reaches the building skin, thereby reducing the exterior surface temperature and the amount of heat conducted into the interior.
- In winter, the boundary layer created by an evergreen leaf blanket serves an insulating function that restricts heat losses.

☒

- As is mentioned by Watson and Labs,[4] a summer disadvantage of planted facades is the fact that warm air is trapped around the shell of the building. The beneficial effect that a cool current of air could have, in driving away this belt of warm air, is reduced. This disadvantage does not exist in cases where the air current is strong enough to cool the external surfaces of the building.
- Do not decide to put plants next or onto the building skin if you are not sure whether you are going to keep them for many years. Many climbing plants destroy the colour coating and the plaster and they may be difficult to remove.

Figure 16.66. A planted facade

Planted roof. Big urban building complexes satisfy a lot of needs, but at the same time destroy the natural aspects of cities. This has many negative effects that should be kept under control as far as possible.

A garden on the roof of a building is one of the positive interventions the urban environment needs. Such gardens can be used by the residents as rest or recreational spaces, improving in this way both the quality of life and the microclimatic conditions.[15] Some construction detail is shown in Figure 16.67.

Figure 16.67. Planted roof – construction detail

Positive and negative considerations relating to planted roofs are as follows.

☑

- Viewed in terms of energy saving, a planted roof positively influences the interior climatic conditions of a building, since the conditions are considerably improved both in winter (higher internal temperatures) and in summer (lower internal temperatures), thus reducing energy consumption.

☒

- The only negative aspect that can be identified in installing a planted roof is the increase in the construction's weight and cost. However, there are many positive aspects. As well as saving energy and environmental benefits, the quality of life in the interior spaces is improved and the value of the property is raised.

Internal gains control

Electric lighting. Positive and negative considerations are as follows.

☑

- The building and its openings should be properly designed in order to ensure natural lighting of internal spaces, and so reduce the need for artificial lighting and therefore the cooling load.
- Natural lighting of the internal spaces through openings should be designed in combination with sunshading devices so as to avoid solar heat gains. It should also be combined with the use of high-efficiency artificial lighting. These measures help reduce the need for artificial lighting, as

☒

- Do not design an entirely 'blind' building with absolutely no external windows. Even if visual contact with the exterior is not required by its use (theatres, store-houses etc.), openings are necessary, mainly for natural ventilation.
- It is not preferable to use artificial lighting of low efficiency (Figure 16.68). It contributes to increased electricity consumption and cooling-load demands. The heat produced by artificial lighting adds to the casual gains, increasing the indoor temperature.

well as appreciably reducing the cooling load.

- In urban office buildings and especially commercial buildings, the effect of artificial lighting on the cooling load is significantly high, because of the increased lighting needs and therefore increased heat generation. In these cases it is essential to look for alternative solutions, such as roof openings, skylights etc., in order to increase natural lighting and consequently reduce the need for artificial lighting.

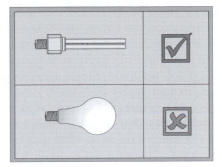

Figure 16.68. Artificial lighting of high and low efficiency

Electrical appliances and equipment. Positive and negative considerations are as follows.

☑

- Use domestic appliances and office equipment (Figure 16.69) of high efficiency to avoid increasing the cooling load.
- The boilers and the hot-water tanks should be carefully insulated. In this way, not only is their energy efficiency improved, but heat flow to the internal spaces is minimized as well.

☒

- Do not place electrical appliances and equipment in positions in the building from where it will be difficult to expel the accumulated heat through natural ventilation.

Figure 16.69. Office equipment

Inhabitants. The inhabitants themselves are a significant source of thermal gains. The heat produced by a human body depends on the person's activity. A seated adult emits about 80 watts of energy to the environment, an amount that is increased by more intense activities. Especially in urban buildings or buildings for special use, e.g. administration and public buildings, the casual gains caused by the occupants are the key factor for determining the cooling (and heating) loads. In these cases population is an inelastic factor and it is therefore essential to focus on passive cooling techniques (extended cross-ventilation) and to control overheating in order to decrease the cooling load demands and to improve the sense of thermal comfort before using air-conditioning systems.

Natural cooling

Ventilation strategies

It is well known that air movement is caused by the difference of pressure distribution between the interior and exterior of a building. Thus the air moves from areas of high pressure to areas of low pressure. This airflow, which is directly connected with the ventilation of the interior, has a cooling effect, because it drives away heat both from the building and the human body itself.

If the difference in pressure is insufficient to cause the movement of air naturally (wind and stack effect), then this action can be reinforced by mechanical means (fans). Forced air movement should generally be avoided and should be resorted to only when all the natural ventilation techniques presented below have been exhausted and when these, in spite of everything, have proved to be inadequate for the natural ventilation of the building or if it is discovered that opening the windows causes noise disturbance or security problems.

Ceiling fans are generally good choices, since by increasing the air speed they increase the heat convection exchange, thus markedly improving the sense of thermal comfort. Measurements have shown that as air movement in the interior speeds up, the acceptable indoor air temperature increases by about 2 K.[16]

Openings. By the frequent ventilation of the spaces throughout the day, and especially by night-time ventilation, natural cooling is achieved, as warm air is removed from the building and from the human body itself as well. Positive and negative considerations are as follows.

☑

- The arrangement of the openings in the building elevations is the determining factor for efficient natural ventilation (Figure 16.70).
- In urban buildings, if the wind flow is parallel to the external openings or if there are windows on adjacent walls, wingwalls that are vertical to the wind-exposed walls can improve the ability of the existing openings to capture the incoming wind or

☒

- Do not place the openings in a facade parallel to the direction of the wind. Air will flow around the building without entering the living spaces and natural cooling will not be achieved.
- Avoid single-sided ventilation. It is an inefficient cooling strategy compared to cross-ventilation.
- It is not recommended that the openings (inlet–outlet) be placed in the upper part of

to function as air exits and thus assist cross-ventilation.

- It is advisable to place air inlets in high positions, if by the use of night-time ventilation you seek to discharge the thermal load of a compact roof that works as a heat storage element during the daytime.
- External openings close to living areas increase a sense of hotness during summer as a result of the evaporation of perspiration from human skin.
- To ensure a feeling of coolness for the inhabitants through cross-ventilation, position the openings at body height.
- In urban buildings where it is impossible to form a cross-ventilation flow because of the lack of opposite openings, an inappropriate internal space arrangement or the disproportionate volume of the building (too great a depth or length), the roof outlets and the central staircase could increase a cross and upward airflow.
- It is better to have two openings placed as far apart as possible on the same elevation of a building, rather than one single opening in the same surface area. In this way the ventilation of the building is noticeably improved.
- Always try to ensure the cross-ventilation of the spaces by placing openings in opposite walls. This is the most effective strategy for natural cooling.
- It is important to place an opening diagonally opposite the main direction of the wind, so as to allow airflow to circulate throughout the entire building.
- Ceiling fans, oscillating fans or box fans can be used as hybrid systems in order to increase the airflow speed and to improve the sense of thermal comfort and the convection thermal exchange.

the space. If in the range 1.00 to 1.50 m in height, they will not generate a high air velocity in the living or working zone.

- Do not use forced air movement (mechanical ventilation) inside the living spaces when natural ventilation can be achieved, unless noise and security problems are created when the windows are opened, as is often the case in urban environments.

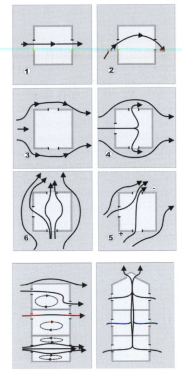

Figure 16.70. Ventilation strategies by the arrangement of openings.[17]

Wind tower. Positive and negative considerations are as follows.

☑
- The construction of a wind tower is quite simple and feasible. If it is not possible to have openings on the windward side of an

- It is not recommended that the operation of the ventilation system be interrupted during the night, when

urban building, a wind tower can be constructed in the case of a new building, or an existing chimney can be used in the case of an existing building, forming wind-scoop inlets in the upper side (Figure 16.71 (left)).

- The wind tower should have adequate openings in its upper side in order to direct the air captured by the inlets into the inner spaces.
- Openings on the leeward side of the building are necessary for air circulation and the cooling of the internal spaces (Figure 16.71 (left)).
- A wind tower system can also work in reverse. If it is impossible to have the openings on the leeward side and they are therefore placed on the windward side, then the air can enter through common windows. The wind tower should be constructed in the low-pressure region, in order for the hot-air suction to start the airflow up the tower (Figure 16.71 (right)).
- If it is not possible to have openings on both sides (the windward and the leeward), then a combination of two wind towers would be effective. One wind tower would be used as a cool-air inlet and the other as a warm-air outlet.

it is more effective, because the external air is cold and heat is expelled from the building elements much more easily. The use of automatic control of natural ventilation by flap valves can be a feasible option.

- Do not obstruct the airflow in the interior of the building by shutting the inlets of the wind tower.
- Do not forget to open the windows on the leeward side of the building and, in particular, do not obstruct the airflow towards this side by placing obstacles in front of the opening (walls, furniture, decorative elements, etc.).
- Placing the wind tower's internal openings at a higher level in the internal spaces is not recommended, because natural cooling at the living and working level would not be achieved.

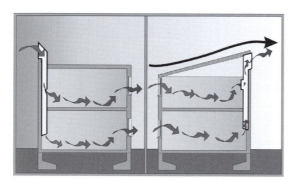

Figure 16.71. Wind tower operation: inlet of cool air is shown on the left and outlet of warm air on the right

Solar chimney. Positive and negative considerations are as follows.

- In urban buildings where there are fixed west-oriented facades, the buoyancy effect can be exploited by implementing the solar-chimney technique.

☒

- Side openings in the chimney should be avoided in order to prevent potential backward flow.

- Construct a chimney on the side that faces the summer sun and leave air outlets at the top of the chimney as well as in the internal spaces of the building (Figure 16.72).
- Cool air inlets can be constructed on the windward side. In this way, buoyancy forces due to the temperature difference between the windward and the leeward sides, when heated by the sun, help to induce an upward flow along the plate.
- Make sure that you can have openings on the windward side before deciding to create a chimney.

- Avoid using reflective materials for the surfaces of the external sides of the chimney. Opt for dark colours that absorb a larger amount of thermal energy, so that higher temperatures are achieved in the interior of the system.

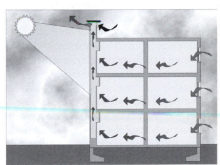

Figure 16.72. A solar chimney – schematic section

Ventilated roof and wall construction. Vent-skin walls (Figure 16.73) or roofs may improve the dissipation of heat stored inside the building shell and intercept solar access to the main internal wall. With this type of wall and roof the condensation of water vapour inside the structure is prevented.

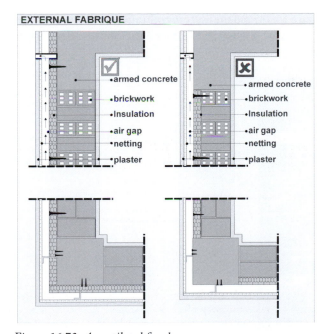

Figure 16.73. A ventilated façade

Positive and negative considerations are as follows.

☑

- Use a ventilated cavity roof or wall construction so as to eliminate the risk of overheating of the building envelope, which is exposed to solar radiation. The heated air column inside the structure will flow upwards by natural convection.
- Provide openings at both the top and the bottom of the wall, or side openings in the vented roof. The building element will effectively cool itself by expelling the excess solar heat gains, thus maintaining an internal temperature near to that of the air outside. This technique has long been applied to roof ventilation and is applicable to walls as well.
- The top opening of a ventilated wall should be protected from rain penetration.

☒

- Do not close the vents or the openings of a ventilated wall or roof because this increases the risk of overheating of the wall as well as of the adjacent internal spaces.
- The lower opening of a ventilated wall should be regularly inspected and obstacles that block air circulation removed. A metal grid could be used for protection.

Cavity shell collector. If the exterior skin of a south-facing cavity wall is a glazing material, the air column will heat faster and the airflow rate of the system will be increased. This configuration features in the three solar heating systems, a mass or Trombe wall (Figure 16.74) and the thermosyphonic panel. Depending on the arrangement of dampers and vent openings, such systems can be used selectively for heating and/or for ventilation.

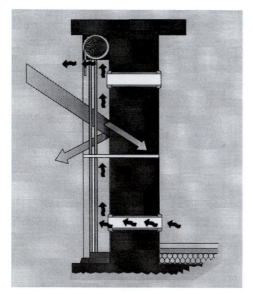

Figure 16.74. Cavity shell collector – Trombe wall

Clerestories and roof vents. Positive and negative considerations are as follows.

- Provide clerestories or roof vents to allow for direct solar access into the interior of the building during winter and to facilitate natural ventilation, and consequently natural cooling, during summer (Figure 16.75).

- Leaving the external openings without solar protection is not recommended. The reduction of the cooling load through ventilation will be cancelled out by the direct solar access to the internal spaces.

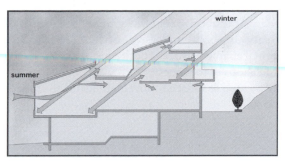

Figure 16.75. Clerestories or roof vents for direct sun access or ventilation of internal spaces

Ground cooling strategies

Direct contact. Positive and negative considerations are as follows.

- There are certain advantages to constructing underground buildings or storeys: protection from noise, dust, storms and solar radiation; limited infiltration; and potentially increased security.
- Underground buildings or storeys can have thermal benefits for both cooling and heating conditions (Figure 16.76). In these buildings maximum utilization of the thermal mass of the ground is achieved.
- The potential for wide-scale construction of such buildings is fairly limited, especially in urban areas, with the possible exception of urban areas in mountainous regions.

- The construction of underground buildings or underground storeys in a building with increased natural lighting demands is not recommended.
- Do not overlook the usual problems arising in underground constructions, such as the increased construction cost, the poor natural lighting and the dampness from which the building will have to be protected.

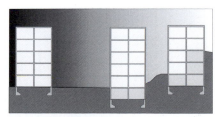

Figure 16.76. Relation of the building to the ground

Buried pipes. The technique of buried pipes (Figure 16.77) should be used for pre-cooling air. Points to note are as follows:

- Plastic or metal durable stainless pipes can be used for the construction of buried pipe systems.
- Place the pipes at a certain depth in the ground (not less than 1.50 m). The entrance of the outside air to the pipes should be in a shaded location, preferably with cool air currents. The air is then introduced into the building through the pipes.
- As reported by Antonucci *et al.*,[17] the temperature drop of the circulated air is a function of the dry-bulb temperature of the inlet air, the ground temperature and the thermal characteristics of both pipes and soil, as well as the air velocity and the dimensions of the pipes.
- Fans should be used to force air movement inside the pipes if the system is not efficient enough.

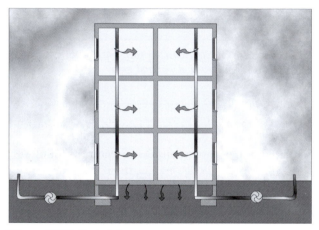

Figure 16.77. Buried pipes

Evaporative cooling systems

Roof spray. In urban multistorey buildings the benefits from this technique (Figure 16.78) mainly affect the top floor.

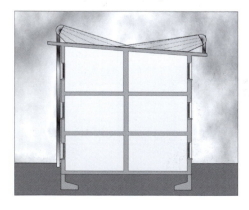

Figure 16.78. Roof spray

Postive and negative considerations are as follows.

☑

- This simple installation that is operable when needed requires no more than a system of external pipes for water supply and the ability to sprinkle the flat roof of the building. Under the influence of water the roof temperature remains constant and, as Watson and Labs report,[4] this is a fairly efficient operation such that over 90% of the solar load upon the roof can be alleviated. The main advantage of this technique is the avoidance of rapid expansion and contraction of the construction elements, which damage the materials of the various layers of the roof.

☒

- It is not recommended that the roof-spraying technique should be used if the slope of the roof is not adequate for the quick removal of water and/or if it is not highly waterproofed.
- Avoid using the roof-spraying technique if the climate in the specific location is humid during the cooling period, because its efficiency is dramatically decreased.

Roof pond. The cooling effect of the roof pond technique (Figure 16.79) is the same as that of spray systems. The main difference lies in the great thermal capacity that a roof pond provides.

The system, if it were converted into a pool on a flat roof, could be applied to urban buildings, such as hotels and sports and fitness centres. However, the high construction cost and the problems arising from the waterproofing requirements of the roof do not allow for extensive application of this specific cooling strategy.

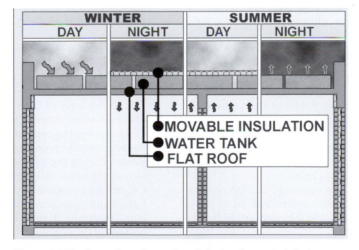

Figure 16.79. Operation of a roof pond during day and night in winter and summer

NATURAL LIGHTING STRATEGIES

Natural lighting plays an important role in urban buildings, and especially in office buildings, hospitals, factories and schools, the lighting needs of which are extremely high (both in quantity and in quality). Therefore, as the natural lighting provided is increased, the need for artificial lighting is reduced accordingly and consequently there is a lower consumption of expensive electric energy.

The window, as has often been mentioned, constitutes a basic component of the shell, allowing a link between the interior of the building and the external environment. This function makes natural lighting and ventilation possible, together with views of the surroundings and passive solar heating of the interior.

The type, size, shape, position and orientation of the openings, in conjunction with various control systems, are basic factors affecting the amount and distribution of light in the illumination space.

Guidelines

Positive and negative considerations are as follows.

☑

- Take into account the requirements of the building in terms of lighting, heating and cooling from the first design stage. Any oversight increases the energy burden of the building, while interventions at a later stage when the building is in use are too difficult and uneconomical to carry out in most cases.
- Choose the position of the spaces and their orientation in relation to their use and the importance of natural lighting for these spaces.
- For purposes of natural lighting, ensure that there are south openings. They provide high lighting levels, if somewhat variable illumination, plus high thermal solar gains in winter and average ones in summer.
- Opt for the division of the total lighting surface into smaller openings. More than one opening produces a more uniform distribution of light.
- Position the openings at high, central points in the outside wall in order to achieve a greater distribution of light into the interior.
- Use of advanced glazing (with low e, low U-values and high light transmittance) is recommended, in order to control ther-

☒

- It is not recommended, as far as natural lighting is concerned, that multistorey buildings should be designed with great depth. Particular attention must be paid to office buildings, teaching rooms, factories etc. and generally to buildings where a specific task is being carried out in a fixed position. Great depth of the rooms results in intensive use of artificial lighting and consumption of electrical energy, often higher than that necessary for the heating of the spaces.
- Do not underestimate the importance of natural lighting (daylight) and think very carefully about replacing it with artificial lighting. Apart from the psychological effect on the users of the building and the effects on visual comfort, the replacement of natural lighting with artificial lighting has negative effects on internal living conditions during the summer period. Particularly in urban office buildings, stores, etc., the heat produced by the use of artificial lighting causes considerable problems, especially during the summer, on account of the increase in internal thermal gains. There is a rise in the indoor temperature, thereby causing the occupants discomfort, and an increase in the cooling

mal losses, thermal gains and high daylight penetration. This measure is considered especially important for areas where natural lighting requirements call for large glazed surfaces.

- Choose the internal surfaces of the walls, ceiling and floor to have a high reflectance value, so as to minimize the amount of light wasted.
- Bear in mind that generally a room will be adequately lit to a depth 2 to 2.5 times the height of the window from the floor, so taller rooms can be daylit to greater depth.[18]
- Provide both sides of the building with openings for uniform lighting when the creative needs of the use of the spaces demand it, as is the case in schools, offices, factories, etc.
- In order to reinforce insufficient lateral lighting, it is recommended that rooflights, which provide the space with a more uniform light, should be used.
- Vertical rooflights with proper sunshading devices are preferable for the lighting of deep spaces. It is also recommended that a south orientation be chosen in order to make use not only of daylight but also of thermal solar gains.
- If you choose to use an internal atrium as a natural lighting technique, remember that it provides passive heating as well.
- Opt for a square-shaped atrium, the height of which should not exceed its width, in order to secure correct lighting levels in all areas. The higher an atrium is in relation to its width, the greater is the reduction in the amount of direct light reaching the lower floors.
- Select high-reflectance materials for the external surfaces that look onto an atrium in order to achieve good daylight penetration into the adjacent spaces. This measure is especially important for the upper floors, in order to maximize the downward reflection of light, thereby improving the lighting of the lower floors.
- Ensure that the structural system of the glass roof of the atrium does not reduce the size of the light-admitting area.
- Ensure that the roof of the atrium is properly ventilated and shaded in order to avoid overheating of the interior during the summer. External adjustable shading is preferable.

load and the amount of energy consumed by the use of artificial cooling.

- Do not place spaces that are in constant use throughout the year or fixed work surfaces (offices for example) on sides with an east or west orientation. The lighting levels on these sides are low. It should be recalled that in these orientations it is difficult to shade the openings from the summer sun, while at the same time glare and overheating are well-known problems that cause the occupants distress.
- Bear in mind that north openings provide a low illumination level, although illumination is constant throughout the day. Thermal solar gains are minimal.
- Be careful to prevent overshadowing of the windows from nearby constructions, trees or other obstructions.
- Avoid horizontal rooflights because, although they improve the natural lighting of the interior, they also increase the danger of overheating during the summer through the almost vertical entry of sunlight.
- Especially in very deep atria, do not use dark colours on the wall surfaces. With the reduction in the reflective power of the interior, the atrium becomes less effective as a natural lighting technique.
- Avoid the use of fixed sunshading devices on the roof of the atrium. Such installations reduce the size of the light-admitting area.
- The use of large openings on the upper floors of an atrium is not advisable. They increase the risk of reducing the quantity of light reaching the lower levels.
- The use of reflective glass is not recommended, especially in south orientations. It obstructs the entrance of natural light, as well as thermal solar gains in winter.
- Avoid the use of prismatic glass in cases where a visual link with the external environment is required. Use it when there is a severe glare problem.
- Do not choose dark colours for the ceiling of the space if you are applying the light-shelf technique. The ceiling must have a high reflectance value.
- Avoid interior shelves. They prevent daylight from entering the space.
- Do not forget to make use of the sunshading devices of the building so as to control

- Reduce the size of the openings on the upper floors of an atrium and increase those on the lower floors. Glass surfaces do not reflect light beams as much as smooth white walls.
- Select flooring materials that are highly reflective for the ground floor of the atrium. A light-coloured type of marble or tile would be an excellent choice. If you think it advisable, use special reflectors as well. By taking these measures, and also designing shallow rooms, the light level of the lower storeys can be increased.
- Protect the glass surfaces on the upper floors of an atrium against the sun in order to avoid glare problems.
- Construct light shelves in south-facing openings. They are more effective here as they reflect more daylight into the back of the rooms.
- In order to avoid irritating glare, the light shelves should be constructed above eye level.
- On the upper surface of the light shelves use highly reflective materials, such as aluminium, mirrors or a highly polished material, so as to direct a large quantity of light into the interior. The ceilings of the individual rooms should be highly reflective too, in order to redirect sunlight to the back of the rooms.
- Make sure that the lower surfaces of the light shelves are reflective only in cases where a system of reflectors has been set up outside, the purpose of which is to redirect light onto the internal floors.
- Opt for moveable louvres instead of light shelves (they have the same function as far as natural lighting is concerned) if you desire greater flexibility in the operation of the lighting system. When the louvres are completely retracted on cloudy days, light is not prevented from entering the interior.
- The light-ducts technique involves the use of a complex and therefore expensive structure. The collection of sunlight requires fixed or moveable devices at the top of the duct, especially designed to redirect the light beams into areas that have a severe lighting problem or are completely

sunlight and thermal gains due to solar radiation in days and seasons when these are not desirable.

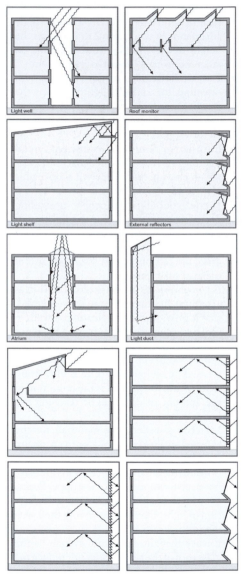

Figure 16.80. Examples of daylighting strategies

dark. The internal surfaces of the shaft should be highly reflective, or possess a system of mirrors in order to direct the light into the desired area.

Figures 16.80 to 16.84 give examples of natural lighting.

Figure 16.81 (above). The main entrance of the School of Technology of the Aristotle University of Thessaloniki. The dark marble of the floor reduces the reflectivity of the surface and therefore the quantity of light lost

Figure 16.82 (above right and right). The dark colour of the floor of an open courtyard reduces the reflectance of the light and so the daylighting level of interior spaces

Figure 16.83 Daylighting with an internal atrium

Figure 16.84. Roof monitors in a sports centre

The author of this chapter (Niobe Chrisomallidou) would like to thank civil engineer Theodore Theodosiou of the Laboratory of Building Construction & Building Physics at the Aristotle University of Thessaloniki for producing the drawings.

REFERENCES

1. Tsanakaki, E. (1994). 'Bioclimatic Architecture' *The World of Buildings*, Vol. 5, pp 30–38.
2. Papadopoulos, M., N. Chrisomallidou, A. Papadopoulos, R. Tsipidou and G. Economides (1996). 'Optimisation of the Design of Low-cost Residences in Order to Improve their Energy Behaviour by Means of Passive Solar Systems'. Research Program Financed by the Greek Workers' Organisation (OEK).
3. Yannas, S. (1994). *Solar Energy and Housing Design*. Vol. 1. *Principles, Objectives, Guidelines*. Architectural Association, London.
4. Watson, D. and Labs, K. (1983). *Climatic Building Design*. McGraw-Hill, New York.
5. Papadopoulos, M. (1978). *Building Physics*. Aristotle University of Thessaloniki.
6. Goulding, J.R., Lewis, J.O. and Steemers, T.C. (1994). *Energy in Architecture*. Batsford, London, for the European Communities.
7. *Material and Building* (1997). Extra Volume. Koskosiolou E, Thessaloniki.
8. Chrisomallidou, N. (1990). 'The Form of the Building as an Energy Design Parameter'. *Proceedings of the ISES Conference on External Perimetral Components in Bioclimatic Architecture, Milan*, pp. 151–155.
9. Chrisomallidou, N. (1987). Design, Construction and Use Parameters Influencing Buildings Energy Behaviour. Economic Evaluation of Energy Saving Measures, Taking into Account the Additional Construction Cost. PhD Thesis, School of Engineering, Aristotle University, Thessaloniki.

10. Santamouris, M., Chrisomallidou, N., Klitsikas, N., Papadopoulos, A. and Tsakiris, N. (1995–96). Save Programme – European Commission, Directorate General for Energy: Energy Rehabilitation of Multi-Use Buildings.

11. Chrisomallidou N. (1990). 'The Influence of Openings in the Building Energy Behavior'. ISES Conference on External Perimetral Components in Bioclimatic Architecture, Milan. Poster Presentation.

12. Schmidt, T., Schulz, M. E., Mangold, D. and Hahne, E. (1995). 'Wärmeschutzverordnung 1995 und dynamische Gebäudesimulation. Ein Vergleich der Heizwärmebedarfsberechnung und Nutzungsmöglichkeiten passiv-solarer Gewinne' OTTI, 7. Symposium. *Therminshe Solarenergie*, pp23–25; April 1997.

13. Colombo, R., Landabaso, A. and Sevilla, A. (1994). Passive Solar Architecture for Mediterranean Area. Institute for Systems Engineering and Information, Joint Research Centre – Commission of the European Communities.

14. Cofaigh, E.O., Olley, J.A. and Lewis, J.O. (1996). *The Climatic Dwelling*. James & James Science Publishers. London.

15. Eumorfopoulou, K. (1987). Planted Roofs. PhD Thesis, School of Engineering, Aristotle University, Thessaloniki.

16. Natural and Low Energy Cooling in Buildings (1994) Thermie Programme Action CRES for the European Commission, Directorate General XVII for Energy.

17. Antinucci, M., Fleuri, B., Lopez, J., Asiain, D., Maldonado, E., Santamouris, M., Tombazis, A. and Yannas, S. (1990). *Horizontal Study on Passive Cooling*. Commission of the European Communities – Building 2000.

18. Energy Research Group – School of Architecture, University College Dublin (1994). Daylighting in Buildings. Thermie Programme Action UCD–OPET for the European Commission, Directorate General XVII for Energy.

FURTHER READING

Baker, N., Fanchiotti, A and Steemers, K. (1993). *Daylighting in Architecture*. James & James Science Publishers. London.

Bayerisches Staatsministerium für Wirtschaft und Verkehr (1993). *Erneuerbare Energien in Bayern*. Munich.

Building Technology (1995). 'Thermie Programme: Promotion of Energy Technology in Europe. *Green Building in Europe*, Issue 5.

Chrisomallidou, N., Papadopoulos, M. *et al.* (1994–95). Research Study – Policy Grants for Energy Saving. Redesign of the Shell of Existing Residences. Financed by Ministry of Environment & Public Works.

Energy Efficient Lighting in Offices (1993). Thermie Programme Action BRECSU-OPET for the European Commission, Directorate General XVII for Energy.

Erhorn, H. and Reiss, J. (1994). *Niedrig-Energie Häuser*. Frauenhofer Institut für Bauphysik, Stuttgart.

Hebgen, H. (1993). *Bauen mit der Sonne*. Energie Verlag, Heidelberg.

Innovative Energy Saving Technologies for Social Housing Programmes. Thermie Programme Action B-112 for the European Commission, Directorate General XVII for Energy.

Papadopoulos, M.,Chrisomallidou, N., Papadopoulos, A., Economidis, G., Kotoulas, K., Zygomalas, M. and Liveris, P. (1988–1992). Monitoring and Evaluation of the Passive Solar Systems in SOLAR VILLAGE III in Lykovrissi – Attica. Financed by the German Ministry of Research & Technology (BMFT) and the Greek Ministry of Industry, Energy & Technology (YBET).

Slater, A.I. and Davidson, P.J.. Energy Efficient Lighting in Buildings. Thermie Programme Action BRECSU-OPET for the Commission of the European Communities, Directorate General XVII for Energy.

Thermie – Building Targeted Projects (1994). *Integrated Clean and Efficient Technologies for the Urban Environment*, Issue 2.

17

Examples of urban buildings

N. Chrisomallidou
Laboratory of Building Construction and Physics,
University of Thessaloniki

છ૭

During the process of designing a building, the designer, usually an architect, takes into account a set of parameters and lays down certain criteria and priorities that have a decisive influence on the 'idea' of the building. Thus, starting with the legal framework (laws and regulations), the building programme, the particular requirements of the client, the available land and the size of the building, the designer moves on to consider the characteristics of the microenvironment (built environment, morphology of the ground, view), the financial factors, etc. In gathering all this information, the designer develops the 'main idea' of the building, putting his/her first thoughts down on paper. Through this process the building gradually begins to take shape, assuming a three-dimensional form (plans, elevations, cross-sections) and becoming integrated into its surroundings.

In recent years, of course, the concept of energy design has come to the fore as an important factor in the process of architectural composition. This tendency has been questioned, strongly criticised and not infrequently rejected by a large number of architects. The problem is believed to stem from the time when energy issues were still at an early stage of scientific inquiry and development and examples of their application in the building sector were by no means ideal. Perhaps the overriding concern in the early years of the development of 'bioclimatic architecture' was to show that the various 'techniques' and 'the building' as a whole were energy-efficient. A much lesser concern was to show the harmonious integration of energy-conscious concepts into the architectural design, producing a building that was morphologically sound.

The situation today has changed significantly, as many of the former opponents of bioclimatic architecture have become keen supporters, while at the same time they have been joined by new professionals who warmly support this new tendency. Furthermore, the constant efforts of many researchers around the world have solved many of the problems involved, advanced knowledge and provided designers, it can be said, with the appropriate computational tools so that they check their decisions at the earliest stage of design. At the same time, technology in the energy sector has attracted the interest of large industries, with the result that today not only is there the technical expertise, but also the means, to design and construct 'buildings with low energy consumption'. Perhaps the only problem that continues to exist is the fact that there is still no wide awareness of the new 'energy logic', either amongst designers or, to a greater extent, amongst the users of buildings; such an awareness is necessary for the application of energy-saving techniques in the construction sector to constitute the rule and not an exception.

From the group of designers, therefore, who use and implement energy-conscious concepts in their designs, we have taken a very small number of urban buildings – seven in total – which are presented in this chapter. The aim here has not been to present all the possible energy-saving strategies that might be applied in an urban building, but rather to show through these examples the effectiveness of a few specific measures and, more generally, the application of the energy logic in urban buildings.

The buildings have been selected on the basis of the following criteria:

- *Climate*. One building from Thessaloniki and two from Athens in Greece are presented, together with one from Nice in France and two buildings from Neckarsulm in Germany.
- *Type of use*. The examples include one multi-use building, two single houses, an apartment house, an apartment and office building, a primary school and a gymnasium.
- *The energy-saving techniques used in their design and construction*. These include the heat insulation of the shell of a conventional building and passive solar systems for heating and cooling of the interior by natural means, or even by active systems.
- *The availability of data and information*. This is not only inevitable, but also essential for their presentation.

The results from the study of these buildings, together with the experience gained from dozens of other similar buildings, show that the more 'fuss' one makes of a building in its design and construction stages and the more one invests in terms of 'time and knowledge of energy issues', the more the building fulfils the energy goals of the designer. In this way, one can eliminate or minimize the adverse effects of the external climatic conditions, the intensive use of the building and its form, and even the negative impact on it of the surrounding built environment.

The issues relating to the 'energy logic' could be said to be quite simple – as long as the basic principles of bioclimatic design are not dismissed out of hand, out of either ignorance or fear of innovation. It should be generally understood that the benefits are truly great – for the average consumer, the national economy and the environment. The most important thing worth stressing, of course, is the fact that these benefits are long-lasting and will continue to be reaped for as long as the building itself continues to stand.

If the application of energy-saving techniques in buildings in suburban and rural environments presents no particular problems, in urban buildings dealing with the energy problem requires greater thought and ingenuity in order to achieve the correct results for the thermal behaviour of a building throughout the year. As has already been mentioned elsewhere, the difficulties stem from the already existing urban fabric of cities, in the development of which the energy parameter was never taken into consideration.

What one can do and what one should avoid in designing an urban building to behave correctly throughout the year and to be able to adjust to the different seasons is the subject of Chapter 16.

Here it is worth summarizing just a few specific energy-saving techniques and pointing out a few approaches that might help us to overcome certain problems that arise in urban buildings. This is backed up by reference to the examples of buildings presented in this chapter.

The energy-conscious design of buildings, bioclimatic design and the rational use of energy – all virtually identical concepts – have one sole aim, to ensure acceptable internal climatic conditions by achieving the correct thermal behaviour of the building through-out the year and so reduce energy consumption (with all the benefits this entails – in terms of financial savings, quality of life, reduction of CO_2 emissions, etc.). In the case of bioclimatic design, the above aim is achieved purely through design processes, or various techniques in the construction of the building, thus reducing dependence on mechanical heating and cooling equipment.

To achieve a reduction in the consumption of energy during the winter period, it is clear that on the one hand it is necessary to reduce the thermal losses of the building (losses through heat conduction and ventilation), while on the other to maximize the gains, mainly thermal solar gains. During the summer, of course, the aim should be to cool the building naturally by minimizing the thermal gains and to reduce the tempera-ture through ventilation and other related means. The above two types of heat flow from and to the building (thermal losses and thermal gains) in practice make up its thermal balance. In cases where the thermal gains in winter are not great enough to cover the thermal losses – and this is largely true of conventional buildings without thermal insulation – heat is produced in the interior of the building by the installation of a heating system to correct the imbalance. Consequently, the aim in this case is to design and construct a building in which the above imbalance is minimal.

The question, therefore, is what can be done to minimize the thermal losses of the building, while at the same time maximizing the thermal solar gains. During the design and construction stages, then, the following measures can be taken:

- The building can be positioned on the north side of the plot to increase the distance between it and the buildings opposite and to avoid as far as possible the risk of shading, which cancels out any possible solar gains. In addition, plants can be grown on the south side in the best microclimatic conditions, providing both a desirable amount of shade and evaporative cooling in summer.
- If the main facade of the building deviates from a southward orientation, solutions can be devised to tilt the axis of the building back towards the south, or at least the facade of the building or its openings. A typical example is the bioclimatic urban building in Athens (Example building 2).
- To allow sunshine to illuminate areas located some distance away from the south side, light ducts or roof skylights can be used, such as those in passive solar buildings (Example buildings 3 and 4). Furthermore, in urban buildings another possible solution is light wells, used in conjunction with reflectors to direct the sunlight into specific areas.
- In the design of the ground plan the various areas should be arranged so that those that are used frequently and require high internal temperatures are positioned on the south side. The remaining ancillary areas, which are infrequently used and require low internal temperatures, should be sited on

the north side so as to reduce the thermal losses from the main areas to the cold external environment.

- In the case of urban buildings where it is impossible to secure a southward orientation, the thermal insulation of the building's shell should be greater than normal, the openings should be reduced to the minimum size required for natural lighting and ventilation of the various areas and advanced windows should be used to offset the lack of thermal solar gains. As can be seen in the first example of the conventional buildings in the Thessaloniki area, the heat insulation of the external building elements itself reduced energy consumption by about 60% in comparison with similar urban buildings. The results would have been even better if more effective insulation had been placed in the external building elements.
- For the storage of thermal solar energy, building elements with a high thermal capacity should be selected. This measure plays an important role, particularly in bioclimatic buildings and areas of constant use, as well as in locations that experience high temperatures in summer.
- In bioclimatic buildings, in particular, all the strategies applied to reduce energy consumption can come to nought if there are increased thermal losses due to extensive ventilation or infiltration of air through openings. Thus, an airtight shell should be created, and, more generally, the ventilation of the various areas should be reduced and controlled, in accordance with their use, within the limits set for reasons of hygiene.
- If the building plot faces south and there is no problem of shading by adjacent buildings, it is advisable to develop the building along the east–west axis in order to make the south facade as large as possible. In this case, serious consideration should be given to the possibility of applying passive solar systems in order to satisfy the second requirement of achieving the maximum amount of inexpensive thermal solar gains. Systems that can be easily applied in practice, with conventional materials and at no great cost, are the following well-known ones:
  – direct solar gains from south openings;
  – mass and trombe walls;
  – sun spaces;
  – thermosyphonic panels.
  The first three of these systems were used successfully in the three urban buildings with passive solar systems in the Athens area (Example buildings 2, 3 and 4).

Which of the above systems is finally chosen depends on a set of parameters, including aesthetics, thermal efficiency, the use of the building (residence, school, hospital, office block, etc.), the cost of construction and maintenance, the system's ease of use by the occupants (summer and winter, day and night), security, shading, privacy, etc.

As well as the above, in order to ensure that the building that has been designed will behave correctly during the summer period and show no signs of overheating, and also to ensure that it will be naturally cooled, the following measures are considered essential:

- The provision of effective forms of shading for the facades through the construction of the most appropriate solar protective structures for each orientation (internal, external, fixed, moveable, horizontal or vertical, etc.) in order to prevent sunlight penetrating the indoor areas. In addition, in office blocks, light industrial premises, etc. the use of reflective or other special types of glass would be an additional effective measure.

- The planting of the outdoor areas with suitable greenery and the creation of areas of water in the vicinity of the building to increase natural cooling through evaporation.
- The provision of cross-ventilation to increase natural cooling. If this measure cannot be achieved with the existing openings in the facades of the building, then the use of openings in the roof of the building, or the construction of a solar chimney to accelerate the removal of warm air from the building, or a windtower to create the necessary air circulation, would be a tried and tested solution (see Example building 4).
- The painting of the external surfaces of the building in a light colour, or the use of materials that will reflect sunlight on road and pavement surfaces.
- The reduction of internal thermal gains by improving the daylighting of the various areas, using high-performance electrical and lighting devices, etc. This measure is particularly important in urban buildings (offices, schools, shops, etc.), in which – because of the high levels of use – the intensive use of artificial lighting is required or in which numerous electrical appliances, which increase the thermal load of the building, are used, resulting in overheating.

The measures mentioned above constitute a basic precondition for the correct behaviour of buildings in terms of energy efficiency. Where emphasis is laid in developing strategies for the summer or winter periods clearly depends on the climatic characteristics of the area in which the building is located. By and large, however, a bioclimatic building should behave correctly during both periods. The techniques that appear advisable for one period do not exclude the application of techniques for the other period as well.

EXAMPLE BUILDING 1 – A CONVENTIONAL MULTI-USE BUILDING IN THESSALONIKI, GREECE

Figure 17.1. The main facades of the building

The architects were G.H. Hatzigeorgiou & P.Zografos and the building is at Egeou 63–65, Thessaloniki, Greece.

Project background

The project is actually concerned with two adjacent, but independent, conventional urban buildings that have features leading to low energy consumption; they are situated in the eastern part of the city.

This is a continuously developing area, with many new luxury buildings. It has attracted the interest of large companies, which have their offices there, either as parts of mixed-use buildings or in their own separate buildings.

The advantageous position of the buildings on the corner of two main streets (Figure 17.1), where there is heavy commercial traffic, was the reason why the company that has its offices on the first floor calls itself 'Junction'. It is interesting to note that the same word in Greek also means 'Contribution'.

The buildings in the area are semi-detached (Table 17.1) and, despite all the positive aspects that have been mentioned, there is an obvious lack of greenery in an area of about one square kilometre and there is a high level of noise during the day.

Site and climate

The latitude of Thessaloniki is 40°40′. The climate of the city is strongly influenced by the sea that it lies alongside. Figures 17.2 and Tables 17.2 and 17.3 present the climatic data for the city for the period 1981–1991.

Table 17.1. Building Data

Uses	Ten apartments/two shops/offices
Number of occupants	45
Year of construction	1995
Scheme of building	Semi-detached
Orientation	NW–SE
Annual energy consumption	93.4 kWh/m^2

Table 17.2. Monthly long-term average data

Month	Solar irradiation on horizontal plane	Degree days	Sunshine hours	Average ambient temperature
	Global MJ/m^2	(base 18°C) °C days	hours	°C
January	49	396	106.3	5.5
February	64	313	120.7	7.0
March	96	268	152.6	9.9
April	135	130	209.4	14.6
May	174	23	268.9	19.7
June	186	0	292.8	24.1
July	205	0	342.4	26.7
August	178	0	306.1	26.3
September	131	0	238.5	22.2
October	87	70	171.1	16.7
November	55	187	119.8	11.9
December	44	338	100.5	7.4

Table 17.3. Annual climatic data

Latitude	40°38'
Longitude	22°47'
Prevailing wind direction (winter)	NW
Average wind speed	1.9 m/s
Relative humidity (winter/summer)	74%/58%
Global irradiation on horizontal plane	117 MJ/m^2
Degree days	1725 days
Sunshine hours	2429.1 hours
Ambient temperature (winter/summer)	8.3°C/25.7°C
Average maximum temperature (annual)	20.9°C

General building description

These are two independent, modern buildings that have been in use since 1995 and so are in excellent condition. Their architecture, as well as the particular nature of their use, justifies the fact that they are separated into two different units with three and four storeys above the ground floor respectively. More specifically, the buildings were constructed at the same time by the same company on two neighbouring plots, with independent entrances, staircases, heating systems, etc. The connecting link is the construction

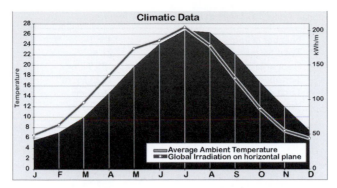

Figure 17.2. Average ambient temperature and global irradiation on the horizontal plane

company called 'Simvoli' that occupies the ground and first floors of both buildings (Figures 17.3 and 17.4).

The buildings have an interesting architecture, with a morphological differentiation for each application, which is achieved by the use of different materials. It is obvious that the designer wanted to draw attention to the ground and first-floor business by using a 'light' facade with dark windows and a distinctive logo on the facade. The apartments, on the other floors, have a more subtle morphology with expensive wooden frames in the openings.

The two independent entrances to the buildings have a special form, with glass bricks and wooden doors so that they are obviously different from the entrance to the shops. As far as the construction is concerned, both buildings are made of excellent materials: the frame is made of reinforced concrete and the fabric consists of a double brick wall, each wall being 7.5cm thick, with a 5 cm thick expanded polystyrene layer in the cavity

Figure 17.3. The main facade on Egeou Street

Figure 17.4. The facade on Adrianoupoleos Street

Table 17.4. Geometrical data on the building

Number of floors	4 + 1
Total height	21 m
Floor area (typical floor)	185.91 m²
Total floor area	1223.55 m²
Total external area	1671 m²
Volume	3890.3 m³
Area-to-volume ratio	0.42
Window area	274.17 m² NW
	11.44 m² NE
	88.76 m² SE
	53.39 m² SW
Total window area	427.76 m²
Vertical wall area: solid	667.2 m²
Opaque-element area	1198.2 m²
Vertical/horizontal (mixed) area	3.20

between the two layers of bricks. All the glazing of the apartments and the first floor offices is double, with wooden frames in the apartments and aluminium ones in the offices. The openings of the shops are covered by single security glazing, which is 1 cm thick (Figures 17.5 to 17.10 and Table 17.4).

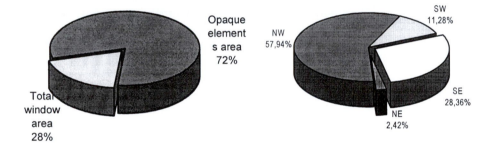

Figure 17.5. Percentages of opaque and transparent building elements

Figure 17.6. Percentage distribution of windows on the sides of the building

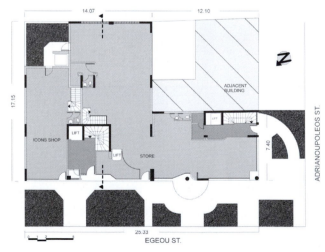

Figure 17.7. Ground floor plan

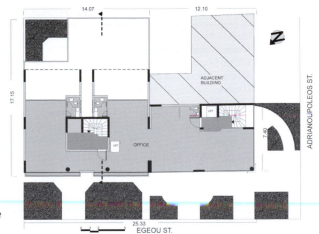

Figure 17.8. First floor plan

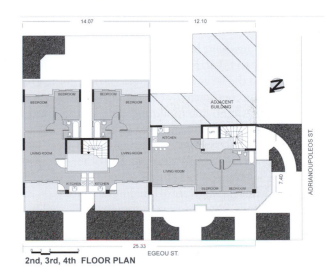

Figure 17.9. Plan of upper floors

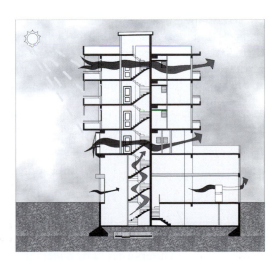

Figure 17.10. Cross section showing shading and natural ventilation

Table 17.5. Thermal characteristics of the building

Heated floor area	1108.32 m²
Heated volume	3890.3 m³
Global heat loss coefficient	2635.407 W/K
Heat transfer coefficient U_m window	3.789 W/m²K
U value	
wall	0.580 W/m²K
floor	0.740 W/m²K
roof	0.500 W/m²K
U_m value	1.317 W/m²K
$U_{m\ max}$ value	0.782 W/m²K
Design temperature	21°C
Space heating load	93.315 kWh/m² per annum
Cooling load	64.4 kWh/m² per annum
Lighting	14.87 kWh/m² per annum

The buildings have been insulated according to the thermal insulation regulations. However, the fact that there is no insulation on the flat roof above the offices of 'Simvoli', together with the large area of single glazing in the ground-floor shops, increases the mean U_m value coefficient by 68% compared to the maximum allowable coefficient of thermal flux. All the other building elements are insulated and have U values that are lower than the permissible values (Table 17.5).

The main facades face south-west and north-west, while the south-east side on Egeou Street faces an empty, unbuilt area. The other sides are in contact with neighbouring buildings. The staircase of building A on Adrianoupoleos Street is external all the way to the top, while the staircase of building B, which is on Egeou Street, is situated centrally and divides each floor into two symmetrical flats. The lighting is achieved through a skylight 1.2 m wide.

A typical flat in building A has a single area that is used as a living room, dining room and kitchen niche, plus a bathroom and two rooms situated on the facade.

The flats on a typical floor of building B are symmetrical in relation to the staircase, cover the same area and have the same rooms. There are two bedrooms and a bathroom that face the empty space at the back of the plot and one area, which faces Egeou Street, that serves as the living room, dining room and kitchen. In both buildings there are large separate balconies, 2 to 3 m wide. The area of the flats is about 60 m² in both buildings. In the ground-floor shops there is a mezzanine and, finally, in the basement there are storerooms and the boiler rooms.

The mean height of each floor is about 3 m and the total heights of the buildings are 18 m and 20.40 m respectively.

Energy systems

Heating in the apartments

The apartments in the two buildings are heated by two independent oil-fired central heating systems. They are in operation for six months every year (20 October to the end of March) for ten hours a day. The boilers are each rated at 100000 kcal/h. There is an

Table 17.6. Heating and cooling system data

Heating system	
Type	Central-A/C-independent
System	Single-pipe
Fuel	Oil/electricity
Power rating	100,000 kcal/h
Hours of operation	10 h/day – office hours
Thermostat temperature	18°C/21°C
Period of operation	October–March
Cooling system	
A/C type	Split units
Power rating (BTU)	5 × 9000 + 4 × 18000 + 3 × 24000
Hours of operation	10:00–22:30
Set temperature	21°C
Period of operation	July–August

internal thermostat in each apartment, which allows independent heating adjustment in each flat. The pipework is insulated and the system is serviced once a year (Table 17.6).

Heating in the 'Simvoli' shop and offices

The company uses air conditioning exclusively for heating and cooling. There are three split units in the shop, which are in operation for about eight hours a day and the thermostat is set at 26°C. In the offices there are:

- one multiple compressor unit with five internal units of 9000 BTU each, which operate during working hours;
- two independent units of 18,000 and 24,000 BTU, which operate only during the morning hours;
- two units of 24,000 BTU each, which are in the conference room of the company and operate occasionally.

The thermostat is set at 21°C for all these units.

Heating in the 'Icons' shop

The shop has an independent heating system, with an internal thermostat that is set at 20°C. The pipes are insulated and the system is serviced once a year.

Ventilation

Natural ventilation is used in the apartments during the day in the summer period. In the shops this occurs all day since the doors are continuously open. Finally, in the office area occupied by 'Simvoli' ventilation is carried out only early in the morning (8:00–10:00).

Shading

The shading of the apartments is achieved by using internal devices like curtains and shutters and external sun protection devices. The offices of 'Simvoli' do not have such devices and dark glazing is the only measure used.

Actual consumption

From the data collected it was calculated that the specific energy consumption for heating is:

- for the apartments 8.1 litres/m^2 per annum or 92.34 kWh/m^2 per annum;
- for the icon shop 8.37 litres/m^2 per annum or 95.418 kWh/m^2 per annum;
- for the company 'Simvoli' 94.42 kWh/m^2 per annum.

The energy consumption for cooling the 'Simvoli' offices and shop was calculated to be 68.4 kWh/m^2 during the summer period. The cooling of the other areas is mainly achieved by natural means. Temperature variations over the winter period are shown in Figure 17.11.

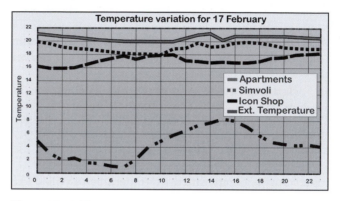

Figure 17.11. Temperature variations for winter period

A set of calculations was carried out in order to study the thermal losses from each construction element (Figure 17.12). The results show, among other things, that the largest percentage of thermal losses is due to the openings, which cover approximately 28% of the total external area of the building, but are responsible for 70% of the total

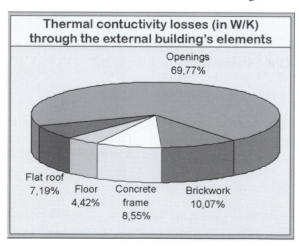

Figure 17.12. Thermal losses by conduction

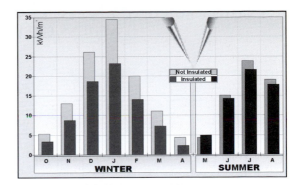

Figure 17.13. Monthly heating and cooling loads for the real situation and for the simulated case of non-insulated building elements

thermal losses. It should be noted, however, that the openings are of excellent quality, double-glazed, except those on the ground floor that are single-glazed but 1cm thick for security reasons. The opaque building elements are responsible for 30% of the total losses. The only non-insulated area is the flat roof above the offices of the company, but the losses through this are rather small as a result of the small surface that it covers (41 m²).

Furthermore, a set of simulations was carried out with TRNSYS in order to study the thermal performance of the building. The simulations were performed for the present situation and for hypothetical case in which the building elements were not insulated. The results of these simulations are presented in Figure 17.13, for both heating and cooling loads. The importance of the thermal insulation is obvious. Although the U_m values of the buildings are greater than the recommended $U_{m\ max}$, one would have expected even better energy consumption values since all the elements are insulated according to the current Greek regulations.

Conclusions

As has been mentioned elsewhere, despite the fact that the individual building elements satisfy the requirements of the current heat insulation regulations, with the exception of the flat roof of the first floor above the offices, in the building as a whole the U_m value of 1.317 W/m²K is greater than the maximum permissible value, i.e. the U_{max} value of 0.782 W/m²K.

This means that in the design of the buildings the surface areas of the openings should have been reduced, mainly on the ground and first floors, while the single-glazed panels on the ground floor should have been replaced with double-glazed ones. Alternatively, the heat insulation of the solid building elements should have been increased, so as to reduce thermal losses from the shells of the buildings.

Thus, if the heat insulation regulations were applied to the letter, one would expect a heat energy consumption rate for these buildings of no more than 60 KWh/m² per annum.[1] In other words, there would be a further 36% improvement in their thermal behaviour, and without doubt higher internal temperatures. In this case, the buildings would certainly fit into the category of buildings with 'low energy consumption'.

Notwithstanding the above observations, the thermal behaviour of these buildings can be regarded as satisfactory, compared to existing urban buildings in the Thessaloniki area with similar use and shape, as studied by Santamouris *et al.*[2]

EXAMPLE BUILDING 2 – A PASSIVE SOLAR SINGLE HOUSE IN ATHENS, GREECE

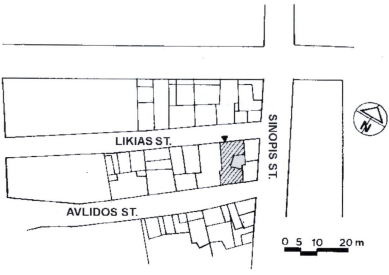

Figure 17.14. Site layout

Project background

The architect was Katerina Spiropoulou and the building is situated in the Ambelokipi district in the centre of Athens, near the Hilton Hotel. It is a four-storey semi-detached house of 210 m² total area and 900 m³ in volume. The area of the building plot is only 99 m² (Figure 17.14). The two main opposite facades are on Likias and Avlidos Streets, while the other two sides adjoin the neighbouring .... buildings (Figure 17.15).

The urban and building regulations, the tall surrounding buildings that form a densely built urban environment and the small plot have strongly influenced the form and organization of the internal spaces of the residence. Table 17.7 summarizes the structure.

Low energy consumption, taking advantage of the mild Mediterranean climate and the building's orientation, were the most important design parameters. The architect and designer Katerina Spyropoulou has said that she also had to solve several problems, such as safety requirements and economic restrictions.

Table 17.7. Building data

Building type	House
Number of occupants	4
Year of construction	1986
Scheme of building	detached
Orientation	S–N
Annual energy consumption	26 kWh/m²

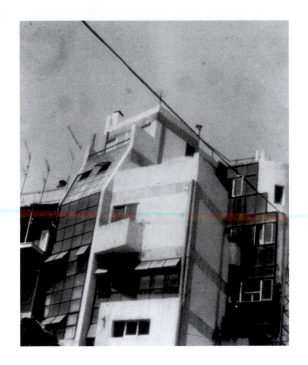

Figure 17.15. General view of the building

Site and climate

The climate of the Athens area is mild and sunny during winter. However, solar radiation is affected by the city's atmospheric pollution. The average monthly temperatures in January and July are 9.4°C and 27.9°C respectively (Figure 17.16). The total annual amount of sunshine is 2818 hours and there is an annual total of 1110 heating degree-days. The prevailing wind direction in the area is northerly during the whole year (Table 17.8).

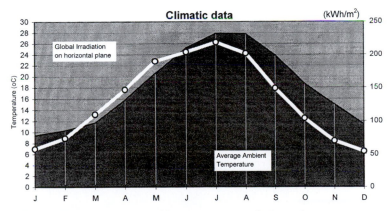

Figure 17.16. Average monthly temperatures and solar radiation

Table 17.8. Average annual climatic data

Latitude	37°58'
Longitude	23°43'
Prevailing wind direction	N
Average wind speed (annual)	2.8 m/s
Relative humidity (winter/summer)	69.4%/52.2%
Global irradiation on horizontal plane (winter/annual)	617/1581 kWh/m²
Degree days (base 18°C)	1110 days
Sunshine hours (winter/annual)	2818 h
Average ambient temperature (winter)	13.2°C/18.1°C
Average maximum temperature (annual)	22.6°C

General building description

The building, a four-storey semi-detached house for a four-member family, was first occupied in 1986. It has a modern architecture compared to the neighbouring urban buildings, characterized by the applied passive solar systems, which are placed on the building's facades.

The limited building plot, only 99 m² in area, the needs of the family and the tall neighbouring buildings rendered it necessary to expand the building in a vertical direction.

The ground floor (Figure 17.17) is 5.5 m in height and is used as a store. It also houses the central heating installations and other auxiliary rooms, since it receives no solar gains because of the height of the neighbouring buildings.

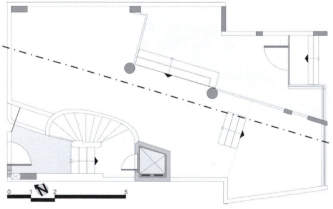

Figure 17.17. Ground floor plan

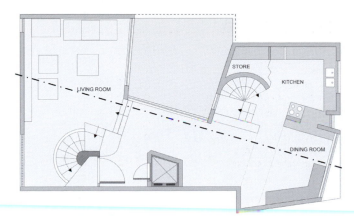

Figure 17.18. First floor plan

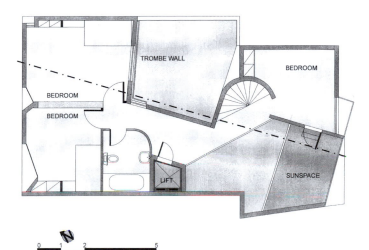

Figure 17.19. Second floor plan

The main residence is therefore raised 5.5 m above this and consists of three successive floors and a small loft between the first and the second floors, 13 m² in area (see Figures 17.18 and 17.22).

The living room, the dining room, the kitchen and a store are on the first floor (Figure 17.18).

The children's bedrooms, a guestroom and a bathroom are placed on the second floor (Figure 17.19).

The parents' bedroom, the necessary auxiliary spaces and an office are on the third floor (Figure 17.20).

The fourth floor, the smallest, includes two study rooms (Figure 17.21).

A sunspace stretches from a height of 9 meters (mezzanine floor) up to the bedroom floors (Figure 17.22).

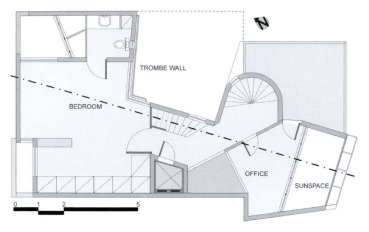

Figure 17.20. Third floor plan

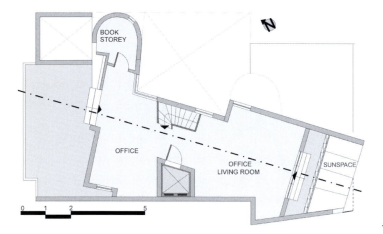

Figure 17.21. Fourth floor plan

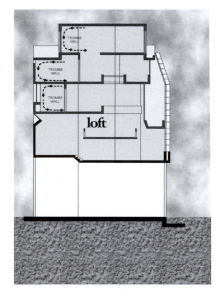

Figure 17.22. Cross section

Energy-saving features

The densely built environment, the narrow streets, the tall buildings and the narrow facades raised problems for passive solar design and led to the following control measures:

- The building's axis has been rotated in a north–south direction, so that the main facades are not parallel with the road's axis, forming an angle of 18°.
- The sunspace (greenhouse) has been placed on the south façade, covering a stretch from the 9.0 m level up to the fourth floor.
- An internal balcony has been formed inside the greenhouse, leading from the living room.
- The loft between the first and the second floors allows deep penetration of the sunlight and free air circulation, unifying the interior spaces.

The architect's intention was to ensure low energy consumption while maintaining a low construction budget.

The building's carefully insulated envelope and the small openings in the north facade ensure satisfactory internal thermal comfort conditions and low energy consumption during winter.

Solar protection during summer is rather inadequate, especially in the greenhouse, owing to the lack of sunshading devices. Summer overheating is reduced by implementing cross-ventilation though north and south external openings.

Passive solar design

The passive solar systems implemented in the building are:

- the direct solar gain through the southern openings;
- the greenhouse/sunspace;
- the Trombe walls in the envelope's alcoves.

Figure 17.23. The inclined glazed surfaces of the main facade

Direct solar gains

The southern openings incline 18° eastwards and have a total area of 19 m² (Figure 17.23). All glazed areas are metal-framed and the glazing is single and reinforced for security reasons, except on the northern facade, where double glazing is used.

External roller shutters without core insulation are used for night-time insulation. Curtains offer extra shading in the interior.

Sunspace

Indirect solar gains are obtained through the south-oriented greenhouse, which is an intermediate space, bridging the external environment and the interior spaces (Figure 17.24). As the reference point of the house, the sunspace can be seen from almost every inside or outside angle. It is planted in order to act as a garden and to give a sense of nature in this densely built urban area, as a substitute for the missing panoramic view (Figure 17.25).

Covering a glazed area of 19 m², the greenhouse extends 2.5 floors in height and consists of a metal-framed construction filled with single reinforced glass panels for security reasons. The surface between the greenhouse and the interior spaces is of single glazing, providing the living room and the staircase with natural lighting, which is valuable for the elongated building shape and the relatively confined facades.

Figure 17.24. Internal view of the greenhouse

Figure 17.25. The sunspace as a substitute for the missing view

The air, as it warms up inside the greenhouse, moves to the upper floors where the bedrooms are. A fan was to be used to recirculate the air, but has never been installed.

Summer overheating is prevented by using a number of scattered openings in the glazed envelope of the greenhouse and dampers at the top of it. The shading of the greenhouse could be considered insufficient and an internal aluminium-coated curtain (safety blanket) is the only solar protection measure.

Indirect gain systems – Trombe walls

Two Trombe walls have been constructed at certain available points in the external envelope. The first one is east-oriented, covering an area of 8.0 m², while the other is south-oriented covering a area of 30 m². A special cost analysis was performed for these Trombe walls, because a significant part of them is shaded by the building itself (Figures 17.26 and 17.27).

The Trombe walls are made of reinforced concrete and solid brick. Their external envelope is metal-framed, filled with single reinforced glass panes and interrupted at certain points by external openings. Air circulation occurs through dampers placed in the upper and lower parts of the Trombe walls.

Auxiliary heating

There is a central auxiliary heating installation, powered by a natural gas boiler. A traditional fireplace placed in the dining room is also used sporadically. Solar collectors have also been used to provide warm water for the users' needs.

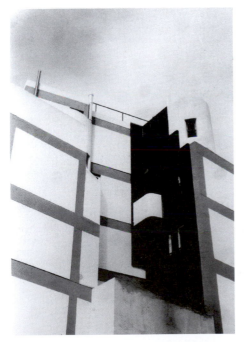

Figure 17.26. The Trombe walls as part of the building's envelope

Figure 17.27. Close-up of the Trombe walls

Figure 17.28. The building's shape assists deep penetration of natural light

Table 17.9. The building's thermal characteristics

Heated floor area	220 m²
Heated volume	670 m³
Ventilation rate	0.8 ach
Global heat loss coefficient	628.5 W/K
Heat transfer coefficient window	4.04
U value	
wall	0.57
floor	1.06
roof	1.757
U_m value	1.61 W/m²K
Design temperature	20°C
Space heating load	25.533 kWh/m² per annum

Figure 17.29. Opaque and transparent building elements

Figure 17.30. The distribution of the windows on the building's sides

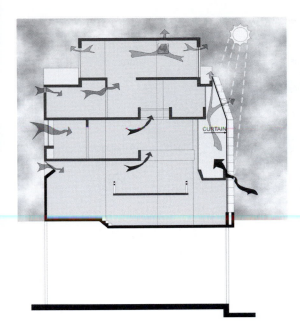

Figure 17.31. Natural cross-ventilation during the summer (cross section)

Thermal behaviour of the passive solar systems

During the heating period, the indoor temperatures are significantly increased as a result of diffused and incidental solar radiation (Figure 17.28). Part of this is stored in the building's thermal mass and particularly in the walls, floors and Trombe walls. The percentage of opaque and transparent building elements and the distribution of the windows in the building's envelope are presented in Figures 17.29 and 17.30 respectively. The overall thermal characteristics of the building are shown in Table 17.9.

During the night the stored heat is returned, keeping indoor temperatures reasonably high, while the external roller shutters and the curtains are kept closed in order to minimize thermal losses to the external environment.

During summer cross-ventilation is the only natural cooling strategy (Figure 17.31) because shading is inadequate, as has been previously mentioned. Overheating of the Trombe walls is eliminated by using the openings in the external glazed envelope, while keeping the dampers closed during the day.

Experimental measurements were taken in this building for a year and a half. A data-logger system was installed to monitor 40 sensors and the measurements were analysed by implementing regression analysis techniques. This was a project carried out by the research team led by Dr E. Angreadaki-Chronaki of Aristotle University of Thessaloniki, while Dr G. Paparsenos and Dr E. Athanasakou performed the regression analysis.[3,4] It emerged that passive solar systems contribute 42% of the building's thermal demand, while internal gains contribute 28% and the other external openings contribute 2% of thermal demand. The auxiliary heating demand is only 28%.

It should be mentioned that temperature stratification phenomena were significant as a result of the open structure of the internal spaces, which helps the buoyancy of the

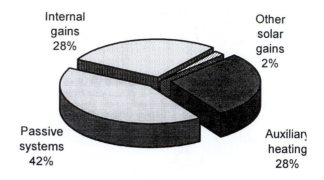

Figure 17.32. Annual energy equilibrium

warm air mass. The building therefore acts as a chimney and the upper floors are warmer. Measurements showed a 3 K temperature difference between the lower and upper floors.

Actual energy consumption

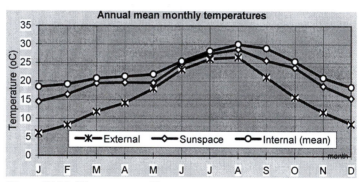

Figure 17.33. The mean monthly temperatures

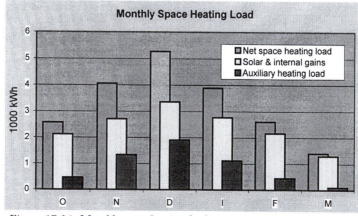

Figure 17.34. Monthly space heating loads

Of the building's heating requirements, 72% were provided by solar and internal thermal gains (Figure 17.32).

The annual requirements for a similar conventionally designed house and the same thermal comfort levels are approximately 12,362.2 kWh, compared with the actual annual auxiliary heating load of 5380.2 kWh (26 kWh/m^2 per annum) in this building with passive solar systems.

The thermal comfort conditions appeared to be satisfactory. The mean monthly temperatures monitored in the sunspace and the internal spaces are presented in Figure 17.33 and the monthly space-heating loads are shown in Figure 17.34.

Discussion

The designers, who were also the inhabitants of the house, ascertained that vertical and horizontal cross-ventilation is very effective for natural cooling during summer.

Indoor temperatures were within the limits of thermal comfort. Only during the summer vacations did the temperature reach 32°C at the top of the fourth floor during August, as a result of lack of ventilation.

From the users' point of view, the frames of the external openings should be more effectively weather-stripped in order to reduce air infiltration losses during winter. Double-glazing could possibly improve thermal comfort and reduce the noise level, if used for the rest of the openings, including the direct-gain passive solar systems. On the other hand, such an improvement would certainly be costly.

EXAMPLE BUILDING 3 – A PASSIVE SOLAR SINGLE HOUSE IN ATHENS, GREECE

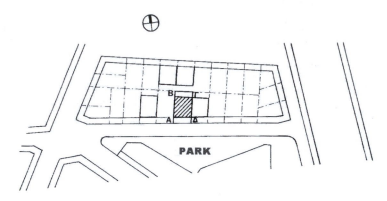

Figure 17.35. Site layout

Project background

The building examined in this case study is a semi-detached house, designed by M. G. Souvatzidis and situated in Athens, in Nea Filothei, a comparatively low-density urban area, with one- and two-storey detached and semi-detached houses. It is on a site of 207 m², inclined towards the south. The main south facade has a panoramic view, facing a park (Figures 17.35 and 17.36). It is therefore suitable for the implementation of passive solar design principles and solar systems without any risk of overshadowing by adjacent

Figure 17.36. General view of the building

Table 17.10. Building data

Building type	Single house
Number of occupants	3
Year of construction	1985
Scheme of building	Semi-detached
Orientation	N–S
Annual energy consumption	71.76 kWh/m²

buildings. The house is inhabited by a three-member family and includes, among other spaces, two studios and the appropriate laboratory facilities. Dimitris Mytaras, a well-known Greek painter, is the owner of the house.

The building data for the house is summarized in Table 17.10.

Site and climate

The climate of the Athens area is mild and sunny during the winter, although the solar radiation is affected by the city's atmospheric pollution. The monthly average temperatures in January and July are 9.4°C and 27.9°C respectively (Figure 17.37). The total amount of annual sunshine is 2818 hours and there is an annual total of 1110 heating degree-days. The prevailing wind direction is northerly to north-easterly (Table 17.11).

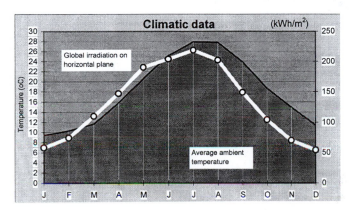

Figure 17.37. Monthly average temperatures and solar radiation

Table 17.11. Annual average climatic data

Latitude	37°58′
Longitude	23°43′
Prevailing wind direction	NE
Average wind speed	2.8 m/s
Relative humidity (winter/summer)	69.4%/52.2%
Global irradiation on the horizontal plane winter/annual	617/1581 KW/m²
Degree days (base 18°)	1110 days
Sunshine hours (winter/annual)	1115/2818 h
Ambient temperature (winter/annual)	13.2°C/18.1°C
Average maximum temperature (annual)	22.6°C

General building description

The building's unique architectural morphology is in sharp contrast to that of the buildings in the surrounding area, which is rather undistinguished.

The house is a semi-detached three-storey building, including a pilotis and a basement. There are northern and a southern facades, while the two remaining opposite sides form the partitions with the adjacent houses, as a result of urban building regulations (Figures 17.38 and 17.39).

There are special needs in this particular house since it is intended for a three-member family, as well as for two professional painters, the two parents. Two separate studios were therefore constructed to fulfil these needs.

The rooms located on the first floor are an office, a studio (atelier), a guest room and a bathroom (Figure 17.40). The two main bedrooms for the couple and their child are

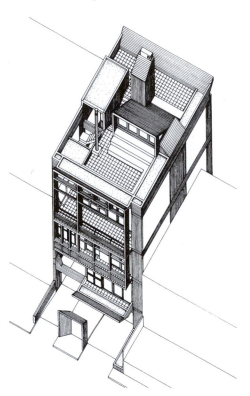

Figure 17.38. Aerial view

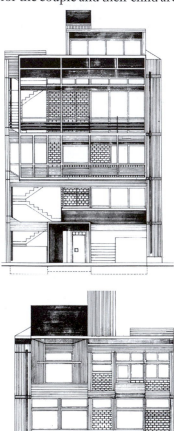

Figure 17.39. The northern and southern facades

located on the second floor, as well as a second studio and a bathroom (Figure 17.41). The rooms located on the third floor are the dining room, the living room (reception) and the kitchen (Figure 17.42).

The pilotis area (i.e. the area beneath the house, which is built on stilts) is appropriately formed in order to serve multiple housing needs; part of it has been modified to serve as a small garden, while a second part serves as an entrance hall and the rest is used as a parking area (Figures 17.43 and 17.44). The central heating installation and other auxiliary rooms are located in the basement.

Part of the building's roof is planted to improve both thermal performance and the building's aesthetics (Figure 17.45).

The passive solar design features used to reduce the heating and cooling loads mainly determine the configuration and architectural characteristics of the south facade. The passive solar systems implemented consist of a sunspace (greenhouse), three Trombe walls and direct solar gain through extended glazed areas.

Figure 17.40. First floor plan

Figure 17.41. Second floor plan

Energy-saving features

The south facade is obviously influenced by passive solar design, which attempts to take full advantage of the rather limited capabilities of this restricted urban site. Extra effort was put into supplying the rooms of the north facade with adequate natural lighting.

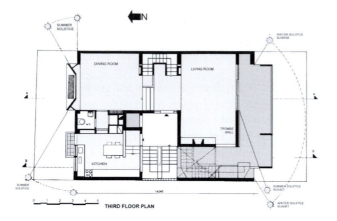

Figure 17.42. Third floor plan

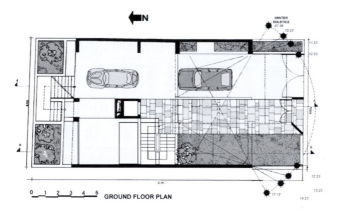

Figure 17.43. Pilotis floor plan

Figure 17.44. View of the pilotis area

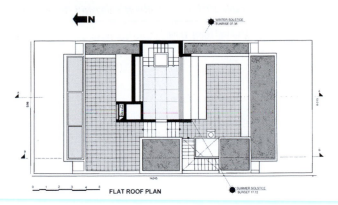

Figure 17.45. Roof floor plan

FLAT ROOF PLAN

The two studios were located on the northern facade in order to be provided with the adequate special lighting that is needed. The studios also act as thermal containment zones, because their energy demands and occupation time are significantly smaller than those of the bedrooms, which were placed on the south facade, in order to take full advantage of solar gains. The majority of the openings (49%) are therefore placed in the south facade (Table 17.12).

Table 17.12. Geometrical data for the building

Number of floors	3 + 1
Total height	11.9 m
Floor area	140.94 m²
Total area	417.86 m²
Total external area	661.00 m²
Volume	1745.00 m³
Area-to-volume ratio	0.379
Window area	
N	30.20 m²
E	1.20 m²
S	40.50 m²
W	10.80 m²
total	82.70 m²
Wall area: solid	437.36 m²
Opaque element area	578.30 m²

The building's envelope is not heavily insulated (Table 17.13). The construction materials are the rather common ones used in modern Greek buildings. The building's frame is of non-insulated reinforced concrete. The external walls consist of double brick walls, 10 cm thick, with 5 cm expanded polystyrene core insulation.

The glazed area covers 12.5% of the envelope's surface, consisting of double-glazed metal-framed side-hung fenestration.

The external surfaces are painted with dark colours to increase heat absorption, while pale colours have been used for the internal surfaces.

The studios' floors are covered with rubber tiles, while granite tiles have been used for the bedrooms and ceramic tiles for the living room, the kitchen and the dining room.

The building's total area is 418 m² and total volume 1745 m³.

Table 17.13. The building's thermal characteristics

Heated floor area	419.86 m²
Heated volume	1745 m³
U values	
window	3.2 W/m²K
wall	0.5 W/m²K
concrete walls	3.3 W/m²K
floor	2.5 W/m²K
roof	0.5 W/m²K
Design temperature	21°C
Space heating load	71.76 kWh/m² per annum
Lighting	14.87 kWh/m² per annum

Passive solar design and systems

Direct gain

Direct solar radiation comes through the glazed area of the south openings and the roof skylight, which covers about 50% of the total horizontal area and is double-glazed. The shading devices used are the overhangs and the side fins formed by the balconies and the facade configuration (Figs. 17.46 and 17.47). Night-time thermal insulation in winter consists of insulated aluminium rolling shutters.

Attached sunspace

The sunspace is attached to the south facade on the second floor and has a total vertical south surface of 8 m². It is connected to one of the bedrooms through a 4 m² double-glazed metal-frame opening, and to the atrium through a 7.5 m² single-glazed metal-framed opening (Figure 17.48).

Natural ventilation is achieved through the openable window area, which is half of the total window area. Shading, provided by the third-floor balcony, prevents overheating.

Figure 17 46. Solar radiation coming in through the south openings (cross section)

Figure 17.47. The roof skylight

Figure 17.48. Interior view of the attached sunspace

Trombe walls

Three Trombe walls were constructed, one on each floor. The one on the first floor has a vertical surface of 4 m² and is part of the external wall of the office. The second one, on the second floor, is part of the bedroom's external surface and has a total vertical area of 1.5 m². On the third floor, the Trombe wall has an area of 7.5 m² and is in contact with the living room (Figure 17.49).

Figure 17.49. The Trombe wall on the third floor

All three Trombe walls are made of 20 cm thick, dense, dark-grey concrete bricks and have 10 cm diameter dampers on their upper and lower sections for air circulation. The outer surface is single-glazed and metal-framed, forming a 10 cm air gap. Fixed concrete elements were used for shading.

The building's thermal behaviour

The building's thermal mass has been increased by the concrete construction elements, assisting the thermal storage process and reducing indoor temperature variations during the winter and summer periods (Figures 17.50 and 17.51).

In the heating period, part of the direct and indirect solar gains is stored during the day and released during the night, decreasing the temperature variations. The Trombe walls and the attached sunspace contribute additional thermal gains by conduction and convection, as well as by thermocirculation through the dampers of the Trombe walls. Thermal losses through the windows are decreased by using insulated roller shutters.

During the summer all the passive solar systems are externally shaded by the concrete side fins and horizontal overhangs, which allow unobstructed solar radiation and natural

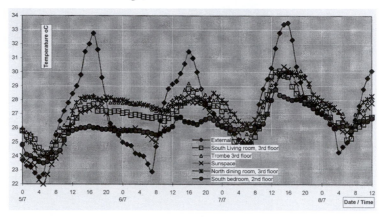

Figure 17.50. Ambient and indoor temperature variations during the summer

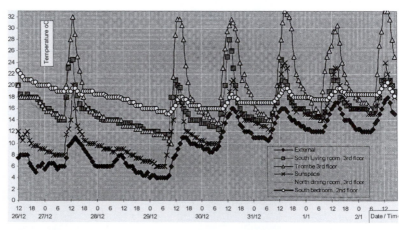

Figure 17.51. Ambient and indoor temperature variations during the winter

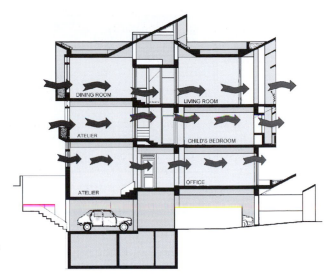

Figure 17.52. Natural cross-ventilation during the summer (cross section)

lighting during the heating period. The dampers of the Trombe walls are always kept closed during the summer.

The house is naturally cooled, during the summer, by cross-ventilation achieved between the north and the south openings (Figure 17.52). This cooling process is assisted by the prevailing north winds, which form a cross airflow, penetrating the north openings and extracting the heat through the south openings. During the night, extra thermal discharge is achieved and overheating is reduced by means of night-time natural cross or single-sided ventilation.

Energy consumption

The total annual energy demand of the building is 47,428 kWh. This demand is partially covered, to the order of 36.5% (17,299 kWh), by thermal gains due to passive solar design, while the rest – 63.5% (30,129 kWh) – is provided by auxiliary heating (Figure 17.53).[5,6]

Passive solar gains are mainly due to direct solar gains (67%), followed by thermal gains from the three Trombe walls (22%) and the attached sunspace (11%) (Figure 17.54).

Figure 17.53. Percentage distribution of annual heating load supply.

Figure 17.54. Percentage distribution of annual passive solar gains

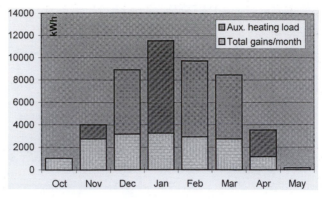

Figure 17.55. Monthly auxiliary heating demand

The maximum monthly auxiliary heating demand is in January, when the total energy load reaches 11,500 kWh (Figure 17.55).

Discussion

The implementation of passive solar design resulted in a 5% increase in the total building's construction budget. This additional cost could be covered in a few years by the energy gains from passive solar systems, which account for 36.5% of the total annual energy demand. The energy results would be more satisfactory for the mild urban climatic conditions of Athens if all the building's elements had stronger insulation, especially the uninsulated concrete walls.

As is shown in Figures 17.50 and 17.51, interior temperature variations appear to be small, compared to the external temperature variations, as a result of the effect of thermal storage. Rooms and spaces beside or near passive solar systems have increased temperatures in winter, while the northern areas of the house are significantly cooler owing to their orientation and to the existence of large openings for studio lighting purposes.

Shading devices prevent extreme temperatures in the south-oriented spaces in the summer. As shown in Figure 17.51 the temperatures of south- and north-oriented spaces do not display extreme differences.

EXAMPLE BUILDING 4 – A PASSIVE SOLAR APARTMENT HOUSE IN ATHENS, GREECE

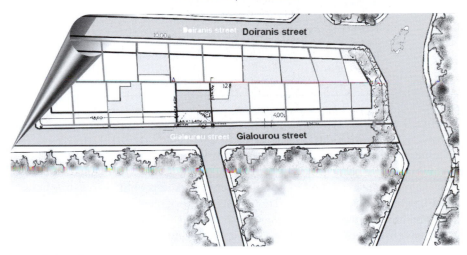

Figure 17.56. Site layout

Project background

The building, designed by M.G. Souratziolis, is situated in the centre of the city, at Gialorou 15, near the Attic Park (Figure 17.56). It consists of three apartments and an

Figure 17.57. Front view of the south facade

Table 17.14. Building data

Building type	Three apartments + an office
Number of occupants	7
Year of construction	1993
Scheme of building	Terraced house
Orientation	SE–NW
Annual energy consumption	53.13 kWh/m²

office built according to the principles of energy-conscious design for passive heating and cooling purposes (Figures 17.57 to 17.58 and Table 17.14).

The nearby buildings are mainly one- and two-storey luxury terrace houses. Attic Park offers improved air quality, thus positively influencing the urban climate, because it creates a green area in the polluted and densely built metropolitan area of Athens.

The bioclimatic architecture and the extensive use of recycled and environmentally friendly building materials have influenced the design. This approach is emphasized by the existence of sculptures and decorative elements (Figure 17.59), as well as by the use of earth products and materials. The bricks used externally and internally are made of loam in order to provide a warm earth look.

According to the architect, '...creating our own home is like building ourselves'; he has also tried to realize his own childhood memories and feelings about the nearby landscape, history and civilization, events, air and rain, light and view, trees and climate.

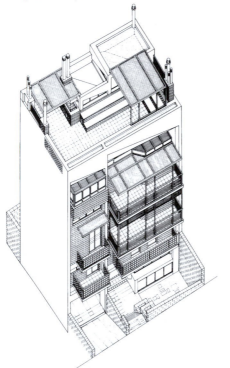

Figure 17.58. Axonometric view of the building

Figure 17.59. One of the decorative elements

Site and climate

The climate of Athens is mild with increased solar radiation during the winter, reduced only by the atmospheric pollution. The average monthly temperatures are 9.6°C in January and 27.9°C in July (Figure 17.60). The total annual amount of sunlight is 2818 hours, well distributed throughout the year and there is an annual total of 1110 degree days (Table 17.15).

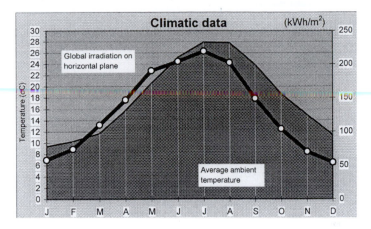

Figure 17.60. Average monthly temperatures and solar radiation

Table 17.15. Average annual climatic data

Latitude	37° 58'
Longitude	23° 43'
Prevailing wind direction	N
Average wind speed	2.8 m/s
Relative humidity (winter/summer)	69.4%/52%
Global irradiation on the horizontal plane (winter/annual)	617/1581 kWh/m²
Degree days (base: 18°C)	1110 days
Sunshine hours (winter/annual)	1115/2818 hours
Average ambient temperature	13.2°C/18.1°C
Average maximum temperature (annual)	22.6°C
Average minimum temperature	11.9°C

The local climate is influenced by the existence of the nearby Attic Park (Attikon Alsos), which is situated near the Acropolis. The park influences air quality, air moisture, temperature, wind flow and ground-reflected radiation.

General building description

The building was designed in 1985 and construction was finished in 1993. The main facade of the building faces Gialourou Street and has a south-east orientation, 35° from south.

Table 17.16. Geometrical data for the building

Number of floors	3 + 1 (Ground Floor) + Basement
Total height	14.7 m
Floor area	101 m²
Total area	455 m²
External area	971.77 m²
Volume	1165.7 m³
Area-to-volume ratio	0.834
Window area	
SE	77.34 m²
NE	1.94 m²
NW	32.40 m²
SW	3.70 m²
total	115.36 m²
Vertical wall area: solid	755.41 m²
Opaque element area	856.41 m²

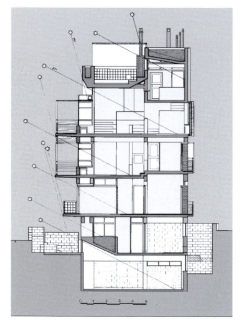

Figure 17.61. Cross section from south to north

The building consists of a basement, a ground floor and three more floors above, with 455 m² total floor area (Table 17.16 and Figure 17.61). There are three apartments with floor areas 190, 100 and 60 m² respectively, which are inhabited by approximately seven people, and an office of 100 m² belonging to the architect, who is the owner of the top apartment. All the apartments are spread out over two floors.

In the basement there is an architect's office, the boiler room and auxiliary rooms (Figure 17.62) The office communicates with a ground-floor office through a separate staircase (Figure 17.63) There is also a central staircase and an elevator for the upper apartments.

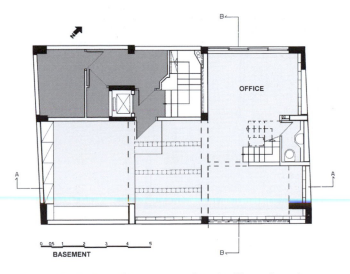

Figure 17.62. Plan of basement, architect's office and auxiliary space

Figure 17.63. Architect's office in the basement with stairs leading to the ground floor

Figure 17.64. Office on the ground floor with the staircase leading to the basement

On the right-hand side of the ground floor there is a separate space, which forms part of the office in the basement (Figures 17.64 and 17.65). On the left-hand side is the first storey of a two-storey 60 m^2 apartment, which has a separate staircase to connect with the rooms on the upper floor.

On the first floor, there are the upper rooms of the ground-floor apartment, including a south-facing living room and a kitchen, as well as the bedrooms of a second two-storey apartment that extends to the second floor (Figure 17.66). The bedrooms of the latter apartment are connected with the rooms on the second floor by a private staircase. Glass bricks are used for the verandas of the first floor as well as for the second-floor bedroom in order to provide privacy.

0 05 1   2   3   4   5   GROUND FLOOR PLAN

Figure 17.65. Ground floor plan with entrance and study

0 05 1   2   3   4   5

FIRST FLOOR PLAN

Figure 17.66. First floor plan

On the second floor there are the living room, the dining room and the kitchen of the two-storey apartment. There are also the master bedrooms and the bathroom of the third-floor apartment (Figure 17.67).

Between the second and third floors a loft has been formed in order to create the two children's bedrooms and a bathroom for the third-floor apartment, which belongs to the architect. The height of the rooms was reduced to the minimum acceptable level and the floor level adjusted in order to create space for this loft.

SECOND FLOOR PLAN

Figure 17.67. Second floor plan with master bedroom and loft with the children's bedrooms of the third-floor apartment

The third floor (Figure 17.68) consists of a living room with a loft and a gallery, a dining room and a kitchen, which all belong to the third two-storey apartment. All the spaces are on several levels and are accompanied by comfortable south-facing verandas with a remarkable view of Attik Park, the Acropolis and Lycavitos hill. The third floor can be accessed via the central staircase and the elevator, as well as by the private staircase of the bedrooms from the floor below (Figures 17.69 to 17.71).

The terrace (Figure 17.72) was built at 14.70 m above ground, using several levels and creating several openings at every level in order to ensure deep penetration of the sun and to allow visual contact between the various levels (Figures 17.73 and 17.74).

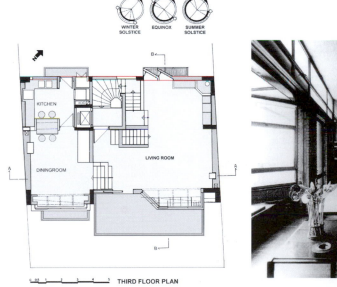

THIRD FLOOR PLAN

Figure 17.68. Third floor plan with living rooms, kitchen, dining room and main entrance to the apartment

Figure 17.69. Living-room and dining area on the third floor with stairs leading to the gallery upstairs and the bedrooms downstairs

Figure 17.70. The south-eastern part of the main living-room with the 3.50 m high solarium, the northern hearth (height 2.40 m) and gallery/library/study upstairs

Figure 17.71. Dining area in the alcove of the south-eastern elevation; objects are made from reused materials

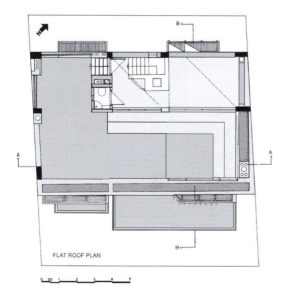

FLAT ROOF PLAN

Figure 17.72. Plan of roof terrace and the gallery/library

Figure 17.73. View of the main living-room from the gallery, showing the sloping gold-painted reflectors on the ceiling

Figure 17.74. Gallery/library/study space with gold-painted sloping ceilings (reflectors) and the exit to the roof terrace

Energy-saving features

The main energy design features of the building include the planted roof, the solarium on the SW facade and the openings in the roof, combined with the direct solar gains, as well as the carefully designed solar shading devices and the gold-painted reflectors in the roof. The separate ventilation ducts for each room, combined with natural ventilation through the openings and the appropriate arrangement of the rooms, are also important energy features.

The terrace roof (Figure 17.75) is formed as a roof garden, with a 1 m thick earth layer to offer cooling in the summer and insulation in the winter. It consists of several levels

Figure 17.75. View of terrace-garden with open-air sitting corner and barbecue

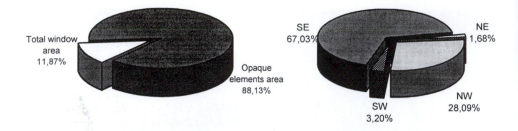

Figure 17.76. Surface area of opaque and transparent building elements

Figure 17.77. Distribution of the windows on the building's sides

Figure 17.78. North-west facade

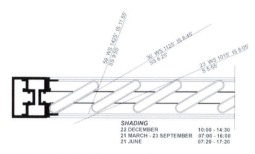

Figure 17.79. Shading devices

and a slightly inclined pediment at the edge. This is painted gold in order to reflect the light and to follow the ancient tradition of gold-painted temple pediments, providing visual contact with the Acropolis.

Large openings are formed in the south-eastern facade to take advantage of the sun and smaller openings are placed in the northern facade in order to achieve adequate ventilation and lighting; these openings cover 12% of the total building envelope and including specially designed openings on the roof (Figures 17.76 to 17.78).

The sunspaces (solaria), made of visible aluminium frames and glass or glass bricks, are used as living rooms during the day in winter, while they are kept closed at night in order to reduce thermal loss.

The carefully designed external shading elements allow the sunlight to enter in winter and block the extensive summer sunlight (Figure 17.79). On the verandas of the upper floors lightweight banisters have been placed, in order to provide adequate shading during the summer days as well as an unobstructed view and privacy (Figure 17.80). In

Figure 17.80. Detail of balcony on the front facade

Figure 17.81. South-western view of terrace-garden with solar pergolas of fixed aluminium louvres, the trellis of aluminium sheeting and the ventilation shafts

this way the bioclimatic design offers a pleasant indoor climate with the minimum possible energy demand.

Every room of the house is equipped with vertical air ducts for passive cooling, which on summer nights is combined with single-sided and cross-ventilation through the external openings (Figure 17.81).

The external walls are insulated with 5 cm expanded polystyrene and an extra 10 cm wall made of insulating bricks (thermoblock) located between the double brickwork constructed with decorative bricks (U value 0.507 W/m^{2}K) (Figure 17.82). The concrete building elements remained without plaster to allow the original materials to be seen. They therefore have no thermal-insulation layer, resulting in U value of 3.249 W/m^{2}K. The flat roof of the building, in the non-planted areas, has a 5 cm thick polystyrene layer plus 10 cm of insulation bricks and 10 cm (mean thickness) of lightweight concrete, which also works as a protective layer. The mean U value of the flat roof is 0.513 W/m^{2}K.

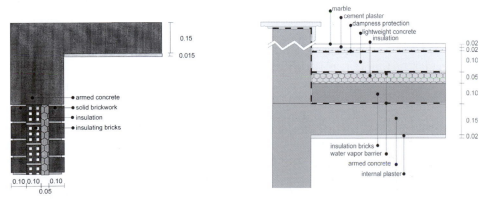

Figure 17.82. The structure of the external walls and the roof

Heating system – natural cooling

Heating system

The building is heated by a central heating system with a boiler rated at 60,000 kcal/h. It is equipped with thermostats in each apartment set at 21°C. The system was installed in 1993 and is serviced once a year. The boiler is used to provide hot water for domestic use.

The annual consumption for space and water heating is approximately 25,000 KWh.

Natural cooling (summer period)

The building is cooled by natural ventilation during the summer period. Overheating is avoided as a result of the planted roof and the shading elements of the building.

Ventilation. Natural ventilation is mainly provided by cross-ventilation through the openings in the two main facades, especially on summer nights. The vertical air ducts for passive cooling, which take advantage of the temperature effect in the summer, carry out the warm air on summer nights.

Shading. The building is equipped with a variety of fixed shading elements. The balconies of the first floor and the bedroom on the second floor are shaded by glazed bricks and the other balconies are equipped with shading blinds. The openings in the south facade that have no balconies above them are equipped with overhangs.

Actual energy consumption

The specific heating energy consumption of the building is 53.13 KWh/m² per annum. It was calculated that the total annual energy saving of the biggest apartment on the second and third floors is about 56%. The energy saving is 41.4% for heating and 75.5% for cooling (Figures 17.83 and 17.84).

Conclusions

The building described is an example of the energy benefits of the passive solar systems incorporated in an ordinary building.[7,8] The aim of the architect was not only to reduce

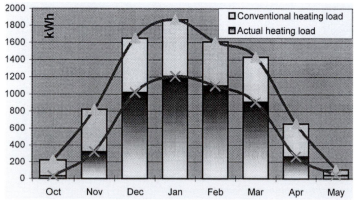

Figure 17.83 Heating load

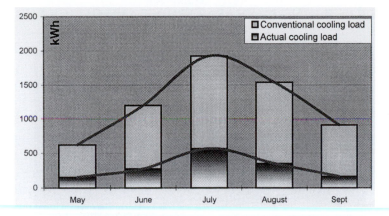

Figure 17.84 Cooling load

energy consumption but also to integrate passive solar design into the building's architecture and decoration. The heating and cooling energy consumption of the examined building were found to be remarkably low, compared with similar conventional urban buildings.

EXAMPLE BUILDING 5 – AN APARTMENT AND OFFICE BUILDING IN NICE, FRANCE

Figure 17.85. A view of Nice

Project background

Le Presbytère de Ste Reparate, 3 Rue Ste Reparate, is located downtown in the old city of Nice (Figures 17.85 and 17.86). It was built during the XVIIIth century (after 1650 and before 1780) in a densely built area. It is attached to the Ste Reparate church and was built as its presbytery.

Le Presbytère consists, like most of the surrounding buildings, of four floors and an attic. It is used permanently as an apartment and office building and for occasional music-school activities (Table 17.17). The floor area is 400 m^2 and the total volume 1500 m^3.

Summer comfort was one of the main objectives in the design of the building, which takes advantage of the urban structure as well as of building details, to achieve a

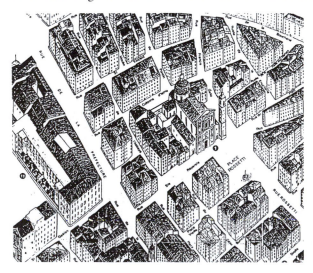

Figure 17.86. Perspective of the building and its surroundings

Table 17.17. Building data

Building type	Apartment/office/school
Year of construction	XVIIIth century
Scheme of building	Attached
Orientation	N–S

maximum comfort level: weak solar penetration, a large surface area open to air circulation (35 times that of modern city buildings) for an efficient convective exchange at night (a dry and cold earth breeze), all-pervasive ventilation, a uniform air temperature and hygrometry to avoid condensation, internal and external shutters, numerous windows that are weakly sunlit to achieve uniform lighting, north openings, interior 'clairoirs' at stairwell level for a very efficient air circulation at night, plus slate paving in areas of common use. Figure 17.87 shows various views of the building and its surroundings.

Figure 17.87. External views of the building and its surroundings

Site and climate

The latitude of the site is 43.8°N and the longitude is 7°E. The elevation is about 5 m above sea level. The climate of the city is strongly influenced by the adjacent sea. Microclimatic conditions are created by adjacent hills and mountains, the Paillon stream and the urban structure (Figure 17.88 and Table 17.18).

The narrow north–south streets, protected by large Italian penthouse caps minimize the direct solar incidence at midday. The east–west streets are slightly ascending, which increases the air velocity as a result of the Venturi effect. The building density (mean street width 3.8 m) and the height (street depth 20 m) restrict the sunlit areas to roof level, preserving only the necessary visual light.

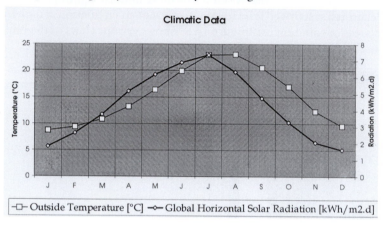

Figure 17.88 Average daily temperature and solar radiation

Table 17.18. Average climatic data

Prevailing wind direction	N
Average wind speed	3.9 m/s
Precipitation	64 mm/month
Relative humidity	72%
Heating degree days	1445 days
Cooling degree days	68 days

General building description

The building has four storeys and a huge attic (Table 17.19, Figures 17.89 and 17.90). The block with the two study rooms faces Abbey Street. The stairwell, the entrance and two other blocks are recessed, inserted between buildings and oriented towards the north and the east (before the French Revolution, all the buildings adjoining the cathedral constituted the bishop's residence). Both blocks and floors are independent, connected only by the staircase. The ceilings are high (>3 m) and the main walls thick (0.6 m).

Windows are numerous to distribute light everywhere and ensure cross-ventilation. They are single-glazed and have both internal and external shutters; on the south side the lower parts of the external shutters are hinged at the top so that they can be opened (Figures 17.91 and 17.92)

Table 17.19. Geometrical data for the building

Number of floors	4 + attic
Total height	12 m
Floor area	400 m²
Total volume	1500 m³
Floor-ceiling height	3 m
City building density	66%

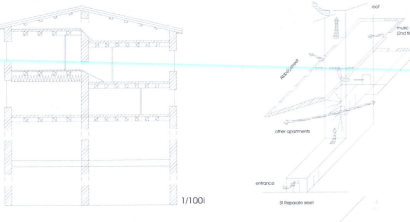

Figure 17.89. General building structure

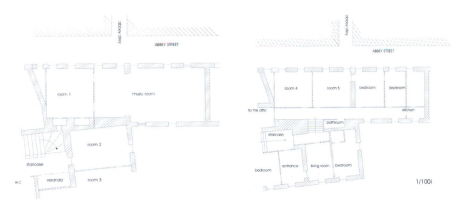

Figure 17.90. Second and third floor organization

The staircase is open at two ends (with 'clairoirs', a sort of guide for ventilation above the entrance door, and a fanlight at roof level, as in many buildings in this city); see Figure 17.93. Here, the building can be viewed as a type of ancient construction and strong thermal gradients are felt when moving upstairs.

Figure 17.91. Building openings and shutters

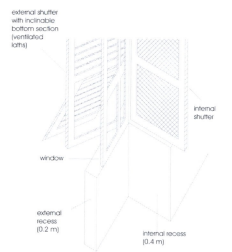

external shutter
with inclinable
bottom section
(ventilated
laths)

internal
shutter

window

external
recess
(0.2 m)

internal recess
(0.4 m)

Figure 17.92. External and internal shutters

*Figure 17.93.
Indoor views of
the music room
and the stairwell*

Architectural elements

The main walls are constructed of masonry and have a U value of 1.9 W/m²K, while the floor and the ceilings over unheated rooms have a U value equal to 1.3 W/m²K. Some elements of the construction are shown in Figures 17.94 and 17.95.

As noted above, the windows are equipped with external and internal wooden shutters with an inclinable bottom section. The U value of the windows is about 5.2 W/m²K.

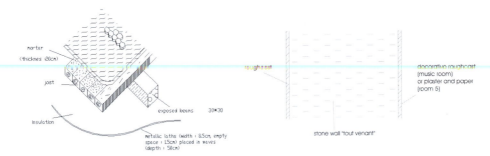

Figure 17.94. Ceiling of the music room   Figure 17.95. Constitution of the main walls

Building behaviour

The occupants feel that the building is fresh in summer and not too cold in winter. This is indirectly confirmed by the reduced amount of electrical and gas heating equipment, which, according to the occupants, is only used occasionally, usually to lower the feeling of humidity.

Windows closed in the daytime and solar shading and thick walls are responsible for comfortable conditions in summer. The maximum inside air temperature is 4 degrees lower than that outside.

Moreover, the stairwell offers the opportunity for increasing, through the chimney effect, the ventilation of the upper levels by bringing up cold air from the lower levels.

Simulations have been conducted in order to demonstrate the contribution of the passive cooling techniques used in the building to maintaining comfortable conditions. Figure 17.96 to 17.98 show the results of these simulations (with and without imposed convective coefficients) compared to the actual monitored results.

Figure 17.99 to 17.101 show the efficiency of different techniques through different simulations:

- reduced ventilation for all zones all day long (sim. 2);
- lower-inertia materials in place of the traditional materials (sim. 3);
- reduced heat gains through the insulation of the floor in the attic (sim. 4).

Conclusions

In fact, as a result of the design of the buildings, it is the city itself that constitutes a climatized zone: rooms, courtyards and staircases open at two ends can only be

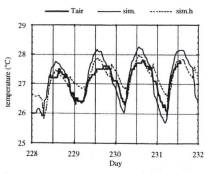

Figure 17.96. Air temperature inside the
music room

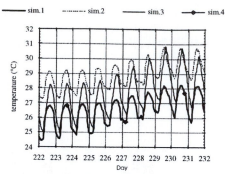

Figure 17.99. Sensitivity analysis for the
music room

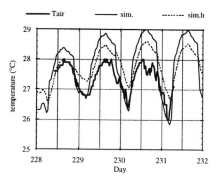

Figure 17.97. Air temperature inside
the office

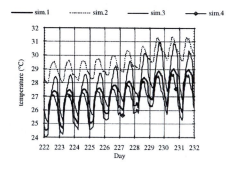

Figure 17.100. Sensitivity analysis for the
office

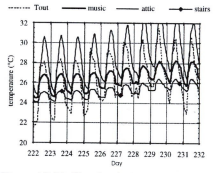

Figure 17.98. Simulated air temperature
differences between various zones of the
building

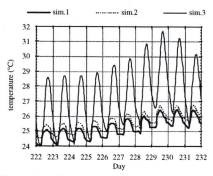

Figure 17.101. Sensitivity analysis for the
staircase

differentiated from the streets by their use. Weak gradients (not only of air temperature,
but also of humidity and air velocity) and common life (odours, noises…) add their
contribution to the feeling of thermal comfort.

EXAMPLE BUILDINGS 6 AND 7 – A PRIMARY SCHOOL AND A GYMNASIUM WITH A CENTRAL SOLAR HEATING PLANT HAVING SEASONAL STORAGE IN NECKARSULM, GERMANY

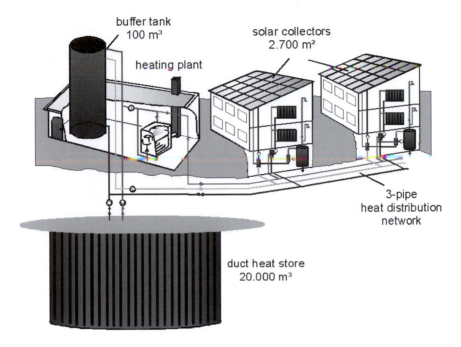

Figure 17.102. Layout of the active solar system with seasonal storage in Neckarsulm, Germany

Project background

Integrated energy approaches to heat provision for urban buildings have been developed in Germany in recent years in order to obtain the maximum saving of fossil fuels at the lowest possible additional cost. These approaches combine insulation measures, rational energy conversion and supply and the active use of solar energy. More than half of the fossil-fuel demand and equally CO_2 emissions of a conventional urban housing estate can be saved with a central solar heating plant with seasonal storage.

In Neckarsulm in South Germany the first building stage of the third pilot project with solar-supported seasonal storage was finished in 1998. Figure 17.102 shows the layout of the central solar heating system.

Solar energy is collected by the collector fields and fed into the seasonal heat store via the three-pipe heat distribution network. The seasonal heat store itself is a duct heat store, where heat is fed directly into the ground by ground heat exchangers of polybuthene. During the summer the storage is charged, while during the winter the heat is discharged to supply the conventional heat generation system in the central heating plant. Each year about 50% of the total heat energy demand of the entire housing estate is supplied by the sun.[9]

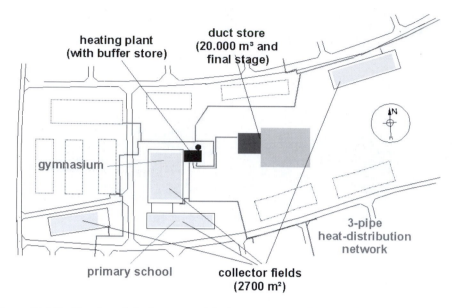

Figure 17.103. Site plan of the first stage of the housing estate in Neckarsulm, Germany

A site plan of the housing estate with the solar installations of the first building stage is given in Figure 17.103. The collector fields are mounted on the roofs of the buildings. The size of the duct store that is connected to the already installed collector area of 2700 m² is 20,000 m³.

The two buildings that are described below are the primary school and the gymnasium. Both have been built by the city of Neckarsulm in the awareness that they represent the uniqueness of the solar-supported central heat supply.

The city of Neckarsulm commissioned a consultant to develop a low-energy concept for each of these buildings, taking into account heat insulation, daylighting, use of and protection from passive solar heat, ventilation, room heating, rational use of electric power, ecology of building materials and usage of rain water.[10] For all energy-saving measures, the energy-saving effects and their profitability were calculated within a comprehensive energy concept.[11]

The only item that had to be discounted was the usage of rainwater within the buildings because in primary schools and gymnasia the potential for rainwater application is very small.

EXAMPLE BUILDING 6 – A SOLAR PRIMARY SCHOOL IN NECKARSULM, GERMANY

Figure 17.104. South-west view of the primary school

Table 17.20. Characteristic values for the primary school

Ground area	1196 m²
Opaque wall area	
north	458 m²
east	175 m²
south	230 m²
west	141 m²
Window area	
north	283 m²
east	30 m²
south	386 m²
west	80 m²
Roof area	1217 m²

Table 17.21. Geometrical data for the building

Heated volume	9612 m³
Surface of heated volume	4191 m²
Area-to-volume ratio	0.44
Heated floor area	1910 m²

Building description

The school is in the Amorbach area of Neckarsulm and was built in 1996/97. The architect was N. Parvanta, Fellbach, and the energy concepts were developed by Steinbeis Transferzentrum RES.

The primary school consists of two blocks each with two floors. The block on the west side, most of which is mainly visible in Figure 17.104, consists of classrooms to the south and staircases to the north, connected by a foyer whose facades are made of glazing in a light wooden structure. The second block located to the east, contains classrooms facing due south and workrooms and adjoining rooms facing due north, connected by a simple corridor.

Main features of the building

The characteristic values listed in Table 17.20 show that the design of the building shape took into account the fact that a small area-to-volume ratio helps to save energy for room heating. The geometrical data for the building is given in Table 17.21. The opaque facades consist of different constructions: while the south facade is of light construction, including the windows, all other walls consist of concrete or bricks with outside insulation. The roof is partly tilted to the south to carry the solar collectors.

Use of and protection from passive solar energy

The building is orientated directly south. Because the street to the south is quite broad, the surrounding buildings do not shade the south facade, shown in Figure 17.105. The west facade is shaded on average 30% by overhangs from the building itself and by a shopping centre that is located on the west side of the school.

The design of the outside planting took care of the shading of the glass facades. Trees are placed so that they protect the south facade from the morning sun and prevent glare in the classrooms in summer. In winter, the trees lose their leaves and the sun is able to shine into the classrooms to support room heating.

For overheating protection in summer, the roof overhang was first thought to be sufficient. A simulation of the indoor temperatures with TRNSYS showed the

Figure 17.105. View of the south facade

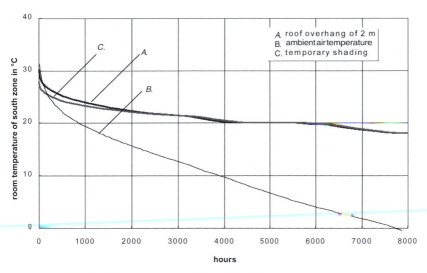

Figure 17.106. Overheating protection by shading devices

necessity of an outside shading device;[12] even with a roof overhang of 2 m the number of hours per year with an indoor temperature over 25°C amounted to 550 (Figure 17.106). With temporary outside shading the maximum temperature can be reduced from 30°C to 28°C and the number of hours per year over 25°C reduced to 264. To admit daylight, a blind was chosen so that its panels could be tilted to keep out glare but guide the light up to the ceiling. The ceiling was therefore designed to be white in colour.

Energy concept

Within the comprehensive energy concept that was prepared by the Steinbeis-Transferzentrum, all possible energy-saving measures were optimized with their energy-saving effects, the additional investment costs and the feasibility of the architectural design itself taken into account. The targeted energy consumption is shown in Table 17.22.

Table 17.22. Targeted energy consumption of the building

	Final energy use (kWh/m² per annum)	Prime energy supply (kWh/m² per annum)
Electric power consumption	12	36
Hot water preparation	5	5
Room heating	75	75

Heat insulation

The national building code in Germany, called *Wärmeschutzverordnung*,[13] restricts the yearly energy demand for room heating. The heat insulation concept, as shown in Table

Table 17.23. Insulation measures

Outside concrete wall to air	16 cm	Mineral wool, k = 0.04 W/mK
Outside light construction to air	14 cm	mineral wool, k = 0.035 W/mK
Outside concrete wall to ground	12 cm	XPS-polystyrene, k = 0.035 W/mK
Outside foundation	5 cm	XPS-polystyrene, k = 0.035 W/mK
Tilted roof	30 cm	Mineral wool, k = 0.04 W/mK, between crossed rafters
Flat roof	6 cm	EPS-polystyrene, k = 0.04 W/mK
	20 cm	EPS-polystyrene, k = 0.035 W/mK
Basement to ground	2 cm	Mineral wool, k = 0.04 W/mK
	10 cm	EPS-polystyrene, k = 0.035 W/mK
Windows:		
only glass		U value: 1.1 W/m²K
glass and framing		U value: 1.3 W/m²K

Table 17.24. Heat demand of the building

Limit in the national building code	21.4 kWh/³ per annum
Room heating demand obtained	14.8 kWh/³ per annum
Savings	31%

17.23, restricts the yearly space heating demand to 31% lower than the required limit (Table 17.24).

The insulation measures required an architectural design concept that avoided heat bridges and ensured an airtight outer surface. Inaccurate planning and particularly implementation would have destroyed the low-energy approach of the building, because heat bridges and outer surfaces that are not tight can heighten the room heat demand of a building to a great extent, easily by over 30%.

Natural ventilation

Schoolchildren do not only need fresh air in breaks but also during classes. Opening windows during class would not be an acceptable measure for fresh air ventilation because of the noisy school surroundings and especially because the children sitting close to the windows would be subjected to cold air.

Thus a controlled ventilation system has been included in each classroom, using the natural lift of heated air to ventilate the whole building: through controlled inlet flaps, the outside covers of which are visible in Figure 17.107, fresh air flows into the classrooms. Directly behind the inlet device the air flows through the radiators to be preheated, supplies the children with oxygen, is heated, rises up to the ceiling and out of the classroom into the corridor through controlled flaps. From the corridor an exhaust chimney leads the air out. On the first floor, the exhaust air from the classrooms flows through north windows directly to the outside (see Figure 17.108).

Heating system

Radiators supplied with hot water heat the school. The supply temperature is adapted to the outside air temperature. The maximum supply temperature is 70°C for outside

Figure 17.107. Close-up of the south facade with openings for natural ventilation

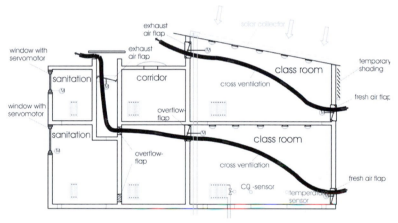

Figure 17.108. Cross section of the building with natural ventilation

temperatures of –12 °C and lower, while the return temperature is 40°C or lower. The heating system can be individually controlled for each room, both directly inside the room and from the control centre that is supervised by the caretaker.

Integration of solar collectors

The primary school carries a part of the collector field for the central solar heating plant with seasonal storage that has been erected nearby. The solar collectors had to be integrated into the architectural concept and into the building structure itself. The tilted roof of the building was covered with a watertight, but not weatherproof, liner, on which wooden rafters were mounted to take the collectors. The collectors were prefabricated by the supplier so as to cover the entire length of the roof. Such collectors are called solar roofs.

*Figure 17.109. Mounting
the collectors*

*Figure 17.110. The entire collector area as
the roof of the primary school*

With a movable crane and a mounting tool the collectors were mounted onto the rafters (Figure 17.109). The entire collector area, covering about 600 m², as shown in Figure 17.110, took a few days to mount. The collector replaces the roof tiles and protects the building just like an ordinary roof. To connect the collector field to the heating network, two pipes, each with a diameter of about 6 cm, were integrated into one of the installation wells, as shown in Figure 17.108.

Electric power consumption

In school buildings, the main consumers of electric power are lighting and ventilation. However, the ventilation of the school needs very little energy because it is driven by natural forces and not by fans.

To reduce the power consumption for lighting, emphasis was placed first on designing the classrooms and the temporary shading devices in such a way that daylight could be used. Second, artificial lighting could be added to provide the extra light needed within every room. The control of lighting and other electric power consumers is effected by an EIB (European Installation Bus). This allows control of every electric power consumer directly from the central control and of the automation energy-saving measures; for example all lights are turned off at 10 p.m. Another advantage of such a bus system is that later renovations are very easily to realize. If a new switch needs to be built in, it only has to be added to the bus. There is no need for further cable installation.

The total electric power consumption of the primary school is targeted to be about 10 to 12 kWh/m² per annum. A current average figure for German schools is about 30 to 35 kWh/m² per annum.

EXAMPLE BUILDING 7 – A SOLAR GYMNASIUM IN NECKARSULM, GERMANY

Building description

The gymnasium, which is divisible into three parts, is, like the primary school, in the Amorbach area of Neckarsulm and was built in 1996/97. The architect was N. Parvanta, Fellbach, and the energy concepts were developed by Steinbeis Transferzentrum RES.

The gymnasium stands to the north of the primary school and is connected to it by an underground corridor. As shown in Figure 17.103, the gymnasium has a north–south orientation, with the main glazing on the west facade. The north and south walls are mainly closed. On the one hand, this partly hinders passive solar energy use, while on the other the design supports the usage of the gymnasium by integrating rebound walls into the north and south facades; these opposite walls are the ones that will be mainly exposed to ball games.

The gymnasium is used by different schools and local sports clubs. For tournaments, terraces and a small foyer with reception facilities are also integrated into the gymnasium.

The greatly varying numbers of people in the gymnasium, from about 10 persons up to 400, in the case of a tournament, calls for a very flexible ventilation system that should also be a target for low-energy building concepts.

Main features of the building

The characteristic values of the gymnasium are given in Table 17.25. Because a gymnasium can only be built as a cube, the area-to-volume ratio is very low.

The geometrical data for the building are given in Table 17.26. All the opaque facades consist of concrete walls, whereas the south and north facades are panelled. Half of the east walls are in contact with the ground, the rest being mainly glazing. The west facade, too, mainly consists of glazing and only a small pedestal is covered by insulation and plaster.

Table 17.25. Characteristic values of the gymnasium

Heated volume	14,786 m³
Surface of heated volume	5390 m²
Area-to-volume ratio	0.36
Heated floor area	2270 m²

Table 17.26. Geometrical data for the building

Ground area	1954 m²
Opaque wall area	
north	252 m²
east	236 m²
south	256 m²
west	119 m²
Window area	
north	93 m²
east	159 m²
south	107 m²
west	260 m²
Roof area	1954 m²

The roof is totally flat with the supporting structure outside. This reversal of ordinary roof construction gives the gymnasium a flat inside ceiling, which is an advantage for some ball games and allows a very good integration of solar collectors.

Use of and protection from passive solar energy

The best way to integrate daylighting without the risk of overheating during summer would have been to integrate glazed north-facing panels into the roof. This possibility was rejected by the city of Neckarsulm, which asked for a totally closed roof. As mentioned above, the north and south walls have been designed to be mainly closed to enable the installation of rebound walls inside the gymnasium. Thus only the east and west sides could be used for glazing.

Most glazed western facades do not retain a lot of energy for room heating, but lead to high overheating in summer. The architectural concept therefore put emphasis on the transparency of the building and consequently on a more or less totally glazed west facade. To prevent overheating of the building during the summer season and to enable the use of daylighting at the same time, the west facade is provided with an outside blind at two levels. Therefore, it is possible to shut down only the lower part of the blind and use the upper, unshaded part of the west glazing for daylighting. A second aid against overheating is the use of the natural ventilation system that is built into the gymnasium for night cooling in the summer.

Energy concept

Within the comprehensive energy concept prepared by the Steinbeis-Transferzentrum, all possible energy-saving measures were optimized, as for the primary school. The targeted energy consumption is shown in Table 17.27.

Table 17.27. Targeted energy consumption of the building

	Final energy use (kWh/m² per annum)	Prime energy supply (kWh/m² per annum)
Electric power consumption	12	36
Room heating	78	78

Heat insulation

The heat insulation concept, as shown in Table 17.28, reduces the yearly space heating demand to 40% lower than the required limit (Table 17.29), taking into account the heat recovery system for ventilation. Here, too, the insulation measures required an architectural design concept that would prevent heat bridges and secure a tight outer surface. The glass is of lower thermal quality than in the school because for the gymnasium special security glass had to be chosen, which would have increased the cost of the glazing significantly if the same U-values had been required as for the school building.

Table 17.28. Insulation measures

Outside concrete wall to air	16 cm	Mineral wool, k = 0.04 W/mK
Outside concrete wall to ground	12 cm	XPS-polystyrene, k = 0.035 W/mK
Outside foundation	5 cm	XPS-polystyrene, k = 0.035 W/mK
Flat roof to ground	12 cm	XPS-polystyrene, k = 0.035 W/mK
Flat roof	22 cm	EPS-polystyrene, k = 0.04 W/mK
Basement with sports floor	15 cm	U value: 0.29 W/m²K
Basement to ground	2 cm	Mineral wool, k = 0.04 W/mK
	10 cm	EPS-polystyrene, k = 0.035 W/mK
Windows		
only glass		U value: 1.3 W/m²K
glass and framing		U value: 1.4 W/m²K

Table 17.29. Heat demand of the building

Limit in the national building code	20.1 kWh/² per annum
Room heating demand obtained	12.0 kWh/² per annum
Savings	40%

Heating and ventilation system

The heating and ventilation of the gymnasium form a combined concept in which each technology depends on the other. A scheme of the system is given in Figure 17.111.

Like the school, the gymnasium is heated by hot water, the supply temperature of which is adapted to the outside air temperature. The sports area is heated by ceiling radiators that allow quick reactions to changes of thermal loads. An energy-saving heating plant for the sports area would be ceiling radiators, because they need less energy for room heating than, for example, an air heating system.

The heating system can be controlled for each third of the sports area itself, both directly inside the sports area or from the control centre that is supervised by the caretaker.

The ventilation system has to fulfil different needs; as mentioned above, the occupants can vary from about 10 persons up to 400.

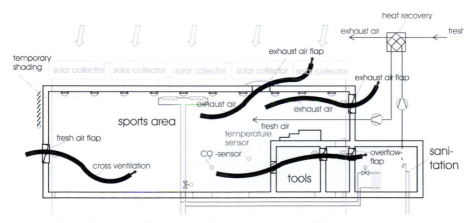

Figure 17.111. Scheme of the heating and ventilation system of the gymnasium

The shower rooms and toilets have to be ventilated at all times because they are underground. The air that is expelled heats up fresh air. This preheated air is blown into the sports area through valves that lead the air to the west facade. If additional fresh air is required, it enters the sports area through the controlled fresh-air flaps in the west facade. The fresh air is heated by the heating system on the ceiling and by those taking part in the sports activities; it is then directed through the tool room and the corridors, through to the shower rooms/toilets and expelled from there. Thus the whole gymnasium is ventilated by one simple system.

It has to be mentioned that, in the overflow flap from the corridors to the shower rooms, radiators heat the entrance air to provide a high indoor temperature in the shower rooms.

This ventilation system with heat recovery is sufficient to maintain a good indoor air quality, if the gymnasium is only used by up to 70 persons. In the case of tournaments, the additional air that is needed to ventilate the sports area enters through the controlled fresh-air flaps and rises as a result of natural forces up and out of the exhaust air flaps, which are also controlled.

In the summer, the fresh-air flaps and the exhaust-air flaps are kept open during the night to ventilate the whole building with cold night air and to cool it down until the sun heats the building up again the next day.

Integration of solar collectors

Owing to the special roof construction with the outside support structure, the collectors were easy to mount on the support structure. Thus, only very few additional costs for the construction of the collectors arose. The collectors were mounted with a movable crane without disturbing the building of the gymnasium (Figure 17.112). Up to spring 1998,

Figure 17.112. Mounting of the collectors on the support structure

only half of the roof area had been covered with collectors. After the next building stage the gymnasium will carry about 1300 m² of solar collectors.

Electric power consumption

As in the primary school, electric power consumption could be kept low by keeping the ventilation system small and by using natural forces. In addition, an EIB, the advantages of which have already been described, has also been installed in the gymnasium.

The author of this chapter (Niobe Chrisomallidou) would like to thank those at the Laboratory of Building Construction and Building Physics (L.B.C.P.) at the Aristotle University of Thessaloniki who collaborated in the project, and in particular civil engineers Theodore Theodosiou and Chrysostomos Kouinakis for their assistance in compiling this chapter.

 The author would also like to thank all those fellow engineers – whose names are mentioned in the text – who designed, constructed or assessed the energy efficiency of the examples of urban buildings presented, for providing the information necessary to present the buildings concerned. Without their contributions it would have been impossible to complete this chapter.

 Finally, the texts and drawings for Example buildings 5, 6 and 7 were produced by ENTPE-LASH, Laboratoire de Sciences de l'Habitat de l'Ecole Nationale des Travaux Publics de l'Etat, Vaulx-en-Velin, France (Example building 5) and the Institut für Thermodynamik und Wärmetechnik I.T.W., Universität Stuttgart (Example buildings 6 and 7), within the framework of the Save Community programme entitled 'Polistudies'.

REFERENCES

1. Chrisomallidou, N. (1994). 'Thermal behaviour of buildings – Passive Solar Systems'. *Proceedings of the CIENE (Central Institute for Energy Efficiency Education) on Solar Energy and Energy Conservation in Urban Buildings – Energy Conservation series.* University of Athens, pp. 19–35.

2. Santamouris, M., Chrisomallidou, N., Klitsikas, N., Papadopoulos, A. and Tsakiris, N. (1995–96). Energy Rehabilitation of Multi-Use Buildings. Save Programme – European Commission, Directorate General for Energy.

3. Andreadaki, E. (1992). *Monitoring of Two Passive Solar Houses.* Final Report of Research Programme financed by the Commission of the European Communities, D.G. XII/B/2. Contact: E. Andreadaki, University Campus, 54006 Thessaloniki, Greece.

4. Andreadaki, E. (ed.) (1992). *Single House in Ambelokipus – Athens. CRES Bioclimatic Architecture. Application in Greece.* Brochure No. 3. *Athens.*

5. Andreadaki, E. (ed.) (1992). *Single House in New Philothei – Athens. CRES Bioclimatic Architecture. Application in Greece.* Brochure No. 4. *Athens.*

6. Soubatzidis, M.G. (1997). 'Bioclimatic Architecture'. *Material & Building*, Vol. 36, pp. 42–47.

7. Souvatzidis, M. (1996). 'Architect's Residence and Office near the Attic Park'. *Architecture in Greece*, Vol. 30, pp 47–55.

8. Souvatzidis, M. (1996). 'Solar block apartment in Attic park'. *National Conference – Institute of Solar Techniques –Proceedings*, Vol. A , pp. 226–229.

9. Seiwald, H. (1997). 'Die solar unterstützte Nahwärmeversorgung mit saisonalem Erdsonden-Wärmespeicher in Neckarsulm' ('The Solar-assisted District Heating System with Duct

Heat Store in Neckarsulm – in German). *Proceedings of the 3rd Symposium erdgekoppelte Wärmepumpen*, Schloss Rauischholzhausen, Germany.

10. 'Steinbeis-Transferzentrum für rationelle Energienutzung und Solartechnik' (1996). Integrales Energiekonzept, Stuttgart.

11. Mangold, D., Schmidt, T., Wachholz, K. and Hahne, E. (1997). 'Solar meets Business: Comprehensive Energy Concepts for Rational Energy Use in the Most Cost-effective Way'. *Proceedings of ISES Solar World Congress*, Taejon, Korea.

12. TRNSYS: A TRaNsient System Simulation program (1996). Solar Energy Laboratory, University of Wisconsin, Madison, USA and Transsolar, Stuttgart, Germany.

13. Wärmeschutzverordnung 1995 (1995). *Verordnung über einen energiesparenden Wärmeschutz bei Gebäuden - Wärmeschutzverordnung 1995* (German national building code). Bundesgesetzblatt Teil 1, Bonn, Germany.

Appendix

Appendix

User's manual for POLISTUDIES: a multimedia tool for studying the energy efficiency of buildings in urban environments

N. Klitsikas

Department of Applied Physics, University of Athens

෨

POLISTUDIES is a complete multimedia educational package, which includes training material on the optimum integration of energy conservation techniques in urban buildings and on the appropriate consideration of the urban environmental issues.

The electronic tool was developed in CD-ROM format using the authoring software TOOLBOOK II PUBLISHER, copyright of Asymetrix company, in the framework of the SAVE programme, funded partially by DG XVII of the European Commission. The institutes participating in the project are shown in Table 18.1.

Table 18.1. Information about POLISTUDIES

European Commission
DG XVII
A multimedia tool for buildings in urban environments, *Version 1.0, October 1998*

Participating institutions
Central Institution for Energy Efficiency Education, University of Athens, Greece
Institut für Thermodynamik und Wärmetechnik Universität Stuttgart, Germany
Laboratoire Sciences de l'Habitat, Ecole Nationale des Traveaux Publics de l'Etat, Lyons, France
Laboratory of Building Construction & Physics, University of Thessaloniki, Greece

The material included on the CD is structured around seven main books. Each book includes one or more chapters.

The following pages contain instructions on how you can use the tool, browse through its chapters and books and use the hyperlinks provided.

Cities are increasingly expanding their boundaries and populations and as stated 'from the climatological point of view, human history is defined as the history of urbanization'.

INSTALLATION

Recommended configuration

- Pentium PC with Windows 95;
- 16 MB RAM;
- 15–20 MB free space on hard disk for temporary cache files;
- VGA graphics card with 16-bit colour;
- CD-ROM drive;
- Soundblaster card (optional).

It is possible that the tool will run on a PC with a less advanced configuration, but it has been tested and runs optimally when using the above configuration.

Setup

The software runs directly from the CD-ROM and thus no installation is required. The CD-ROM is 'autorun'. Thus, when you insert it in the CD-ROM drive the execution of the tool will start automatically.

If the autorun feature is disabled in your PC, you can run the program by executing the file POLIS.BAT, using either Explorer or the Run command in the Start menu.

Settings

Be sure that your monitor resolution is set at 640 × 480 with a colour depth of 16 or 24 bits. The software is optimized to run with this configuration. If it runs with a different configuration, the visual output may not be appropriate.

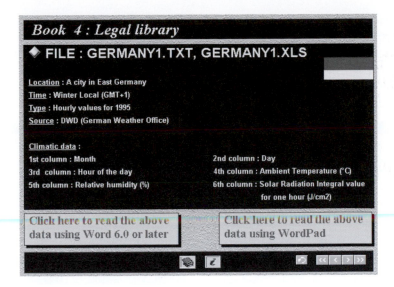

THE STRUCTURE OF THE TOOL

The tool is structured around seven 'knowledge blocks', which contain all the educational and training material that has been developed. Each of these seven blocks contains appropriate hyperlinks to the other blocks.

The contents of the each 'knowledge block', as they were developed, are described below.

Book 1. Computerized Handbook Encyclopedia

Chapter 1 : Urban building climatology

- The urban climate
- Wind distribution
- Vertical distribution
- Air flow in canyons
- Prediction of air speed in urban canyons
- Measurements of wind speed in urban canyons
- The urban temperature field
- Heat island studies
- Summer heat island in Athens
- Urban heat island models
- Simplified or correlation heat island and urban canyon expressions
- The impact of the main design parameters
- The role of the surface albedo
- The role of green spaces
- The role of the street layout

- The role of anthropogenic heating
- Energy consumption in urban spaces
- Increased electricity consumption due to high urban temperatures in some USA cities
- Energy studies related to heat island in Athens, Greece
- Other studies

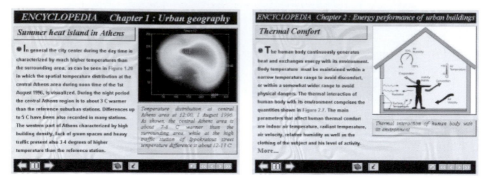

Chapter 2. Energy performance of urban buildings

- Fundamentals of heat and mass transfer
- Prerequisites of urban buildings
- Energy balance of urban buildings

Chapter 3. Applied lighting technology

- Artificial lighting
- Fundamentals
- Light sources
- Luminaires
- Lighting control
- Daylighting
- Designing with daylight in urban spaces
- Glazing materials
- Components and technologies

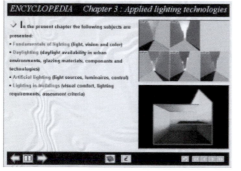

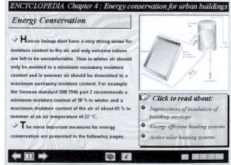

Chapter 4. Energy conservation and passive solar technologies for urban buildings

- Existing energy saving energy techniques in urban buildings
- Active solar heating systems
- Passive solar technologies

Chapter 5. Guidelines to integrate energy conservation

- Design guidelines
- Energy conservation strategies
- Passive solar heating strategies
- Passive cooling strategies
- Users' energy behaviour
- Natural lighting strategies
- Artificial lighting
- Active solar systems
- DHW
- Space heating systems
- Air conditioning
- Waste heat recovery
- Combining systems
- Appliances

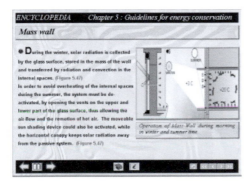

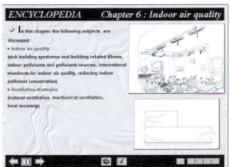

Chapter 6. Indoor air quality and ventilation

- Indoor air quality
- SBS and BRI
- Indoor pollutants and pollutant sources
- Reducing indoor pollutant concentration
- Ventilation strategies
- Natural ventilation
- Mechanical ventilation

Chapter 7. Applied energy and resources management in urban environment

- Central energy generation and supply
- Solar assisted district heating systems
- District cooling systems

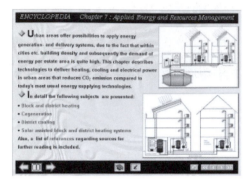
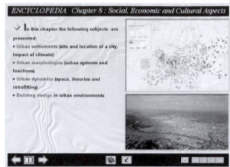

Chapter 8. Social, economic and cultural aspects of urban environment

- Urban settlements
- The site
- The location
- Impact of climate
- Urban morphologies
- Urban system
- Urban functions
- Urban dynamics
- Urban space
- Urban theories
- Urban dynamics
- Urban retrofitting and building design

Book 2. Computerized evaluation tools

Software tools developed by the authoring groups are integrated in the multimedia tool, so that the user is able to evaluate the energy performance of a building located in an urban environment, as well as the effect of the urban microclimate on this performance. Specifically, the tools SUMMER, AIOLOS and CANYON are integrated in the CD-ROM and the user's manuals for these tools are also included. Finally, the book contains detailed information on all the available design and calculation tools concerned with buildings in the urban environment.

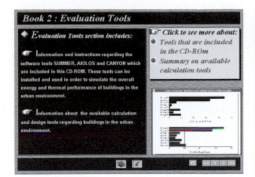

Book 3. Examples of urban buildings

Nine typical and bioclimatic buildings (five from Greece, two from Germany and two from France) are presented in this book as case studies on the performance of buildings in urban environments. The buildings on the CD include those described in Chapter 17 of the present book. The presentation of these buildings is based on the following framework:

- the project background;
- the site and climate;
- a general description of the building;
- an architectural description with the appropriate architectural drawings;
- the construction details of the building with a short description;
- the energy systems or the energy-saving features applied to the building for heating and cooling proposes;
- the passive solar design of the respective systems – if it exists – and its thermal behaviour;
- the actual energy consumption; and finally
- conclusions and comments on the energy behaviour of the building.

A series of drawings, pictures, videos and construction details completes the presentation of the urban buildings, giving a general and global view of the problems encountered and presenting the interventions and the design solutions implemented by the designers of each building, in order to decrease its energy consumption and to improve its energy efficiency. The presented buildings cover a wide range of climatic zones.

Book 4. Legal library

Information on the laws and regulations regarding the construction of buildings, as established in the three participating countries, are presented in this book. The block includes information about Greek, French and German legislation and standards. This information is classified both by regulation and by specific keywords (air quality, insulation, daylight, ventilation etc.).

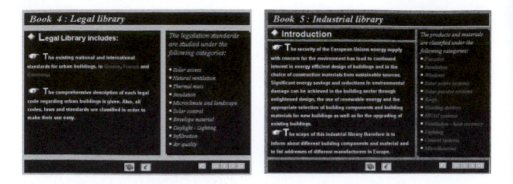

Book 5. Industrial library

Information about the energy efficient materials and systems produced by over 1000 manufacturers all over Europe are included in here., The following data are presented in detail for each product:

- the name of the product;
- visual material;
- properties of the product;
- cost;
- name and address of the manufacturer.

These data are classified and presented in twelve different categories.

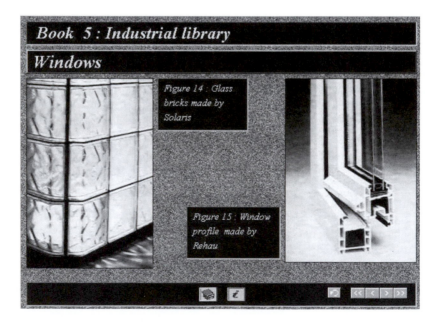

Figure 14 : Glass bricks made by Solaris

Figure 15 : Window profile made by Rehau

Book 6. Economic evaluation

This knowledge block presents and implements methods for cost–benefit analysis and other economic methodologies offering advanced opportunities for global evaluation of energy-conservation scenarios. Basic – discount and no-discount – methods, as well as more accurate and detailed methods for evaluating an energy-conservation investment project are presented. Examples and graphical representations are also included.

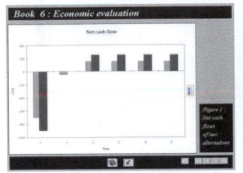
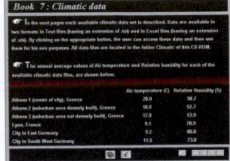

Book 7. Climatic documentation

Climatic data monitored in urban stations are presented for various locations in the three participating countries (three stations in Athens, one in Lyons and two in Germany). These data are appropriate, on the one hand, for performing realistic thermal and energy simulations of buildings in the urban environment and, on the other hand, for evaluating the heat-island effect in urban areas.

USING THE TOOL

- First, an introductory page is shown, giving information about the authoring teams.

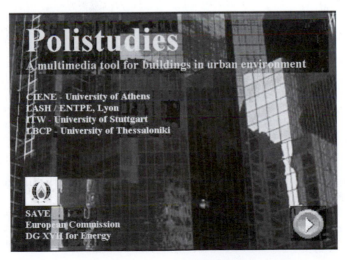

- Click the button at the bottom right to continue to the next page, which is the main contents page of the tool. From here you can go to any of the seven books of the tool, by clicking the appropriate title, or you can access the on-line guidelines by clicking the *Help* button.

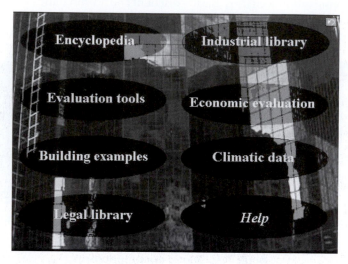

- If you select the first book (Encyclopedia), you will move to another contents page, which presents the titles of the eight chapters of the book. Click on the title to jump to any of the chapters.

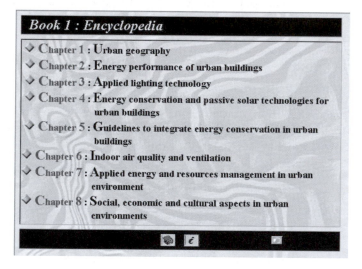

- If you move to any of the other six books you will go directly to this book (as they all have only one chapter).
- In order to navigate through the pages of each book or chapter (in the case of Encyclopedia), use the toolbar at the bottom of the page.

The following figure gives information about the function of each button.

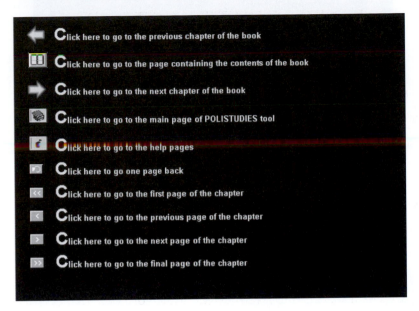

- On almost every page of the tool you will find a **Back** button. Click on this to go one page back, i.e. to the page you were on before entering the current one.

- Click on **hypertext** words (in red) or pictures to get more information about a particular subject, to go to another topic or to enlarge a picture. You can identify the hyperlink objects (either text, buttons or pictures), by the shape (the image of a hand) that the cursor takes when it moves over them:

- In the Encyclopedia, many of the hyperlinks lead to pages containing detailed information about a subject. These pages have a characteristic blue background. The following figure gives an example of such a page:

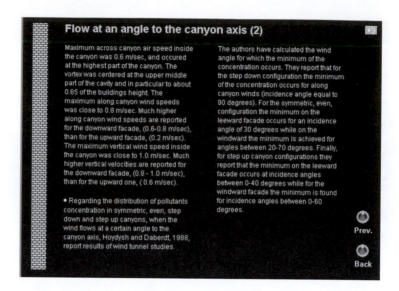

Click on the buttons on the right-hand side to navigate through these pages, as explained below:

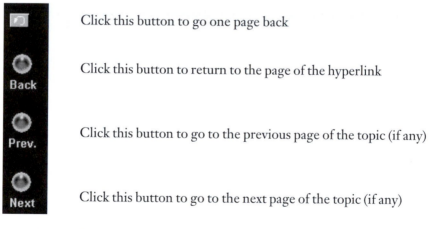

Click this button to go one page back

Click this button to return to the page of the hyperlink

Click this button to go to the previous page of the topic (if any)

Click this button to go to the next page of the topic (if any)

• As mentioned above, you can zoom into most of the pictures and drawings presented.

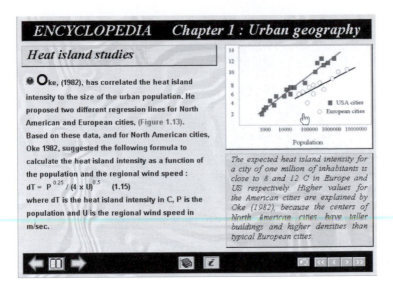

In order to do this, click on the picture (when the hand shaped cursor appears) and you will zoom into a larger size of the picture. Click on the **Back** button in order to return to the hyperlink page.

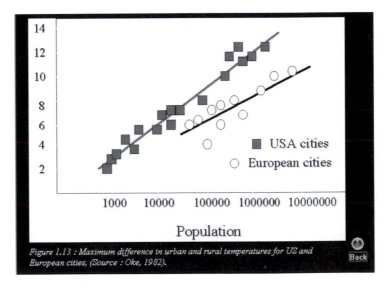

Figure 1.13 : Maximum difference in urban and rural temperatures for US and European cities, (Source : Oke, 1982).

- In order to watch one of the videos, click on the play button. You can use the toolbar at the bottom of the video window to stop, play, pause or move forwards and backwards, in the same way as you would with a normal video player.

 Pause the video

Stop the video

Play the video

Move the slider in order to go forwards or backwards.

- In Book 2 (Computerized evaluation tools), Book 4 (Legal library), Book 5: (Industrial libraries) and Book 7 (Climatic documentation), the tool will call other Windows applications in order either to install software tools or to present information and data. All the external applications that are called by the POLISTUDIES tool run in parallel with it. Specifically:
  - In Book 2, you can install the tools SUMMER BUILDING, SUMMER TECHNIQUES, AIOLOS or CANYON that are integrated in the CD-ROM. When you click the appropriate button, the setup program for each of the tools is executed and then you can install the tools on your hard disk by following the on-line guidelines. Refer to the relevant section of the tool for more information on the installation of the above-mentioned programs and to the user manual for each tool (included on the CD-ROM) for more information about it. You can read the manuals (and of course print them) using either Microsoft Word (version 6.0 or later – if it is installed on the

system) or the Windows Wordpad tool (which is already installed in Windows 95), by clicking the appropriate button.

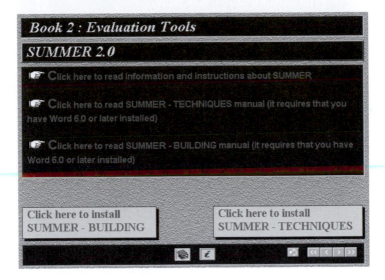

– In Books 4 and 5 you can access all the available information in the Legal and Industrial libraries respectively, using either Microsoft Word (version 6.0 or later – if it is installed on the system) or the Windows Wordpad tool (which is already installed in Windows 95), by clicking the appropriate button.

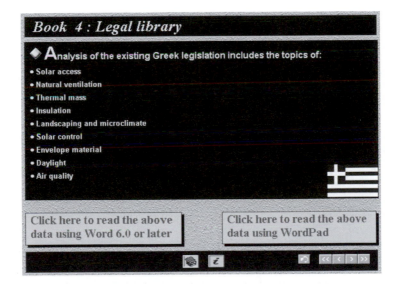

– In Book 7 you can edit the included climatic data files, using either Microsoft Excel (version 5.0 or later – if it is installed on the system) or the Windows Wordpad tool (which is already installed in Windows 95), by clicking the appropriate button.

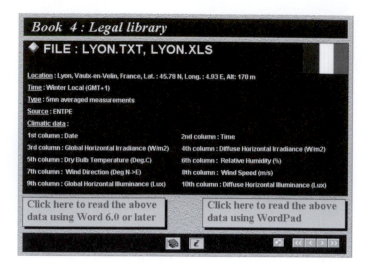

- Finally, note that by clicking the appropriate button you can access the on-line help from every page of the tool.

Index

ℰℜ